# Lecture Notes in Computer Science 12289

More information about this series at http://www.springer.com/series/7407

Marco Gribaudo · David N. Jansen ·
Anne Remke (Eds.)

# Quantitative Evaluation of Systems

17th International Conference, QEST 2020
Vienna, Austria, August 31 – September 3, 2020
Proceedings

 Springer

*Editors*
Marco Gribaudo (ID)
Politecnico di Milano
Milan, Italy

David N. Jansen (ID)
Chinese Academy of Sciences
Beijing, China

Anne Remke
University of Münster
Münster, Germany

ISSN 0302-9743          ISSN 1611-3349   (electronic)
Lecture Notes in Computer Science
ISBN 978-3-030-59853-2          ISBN 978-3-030-59854-9   (eBook)
https://doi.org/10.1007/978-3-030-59854-9

LNCS Sublibrary: SL1 – Theoretical Computer Science and General Issues

This Springer imprint is published by the registered company Springer Nature Switzerland AG
The registered company address is: Gewerbestrasse 11, 6330 Cham, Switzerland

# Preface

This volume contains the papers presented at the International Conference on Quantitative Evaluation of Systems (QEST 2020), held online during August 31 – September 3, 2020, and organized by TU Wien, Austria. Following the path of several previous editions, this conference was part of a greater event, this time called QONFEST. This larger forum brought together the 31st Conference on Concurrency Theory (CONCUR 2020), the 18th International Conference on Formal Modelling and Analysis of Timed Systems (FORMATS 2020), the 25th International Conference on Formal Methods for Industrial Critical Systems (FMICS 2020), as well as QEST 2020. QONFEST also included workshops and tutorials before and after these major conferences.

As one of the premier fora for research on quantitative system evaluation and verification of computer systems and networks, QEST covers topics including classic measures involving performance and reliability, as well as quantification of properties that are classically qualitative, such as safety, correctness, and security. QEST welcomes measurement-based studies as well as analytic studies, diversity in the model formalisms and methodologies employed, as well as development of new formalisms and methodologies. QEST also has a tradition of presenting case studies, highlighting the role of quantitative evaluation in the design of systems, where the notion of system is broad. Systems of interest include computer hardware and software architectures, communication systems, embedded systems, infrastructural systems, and biological systems. Moreover, tools for supporting the practical application of research results in all of the aforementioned areas are also of interest to QEST. In short, QEST aims to encourage all aspects of work centered around creating a sound methodological basis for assessing and designing systems using quantitative means.

Following the tradition of previous years' editions of QEST, a special session on frontier topics in the current research landscape was proposed. The topic selected this year was "Predictive performance by machine learning," and it focused on work considering the combination of machine learning and performance prediction.

Thanks to the organization of the QONFEST event, this year's edition of QEST featured three keynote speakers. Evgenia Smirni (College of William and Mary Williamsburg, USA) gave a talk on "Machine Learning Models for Reliability of Large Scale Systems," Annabelle McIver (Mcquarie University, Australia) spoke about "On Privacy and Accuracy in Data Releases," and Tom Henzinger (IST, Austria) presented "A Survey of Bidding Games on Graphs." We also had two tutorials, specifically addressed to the QEST community: "Flexible nets" by Jorge Júlvez (University of Cambridge, UK) and "Verifying Probabilistic Programs" by Benjamin Kaminski (University College London, UK), Joost-Pieter Katoen (RWTH Aachen University, Germany), and Christoph Matheja (ETH Zürich, Switzerland).

The Program Committee (PC) consisted of 32 experts and we received a total of 42 submissions. Each submission was reviewed by at least three PC members or external reviewers. Based on the reviews and the PC discussion phase, the committee decided to accept 20 submissions, of which 19 papers were included in these proceedings (12 full papers and 7 shorter contributions, which also include tool presentations and demos). The program also included a tool demo "Compact and explainable strategy representations using dtControl" by P. Ashok, M. Jackermeier, J. Křetínský, and M. Weininger, not followed by a paper in this book. Conversely, these proceedings also include the contribution "Machine Learning for Reliability Analysis of Large Scale Distributed Systems" by one of the invited speakers, and the abstracts of the tutorials.

We want to thank a lot of people for their efforts despite the generally difficult situation during which this event was organized. Firstly, we thank all the authors who submitted papers, as without them there simply would not be a conference. We feel that in this particular year, thanking the authors is even more important, since everybody experienced a lot of difficulties in cooperating with colleagues as well as in finding novel and different ways of communicating, while coping with new restrictions and regulations never seen before. We would also like to express the same gratitude to the PC members and the additional reviewers for their hard, timely work and for sharing their valued expertise with the rest of the community, as well as EasyChair for supporting the electronic submission and the reviewing process. In particular, we would like to thank Alfred Hofmann and Anna Kramer for their help in preparing this LNCS volume, and we thank Springer for kindly sponsoring the prize for the Best Paper Award. We thank the QONFEST Platinum Sponsor Interchain Foundation, and also the Vienna Center for Logic and Algorithms (VCLA) and TU Wien for their financial support. Special thanks go to the local organization chair and general chair, Ezio Bartocci, for his dedication and excellent work, and to the publicity chair, Carlos E. Budde, who was of great help during all the stages of preparation of the conference program. Finally, we would like to thank Enrico Vicario, chair of the QEST Steering Committee, and Jane Hillston, former chair, for their guidance throughout the past two years, as well as the members of the QEST Steering Committee.

We hope that you find the conference proceedings rewarding and will consider submitting to QEST 2021.

August 2020

Marco Gribaudo
David N. Jansen
Anne Remke

# Organization

## Program Committee

| | |
|---|---|
| Alessandro Abate | University of Oxford, UK |
| Simona Bernardi | Universidad de Zaragoza, Spain |
| Peter Buchholz | TU Dortmund, Germany |
| Carlos E. Budde | University of Twente, The Netherlands |
| Davide Cerotti | Politecnico di Milano, Italy |
| Florin Ciucu | University of Warwick, UK |
| Yuxin Deng | East China Normal University, China |
| Pedro R. D'Argenio | Universidad Nacional de Córdoba and CONICET, Argentina |
| Marco Gribaudo | Politecnico di Milano, Italy |
| Arnd Hartmanns | University of Twente, The Netherlands |
| András Horváth | University of Turin, Italy |
| David N. Jansen | Institute of Software, Chinese Academy of Sciences, China |
| William Knottenbelt | Imperial College London, UK |
| Jan Křetínský | Technical University of Munich, Germany |
| Andrea Marin | University of Venice, Italy |
| Andrew Miner | Iowa State University, USA |
| Laura Nenzi | University of Trieste, Italy |
| Gethin Norman | University of Glasgow, UK |
| Marco Paolieri | University of Southern California, USA |
| David Parker | University of Birmingham, UK |
| Tuan Phung-Duc | University of Tsukuba, Japan |
| Riccardo Pinciroli | College of William and Mary, USA |
| Pavithra Prabhakar | Kansas State University, USA |
| Daniel Reijsbergen | The University of Edinburgh, UK |
| Anne Remke | WWU Münster, Germany |
| Markus Siegle | Universität der Bundeswehr, Germany |
| Meng Sun | Peking University, China |
| Mirco Tribastone | IMT, Lucca, Italy |
| Benny Van Houdt | University of Antwerp, Belgium |
| Andrea Vandin | Sant'Anna School of Advanced Studies, Italy |
| Verena Wolf | Saarland University, Germany |
| Katinka Wolter | Freie Universität zu Berlin, Germany |

## Additional Reviewers

Ashok, Pranav
Backenköhler, Michael
Brihaye, Thomas
Castro, Pablo
de Boer, Pieter-Tjerk
Demasi, Ramiro
Eisentraut, Julia
Gouberman, Alexander
Gros, Timo P.
Großmann, Gerrit
Hasanbeig, Mohammadhosein
Lal, Ratan
Lu, Yuteng
Meggendorfer, Tobias
Molyneux, Gareth
Putruele, Luciano
Sun, Weidi
Weininger, Maximilian
Wijs, Anton
Wolovick, Nicolás

# Contents

**Tutorials**

# Invited Paper

# Machine Learning for Reliability Analysis of Large Scale Systems

Evgenia Smirni[✉]

Department of Computer Science, William and Mary, Williamsburg, VA, USA
esmirni@cs.wm.edu
http://www.cs.wm.edu/~esmirni

**Abstract.** As distributed systems dramatically grow in terms of scale, complexity, and usage, understanding the hidden interactions among system and workload properties becomes an exceedingly difficult task. Machine learning models for prediction of system behavior (and analysis) are increasingly popular but their effectiveness in answering what and why is not always the most favorable. In this talk I will present two reliability analysis studies from two large, distributed systems: one that looks into GPGPU error prediction at the Titan, a large scale high-performance-computing system at ORNL, and one that analyzes the failure characteristics of solid state drives at a Google data center and hard disk drives at the Backblaze data center. Both studies illustrate the difficulty of untangling complex interactions of workload characteristics that lead to failures and of identifying failure root causes from monitored symptoms. Nevertheless, this difficulty can occasionally manifest in spectacular results where failure prediction can be dramatically accurate.

**Keywords:** Data centers · HPC · Storage systems · Reliability · GPUs · SSDs · HDDs

## 1 Overview

Effective workload prediction hold the answers to the conundrum of efficient management in distributed and scaled out systems. Being able to accurately predict the upcoming workload within the next time frame (i.e., in the next 10 min, half hour, hour, or even week) allows the system to make proactive decisions [14], improve its reliability and performance [11,12], and result in better user experience [16]. Machine learning predictors have been used widely to improve on the operating conditions of data centers, see [11–16] and references within.

Beyond the above efforts that mainly focus on predicting workload peaks and system usage, accurately predicting failures of systems components and

The work was partially supported by NSF grants CCF-1649087, CCF-1717532, and IIS-1838022. The work presented here was done in collaboration with J. Alter, L. Yang, B. Nie, J. Xue, R. Pinciroli, D. Tiwari, A. Jog, A. Dimnaku, R. Birke, and L. Chen.

M. Gribaudo et al. (Eds.): QEST 2020, LNCS 12289, pp. 3–7, 2020.
https://doi.org/10.1007/978-3-030-59854-9_1

identifying the reason behind these failures becomes a pressing problem [6]. As systems dramatically increase in scale and complexity, machine learning predictors become critical for understanding the hidden interaction between systems and workload that leads to failures and can result in their proactive system management to dramatically improve on system robustness. In this talk, I will concentrate on how workloads can affect the reliability of two hardware components of data centers, general purpose GPUs (GPGPUs) and storage devices (HDDs and SSDs) and illustrate how the judicious use of machine learning predictors can lead to better understanding of the workload/hardware interaction and system robustness.

## 2  GPGPU Error Prediction

As GPGPUs are more widely adopted in scale-out computing architectures, GPU soft errors become a critical challenge. Evaluation of application resilience to soft errors has been the focus in many recent works [5,9,17]. Understanding the source of GPU soft errors adds further to the challenge. Past work has shown evidence that indicates a plausible relationship between power/cooling infrastructure and GPU errors [6], but there exists no clear understanding on the exact conditions that trigger faults. We focus on understanding the interplay between workload/temperature/power consumption and GPU soft errors on the Titan, one of America's fastest supercomputers for open science [1].

Workload analysis on Titan traces shows that workload characteristics, certain GPU cards, temperature and power consumption could be indicative of GPU errors, but it is non-trivial to exploit them for error prediction [7,8]. Acknowledging the necessity of an error predictor, we elaborate on the challenges, process, and solutions involved in building effective machine-learning-based prediction models. We show how to systematically select features by categorizing them into spatial and temporal dimensions. We illustrate how to overcome the imbalanced dataset challenge and trade-offs by taking advantage of the inherent features of the dataset. We use the selected features to train various machine learning models, including Logistic Regression (LR), Gradient Boosting Decision Tree (GBDT), Support Vector Machine (SVM), and Neural Network (NN).

Finally, we evaluate the machine learning models via different metrics and under diverse testing scenarios. Our results indicate that the proposed models achieve good prediction quality and are robust. In particular, the GBDT-based prediction significantly outperforms other models and results in conservative predictions in identifying as many soft error cases as possible. Our evaluation also uncovers interesting insights from comparison across different models, training/testing data, and feature combinations. We show that the proposed prediction models impose moderate overhead and are practically feasible for GPU soft error prediction.

# 3   The Life and Death of HDDs and SSDs

Storage devices, such as hard disk drives (HDDs) and solid state drives (SSDs), are among the components that affect the data center dependability the most [4], contributing to the 28% of data center failure events [10]. Accurate prediction of storage component failures enables on-time mitigation actions that avoid data loss and increases data center dependability. Failure events observed in HDDs and SSDs are different due to their distinct physical mechanics, it follows that observations from HDD analysis cannot be generalized to SSDs and vice-versa.

We focus on failures of HDDs and SSDs by analyzing disk logs collected from real-world data centers over a period of six years. We analyze *Self-Monitoring, Analysis and Reporting Technology* (SMART) traces from five different HDD models from the Backblaze data center [3] and the logs of three multi-level cell (MLC) models of SSDs collected at a Google data center. Though we are unaware of the data centers' exact workflows for drive repairs, replacements, and retirements (e.g., whether they are done manually or automatically, or the replacement policies in place), we are able to discover key correlations and patterns of failure, as well as generate useful forecasts of future failures. Being able to predict an upcoming drive retirement could allow early action: for example, early replacement before failure happens, migration of data and VMs to other resources, or even allocation of VMs to disks that are not prone to failure.

Drive failures are triggered by a set of attributes and different drive features must be monitored to accurately predict a failure event. Similar to the GPU study, there is no single metric that triggers a drive failure after it reaches a certain threshold. We show that machine learning models that are trained from monitoring logs achieve failure prediction that is both remarkably accurate and timely, both in the HDD and SSD domains. Random forests are especially efficient in their prediction and can be used for further interpretation: they can provide valuable insights on which errors and workload characteristics are most indicative of future catastrophic failures. The predictors are able to anticipate failure events with reasonable accuracy up to several days in advance. We further show that there exist different ways to partition the HDD and SSD datasets to increase model accuracy. This partitioning is based on workload analysis that was first developed in [2] for SSDs and focuses on the discovery of certain drive attributes. We saw that similar partitioning can be also successfully applied for the case of HDDs. We find that in both HDD and SSD cases, datasets may be partitioned to improve the performance of the classifier. Partitioning SSDs on the *Drive age* attribute and HDDs on *head flying hours* (i.e., SMART 240) dramatically increases model accuracy. Finally, the interpretability of the machine learning models derive insights that can be used to drive proactive disk management policies.

# References

1. Top500 Supercomputer Sites, November 2018. https://www.top500.org/lists/2018/11/

2. Alter, J., Xue, J., Dimnaku, A., Smirni, E.: SSD failures in the field: symptoms, causes, and prediction models. In: Taufer, M., Balaji, P., Peña, A.J. (eds.) Proceedings of the International Conference for High Performance Computing, Networking, Storage and Analysis, SC 2019, Denver, Colorado, USA, 17–19 November 2019, pp. 75:1–75:14. ACM (2019). https://doi.org/10.1145/3295500.3356172

3. Backblaze: Hard drive data and stats. https://www.backblaze.com/b2/hard-drive-test-data.html. Accessed 28 Apr 2020

4. Birke, R., Björkqvist, M., Chen, L.Y., Smirni, E., Engbersen, T.: (Big)data in a virtualized world: volume, velocity, and variety in cloud datacenters. In: Schroeder, B., Thereska, E. (eds.) Proceedings of the 12th USENIX conference on File and Storage Technologies, FAST 2014, Santa Clara, CA, USA, 17–20 February 2014, pp. 177–189. USENIX (2014). https://www.usenix.org/conference/fast14/technical-sessions/presentation/birke

5. Nie, B., Jog, A., Smirni, E.: Characterizing accuracy-aware resilience of GPGPU applications. In: 20th IEEE/ACM International Symposium on Cluster, Cloud and Internet Computing, CCGRID 2020, Melbourne, Australia, 11–14 May 2020, pp. 111–120. IEEE (2020). https://doi.org/10.1109/CCGrid49817.2020.00-82

6. Nie, B., Tiwari, D., Gupta, S., Smirni, E., Rogers, J.H.: A large-scale study of soft-errors on GPUs in the field. In: 2016 IEEE International Symposium on High Performance Computer Architecture, HPCA 2016, Barcelona, Spain, 12–16 March 2016, pp. 519–530. IEEE Computer Society (2016). https://doi.org/10.1109/HPCA.2016.7446091

7. Nie, B., Xue, J., Gupta, S., Engelmann, C., Smirni, E., Tiwari, D.: Characterizing temperature, power, and soft-error behaviors in data center systems: insights, challenges, and opportunities. In: 25th IEEE International Symposium on Modeling, Analysis, and Simulation of Computer and Telecommunication Systems, MASCOTS 2017, Banff, AB, Canada, 20–22 September 2017, pp. 22–31. IEEE Computer Society (2017). https://doi.org/10.1109/MASCOTS.2017.12

8. Nie, B., et al.: Machine learning models for GPU error prediction in a large scale HPC system. In: 48th Annual IEEE/IFIP International Conference on Dependable Systems and Networks, DSN 2018, Luxembourg City, Luxembourg, 25–28 June 2018, pp. 95–106. IEEE Computer Society (2018). https://doi.org/10.1109/DSN.2018.00022

9. Nie, B., Yang, L., Jog, A., Smirni, E.: Fault site pruning for practical reliability analysis of GPGPU applications. In: 51st Annual IEEE/ACM International Symposium on Microarchitecture, MICRO 2018, Fukuoka, Japan, 20–24 October 2018, pp. 749–761. IEEE Computer Society (2018). https://doi.org/10.1109/MICRO.2018.00066

10. Pinciroli, R., Yang, L., Alter, J., Smirni, E.: The life and death of SSDs and HDDs: Similarities, differences, and prediction models, pp. 1–14 (2020). (under submission)

11. Xue, J., Birke, R., Chen, L.Y., Smirni, E.: Managing data center tickets: prediction and active sizing. In: 46th Annual IEEE/IFIP International Conference on Dependable Systems and Networks, DSN 2016, Toulouse, France, 28 June–1 July 2016, pp. 335–346. IEEE Computer Society (2016). https://doi.org/10.1109/DSN.2016.38

12. Xue, J., Birke, R., Chen, L.Y., Smirni, E.: Tale of tails: anomaly avoidance in data centers. In: 35th IEEE Symposium on Reliable Distributed Systems, SRDS 2016, Budapest, Hungary, 26–29 September 2016, pp. 91–100. IEEE Computer Society (2016). https://doi.org/10.1109/SRDS.2016.021

13. Xue, J., Birke, R., Chen, L.Y., Smirni, E.: Spatial-temporal prediction models for active ticket managing in data centers. IEEE Trans. Netw. Serv. Manag. **15**(1), 39–52 (2018). https://doi.org/10.1109/TNSM.2018.2794409
14. Xue, J., Nie, B., Smirni, E.: Fill-in the gaps: spatial-temporal models for missing data. In: 13th International Conference on Network and Service Management, CNSM 2017, Tokyo, Japan, 26–30 November 2017, pp. 1–9. IEEE Computer Society (2017). https://doi.org/10.23919/CNSM.2017.8255983
15. Xue, J., Yan, F., Birke, R., Chen, L.Y., Scherer, T., Smirni, E.: PRACTISE: robust prediction of data center time series. In: Tortonesi, M., Schönwälder, J., Madeira, E.R.M., Schmitt, C., Serrat, J. (eds.) 11th International Conference on Network and Service Management, CNSM 2015, Barcelona, Spain, 9–13 November 2015, pp. 126–134. IEEE Computer Society (2015). https://doi.org/10.1109/CNSM.2015.7367348
16. Xue, J., Yan, F., Riska, A., Smirni, E.: Scheduling data analytics work with performance guarantees: queuing and machine learning models in synergy. Clust. Comput. **19**(2), 849–864 (2016). https://doi.org/10.1007/s10586-016-0563-z
17. Yang, L., Nie, B., Jog, A., Smirni, E.: Practical resilience analysis of GPGPU applications in the presence of single- and multi-bit faults. IEEE Trans. Comput. (2020). https://doi.org/10.1109/TC.2020.2980541

# Predictive Performance and Machine Learning

# Tracking the Race Between Deep Reinforcement Learning and Imitation Learning

Timo P. Gros$^{(\boxtimes)}$, Daniel Höller, Jörg Hoffmann, and Verena Wolf

Saarland University, Saarland Informatics Campus, 66123 Saarbrücken, Germany
{timopgros,hoeller,hoffmann,wolf}@cs.uni-saarland.de
https://mosi.uni-saarland.de, http://fai.cs.uni-saarland.de

**Abstract.** Learning-based approaches for solving large sequential decision making problems have become popular in recent years. The resulting agents perform differently and their characteristics depend on those of the underlying learning approach. Here, we consider a benchmark planning problem from the reinforcement learning domain, the Racetrack, to investigate the properties of agents derived from different deep (reinforcement) learning approaches. We compare the performance of deep supervised learning, in particular imitation learning, to reinforcement learning for the Racetrack model. We find that imitation learning yields agents that follow more risky paths. In contrast, the decisions of deep reinforcement learning are more foresighted, i.e., avoid states in which fatal decisions are more likely. Our evaluations show that for this sequential decision making problem, deep reinforcement learning performs best in many aspects even though for imitation learning optimal decisions are considered.

**Keywords:** Deep reinforcement learning · Imitation learning

## 1 Introduction

In recent years, deep learning and especially deep reinforcement learning (DRL) have been applied with great successes to the task of learning near-optimal policies for sequential decision making problems [1,8,9,13–15]. It relies on a feedback loop between self-play and the improvement of the current strategy by reinforcing decisions that lead to good performance.

Passive imitation learning (PIL) is another well-known approach of deep learning, where a policy is based on data which was labeled by an expert [12]. An extension of this approach is active imitation learning (AIL), where after an initial phase of passive learning, additional data is iteratively generated by exploring the state space based on the current strategy and subsequent expert labeling [6,11]. AIL has successfully been applied to common reinforcement learning benchmarks such as cart-pole or bicycle-balancing [6].

M. Gribaudo et al. (Eds.): QEST 2020, LNCS 12289, pp. 11–17, 2020.
https://doi.org/10.1007/978-3-030-59854-9_2

Here we aim at an in-depth study of empirical learning agent behavior for a range of different learning frameworks. We train different agents on a simple benchmark problem named Racetrack [2–4,10,16], using DRL, PIL, and AIL and study their characteristics. We first apply PIL and train agents represented by linear functions and artificial neural networks. For AIL, we use the DAGGER approach [11] to train agents represented by neural networks. Based on the same network architecture, we apply deep reinforcement learning. More specifically, we use deep Q-networks [9]. We compare the resulting agents considering three different aspects: the success rate, the quality of the resulting action sequences, and the relative number of optimal and fatal decisions.

Amongst other things, we find that, *even though it is based on optimal training data, imitation learning leads to unsafe policies, much more risky than those found by RL*. Upon closer inspection, it turns out that this apparent contradiction actually has an intuitive explanation in terms of the nature of the application and the different learning methods: *to minimize time to goal, optimal decisions navigate very closely to dangerous states*. This works well when taking optimal decisions throughout – but is brittle to (and thus fatal in the presence of) even small divergences as are to be expected from a learned policy. We believe that these finding might carry over to many other applications beyond Racetrack.

This short paper repeatedly lacks some details. For further information we refer to the extended version [5].

## 2    Racetrack

Racetrack has been used as a benchmark in the context of planning [3,10] and reinforcement learning [2,16]. It can be played on different maps. Initially, a car is placed randomly at one of the discrete positions on the start line with zero velocity. In every step it can speed up, hold the velocity or slow down in $x$ and/or $y$ dimension. Then, the car moves in a straight line with the new velocity from the old position to a new one, where we discretize the maps into cells. The game is lost when the car crashes, i.e. the straight line between the old position and the new one intersects with or ends in a wall (which surround the track). The game is won when the car either stops at or drives through the goal line. Given a Racetrack map, the game can be modeled as a Markov decision process.

**States.** The current state is uniquely defined by the position $p = (x, y)$ and the velocity $v = (v_x, v_y)$.

**Actions.** Actions represent the acceleration $a$. As the car can be accelerated with values $\{-1, 0, 1\}$ in the $x$ and in the $y$ dimension, there are exactly $3^2 = 9$ different actions available in every state.

**Transitions.** We assume a wet road, so with a chance of 0.1, the acceleration cannot be applied, i.e. $a = (0, 0)$. Otherwise, with probability 0.9, the acceleration is as selected by the action. The new velocity $v' = (v_x', v_y')$ is given by the sum of the acceleration $a = (a_x, a_y)$ and the current velocity. The new position $p' = (x', y')$ is given by adding $v'$ to the current position.

**Rewards/Costs.** As we consider both planning and learning approaches, we define the following two functions: For planning we consider a uniform cost function, such that an optimal planner will find the shortest path to reach the goal line. For RL we consider a reward function that is positive if the step reaches the goal, negative if the step is invalid and 0 otherwise. As RL makes use of discounting, both functions motivate to reach the goal as fast as possible.

**Simulation.** For a given map, we consider several variants of the simulation. We distinguish between normal start (NS), i.e. starting on the start line and random start (RS), i.e. starting on a random (valid) position on the map. Further we consider starting with velocity zero (ZV) and with a random velocity (RV) up to a given upper bound. Lastly we consider both, a noisy (N) and a deterministic (D) version of the game. In the latter, the wet road assumption is dropped.

# 3 Learning

We train the agents by using imitation learning and reinforcement learning.

Although a state is uniquely given by the car's position and velocity, we provide several other features that can be used as state encoding to improve the learning procedure, such as wall sensors and goal distances.

## 3.1 Imitation Learning

Imitation learning is based on labeled training data. To create datasets that can be used for imitation learning, we consider several different variants. In the base case, we uniformly sample possible states, i.e. positions and velocities, and then label them with an optimal action, i.e. an optimal acceleration, by using an expert. This expert is a Racetrack-tailored version of the $A^*$ algorithm. We further consider sampling through the state space with several other options, namely (1) learning from complete trajectories from the uniformly sampled state until the goal, (2) considering *all* optimal actions instead of a random one, and (3) considering only states with a unique solution. Combined with the (RS)/(NS) and (RV)/(ZV) options of the simulation, this results in several different combinations, of which we consider the six most promising ones to create data sets. All datasets contain approximately $10^5$ entries.

**Passive Imitation Learning.** We consider linear functions and neural networks to represent PIL policies. To train linear functions we use the package `sklearn` to apply both *Linear Discriminant Analysis* (LDA) and *Logistic Regression* (LR). To train neural networks we use the package `PyTorch` [7]. We make use of the MSE as loss function.

**Active Imitation Learning.** To represent the class of active imitation learning algorithms, we consider DAGGER [11]. In contrast to PIL, we here consider neural networks only. To have a fair comparison, DAGGER has the same number of samples as PIL, i.e. $10^5$. While the pre-training is important for sampling within the first iteration of the algorithm, the main idea is to generate further entries, which are more important for the training of the agent. Thus, we pre-trained the agent on each of our data sets and then additionally allowed DAGGER to add $10^5$ samples.

### 3.2 Deep Reinforcement Learning

Reinforcement learning is based on self-play without prior knowledge. We focus on the value-based approach of *deep Q-learning* (DQN) [9]. We apply the deep Q-learning algorithm and make use of *experience replay* and *fixed targets* [9].

To enable a fair comparison, we restrict the agents to (1) $10^5$ entries in the replay buffer (the maximal number of entries an agent can learn from at the same time) and (2) $10^5$ episodes that the agent can play in total. The neural network is not pre-trained but initialized randomly. Besides the given options of either starting on the start line (NS) or anywhere on the map (RS), DRL can benefit from learning while the noisy (N) version of Racetrack is simulated instead of the deterministic (D) one. This gives us four different training modes.

## 4    Results

To have an extensive comparison, we consider six possible combinations given by the simulation parameters as described in Sect. 2.

For each learning method we present the best-performing parameter combination of all those that we trained. We investigate three aspects of the behavior of the resulting agents: the success rate, the quality of the resulting action sequences, and the relative number of optimal and fatal decisions.

### 4.1    Success Rate

We first compare how often the agents win a game, i.e. reach the goal, or lose, i.e. crash into a wall. We compare the agents on $10^4$ simulation runs. For each run, all agents start in the same initial state. The results for two of the settings can be found in Fig. 1.

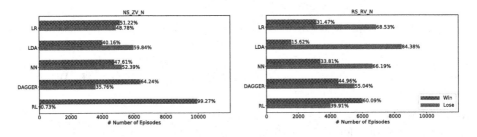

**Fig. 1.** Success rate results for all classes of examined agents.

The linear PIL agents perform worst. Especially with random starting points and velocities, they regularly fail to reach the goal. DAGGER outperforms the passive imitation learning agents, as it has been designed to cope with sequential decision making. Throughout all settings, the DRL agents perform best. They clearly outperform DAGGER, for instance reaching the goal more than 1.5 times more often in the NS-ZV-N setting.

## 4.2 Quality of Action Sequences

We illustrate results for the quality of the chosen action sequences in Fig. 2.

The left plot gives the cumulative reward reached by the agents averaged over all runs (also over those that are not successful). DRL clearly achieves the highest cumulative reward. We remark that the optimal policies computed via A$^*$ give higher cumulative rewards as the goal is reached faster. However, imitation learning achieves lower results on average as it fails more often.

The right of Fig. 2 gives results for the number of steps. When a car crashes, we are not interested in the number of steps taken. Therefore – in this specific analysis – we only report on successful runs. They show that – while reinforcement learning has the most wins and is the best agent considering the reward objective – it is consuming the highest number of steps when reaching the goal. It even takes more steps than linear PIL classifiers.

## 4.3 Quality of Single Action Choices

Next we examine whether the agents choose the optimal acceleration, i.e. an acceleration that does not crash and leads to the goal with as few steps as possible, for different positions and velocities. We distinguish between (1) optimal actions, (2) fatal actions that unavoidably lead to a crash, and (3) secure actions that are neither of the former.

The results are given in Fig. 3. Especially when we start from a random position on the map, we see that (independent of the setting) passive imitation learning with neural networks selects optimal actions more often than active imitation learning or deep reinforcement learning. Interestingly, DAGGER and RL select both secure *and* fatal choices more often than PIL.

## 4.4 Discussion

We found that passive imitation learning agents perform poorly (see Fig. 1) even though they select optimal actions most often. One reason for this is that the data sets from which they learn contain samples that have not been generated by iteratively improving the current policy. Hence, it is not biased towards sequences of dependent decisions leading to good performance. We have observed that DAGGER and in particular DRL sometimes do not select optimal actions, but those with lower risk of hitting a wall. As a result, they need more steps than other approaches before reaching the goal, but the trajectories they use are more

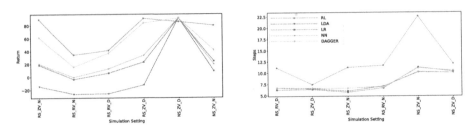

**Fig. 2.** Average reward (left) and average number of steps (right) for all agent classes.

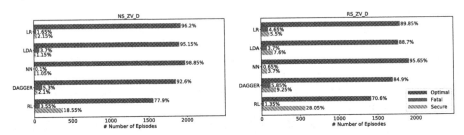

**Fig. 3.** Quality of selected actions.

secure and they crash less often. This is an interesting insight, as all approaches (including PIL) try to optimize the same objective: reaching the goal as soon as possible without hitting a wall.

The fact that both, DAGGER and RL have a relatively high number of fatal actions, but not an increased number of losses, leads us to the assumption that these agents avoid states where they might make fatal decisions, even though those states could help reaching the goal faster.

In summary, DRL performs surprisingly well. In some aspects, it performs even better than active imitation learning, which is not only considered state of the art for sequential decision making [6], but – in contrast to DRL – even has the chance to benefit from expert knowledge.

## 5   Conclusion

We have presented an extensive comparison between different learning approaches to solve the Racetrack benchmark. Even though we provided optimal decisions during imitation learning, the agents based on deep reinforcement learning outperform those of imitation learning in many aspects.

We believe that our observations carry over to other applications, in particular to more complex autonomous vehicle control algorithms. We plan to consider extensions of the Racetrack problem, which include further real-world characteristics of autonomous driving. We believe that, to address the difficulties we observed with imitation learning, further investigations into the combination of expert data sets and reinforcement learning agents are necessary.

**Acknowledgements.** This work has been partially funded by DFG grant 389792660 as part of TRR 248 (see https://perspicuous-computing.science).

# References

1. Agostinelli, F., McAleer, S., Shmakov, A., Baldi, P.: Solving the Rubik's Cube with deep reinforcement learning and search. Nat. Mach. Intell. **1**(8), 356–363 (2019)
2. Barto, A.G., Bradtke, S.J., Singh, S.P.: Learning to act using real-time dynamic programming. Artif. Intell. **72**(1–2), 81–138 (1995)
3. Bonet, B., Geffner, H.: GPT: a tool for planning with uncertainty and partial information. In: Proceedings of the IJCAI Workshop on Planning with Uncertainty and Incomplete Information, pp. 82–87 (2001)
4. Gros, T.P., Hermanns, H., Hoffmann, J., Klauck, M., Steinmetz, M.: Deep statistical model checking. In: Gotsman, A., Sokolova, A. (eds.) FORTE 2020. LNCS, vol. 12136, pp. 96–114. Springer, Cham (2020). https://doi.org/10.1007/978-3-030-50086-3_6
5. Gros, T.P., Höller, D., Hoffmann, J., Wolf, V.: Tracking the race between deep reinforcement learning and imitation learning – extended version. arXiv preprint arXiv:2008.00766 (2020)
6. Judah, K., Fern, A.P., Dietterich, T.G., Tadepalli, P.: Active imitation learning: formal and practical reductions to I.I.D. learning. J. Mach. Learn. Res. **15**(120), 4105–4143 (2014)
7. Ketkar, N.: Introduction to PyTorch. In: Ketkar, N. (ed.) Deep Learning with Python, pp. 195–208. Apress, Berkeley (2017). https://doi.org/10.1007/978-1-4842-2766-4_12
8. Mnih, V., Kavukcuoglu, K., Silver, D., Graves, A., Antonoglou, I., Wierstra, D., Riedmiller, M.: Playing atari with deep reinforcement learning. arXiv preprint arXiv:1312.5602 (2013)
9. Mnih, V., et al.: Human-level control through deep reinforcement learning. Nature **518**(7540), 529–533 (2015)
10. Pineda, L.E., Zilberstein, S.: Planning under uncertainty using reduced models: revisiting determinization. In: Proceedings of the 24th International Conference on Automated Planning and Scheduling (ICAPS), pp. 217–225. AAAI Press (2014)
11. Ross, S., Gordon, G.J., Bagnell, D.: A reduction of imitation learning and structured prediction to no-regret online learning. In: Proceedings of the 14th International Conference on Artificial Intelligence and Statistics (AISTATS). JMLR Proceedings, vol. 15, pp. 627–635. JMLR.org (2011)
12. Schaal, S.: Is imitation learning the route to humanoid robots? Trends Cogn. Sci. **3**(6), 233–242 (1999)
13. Silver, D., et al.: Mastering the game of Go with deep neural networks and tree search. Nature **529**, 484–503 (2016)
14. Silver, D., et al.: A general reinforcement learning algorithm that masters Chess, Shogi, and Go through self-play. Science **362**(6419), 1140–1144 (2018)
15. Silver, D., et al.: Mastering the game of Go without human knowledge. Nature **550**, 354–359 (2017)
16. Sutton, R.S., Barto, A.G.: Reinforcement Learning: An Introduction. Adaptive Computation and Machine Learning, 2nd edn. The MIT Press, Cambridge (2018)

# SafePILCO: A Software Tool for Safe and Data-Efficient Policy Synthesis

Kyriakos Polymenakos$^{(\boxtimes)}$, Nikitas Rontsis, Alessandro Abate, and Stephen Roberts

University of Oxford, Oxford, UK
kpol@robots.ox.ac.uk

**Abstract.** SafePILCO is a software tool for safe and data-efficient policy search with reinforcement learning. It extends the known PILCO algorithm, originally written in MATLAB, to support safe learning. SafePILCO is a Python implementation and leverages existing libraries that allow the codebase to remain short and modular, towards wider use by the verification, reinforcement learning, and control communities.

## 1 Introduction

**Goals and Design Philosophy.** Reinforcement learning (RL) is a well-known, widely-used framework that has recently enjoyed breakthroughs using model-free methods based on deep neural networks [12,14,19]. Notable shortcomings of model-free deep RL algorithms are their need for extensive training datasets, the lack of interpretability, and the difficulty to verify their outcomes. It is data-efficient, which makes it appealing for applications involving physical systems. PILCO [11] (Probabilistic Inference for Learning COntrol) represents a state-of-the-art model-based RL method that relies on Gaussian processes (thus, not on deep neural networks). PILCO does not incorporate safety constraints and comes as a MATLAB implementation. SafePILCO, based on [21], extends the original algorithm with safety constraints embedded in the training procedure and as learning goals, and comes as a concise and clean Python implementation[1].

SafePILCO is underpinned by an object-oriented architecture, enabling code re-use by keeping the implementation short and modular, with the capability to flexibly replace individual components. It takes advantage of available open source libraries, both as building blocks of the core algorithm, and as predefined tasks for evaluating the performance of the algorithm. It uses standard libraries to implement specific sub-tasks and to facilitate extensions, e.g. the GPflow library gives access to an array of models with a consistent interface. Additionally, by using standard scenarios for experimental evaluation, SafePILCO enables users to employ it as a benchmark to easily compare their own methods against.

---

[1] An extended version of this paper is at: http://arxiv.org/abs/2008.03273.

---

**Electronic supplementary material** The online version of this chapter (https://doi.org/10.1007/978-3-030-59854-9_3) contains supplementary material, which is available to authorized users.

© Springer Nature Switzerland AG 2020
M. Gribaudo et al. (Eds.): QEST 2020, LNCS 12289, pp. 18–26, 2020.
https://doi.org/10.1007/978-3-030-59854-9_3

**Related Work.** The original PILCO algorithm [7,11] is a policy search framework [9], employing Gaussian processes (GPs) [22] to learn the model dynamics and to maximise data efficiency. In [10], constraints are incorporated as negative rewards, discouraging the system from visiting parts of the state space. However, these rewards have to be hand-tuned to balance performance and safety. SafePILCO implements a procedure in [21], introduced to synthesise policies satisfying spatial constraints, while retaining safety during training.

Bayesian optimisation has also been used to train policy parameters [2,13,23] towards data efficiency and safety, mapping parameters to the loss/reward directly instead of modelling the system dynamics. Other model-based RL approaches have been proposed recently, describing the system dynamics through probabilistic models based on ensembles of deep neural networks [6,26], or GPs [5,24]. These methods have not been used in combination with safety requirements encoded as spatial constraints. Other GP-based methods either focus on stability [25] or take an approach [15] that provides more conservative guarantees, but which restricts scalability by increasing computational demands.

## 2  Description of the Software Tool

SafePILCO comes as an open source Python package[2]. To make reproduction of the experiments easier we provide additional functionalities (such as logging, post processing results and creating the plots in the paper) in a separate repository[3]. In a standard object-oriented fashion, the main components of the algorithm are organised as objects, following a hierarchy of classes. The main components are:

– the Gaussian process model, providing short-term and long-term predictions;
– the controller, selecting an action based on the state at each time step;
– the parametric reward function, which captures the performance of the algorithm and is also tasked with enforcing safe behaviour;
– scenarios that capture environment dynamics specific to a case study.

Firstly, the environment that the agent interacts with needs to be specified. SafePILCO is designed to seamlessly interface with any environment following the OpenAI gym API. Therefore, gym environments can be directly invoked, as well as user-defined environments equipped with the necessary functionalities.

The PILCO class is the central object of the package, encapsulating the GP model, the controller, and the reward function as attributes. PILCO employs the model and the controller to predict a trajectory, calls the reward function to evaluate it, and uses the gradients calculated through automatic differentiation, to improve the controller parameters. The SafePILCO subclass combines the common additive reward function component used for performance, with a multiplicative component that encodes the risk of violating the safety requirement over any time step of the episode (this is used to enforce safety during training) (Fig. 1).

---

[2] Main package repository: https://github.com/nrontsis/PILCO.

[3] Experiments and figures reproduction repository: https://github.com/kyr-pol/ SafePILCO_Tool-Reproducibility.

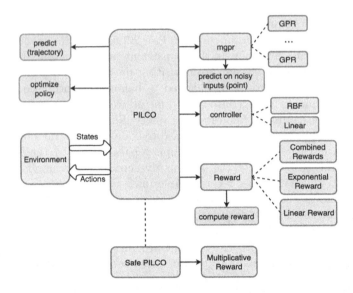

**Fig. 1.** The basic structure of the SafePILCO implementation. Black arrows correspond to object-attribute relationship, dashed lines to inheritance, and wide arrows to data flow. Classes are represented by blue boxes and key functions by green boxes. (Color figure online)

The mgpr class implements the multi-input, multi-output Gaussian process regression that underpins the dynamical model. Specifically, mgpr combines several, multi-input/single-output GP models. These GP models are provided by GPflow, along with standard GP functionalities. Our code provides GP predictions for multiple output dimensions, when the inputs are multi-dimensional and noisy. The mgpr class also allows for priors on the GP hyperparameters.

The policy (or controller) defines how the agent selects appropriate actions at each time step. Policies are implemented as memory-less, deterministic feedback controllers: the control input is directly dependent on the environment state that the agent observes at the current time step. The agent implements a policy $\pi$ of the form $u = \pi^\theta(x)$, where $\theta$ are the policy parameters. The package provides the controller class with two subclasses, one for *linear* controllers, and one for controllers based on *radial basis functions* (RBF). The only extra requirement from the controller is the ability to calculate, for a Gaussian-distributed state (including the predicted states during the planning phase), a similarly Gaussian-distributed control input, so that the state and input are *jointly* Gaussian. The policies are parametric and optimising the values of these parameters $\theta$ constitutes the overall policy search objective.

The final part concerns the specification of the reward function. We note that this is different from some RL literature: in SafePILCO, much like for the original PILCO [11] algorithm, the reward function is known analytically *a-priori*: this is necessary for the GP model to estimate the reward of a proposed policy, without

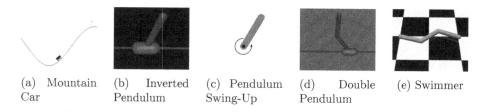

(a) Mountain Car    (b) Inverted Pendulum    (c) Pendulum Swing-Up    (d) Double Pendulum    (e) Swimmer

**Fig. 2.** Snapshots of the Open-AI gym environments used in the case studies.

interacting with the environment. The `reward` class implements the standard reward, while also encoding an adaptively weighted penalty that encourages constraint satisfaction. The class provides, at any given state, scalar outputs that capture the expected reward and the constraint violation probability. A composite reward function is used to train the policy that combines these two components. We further note that it is the choice of reward function, along with the environment, that defines a task: indeed, we can design multiple tasks with a shared environment by varying the reward function. This setup is therefore suitable for multi-task learning [8], as the same model is valid for multiple tasks.

**Libraries.** The tool relies on other Python packages, allowing us to leverage their optimised functionalities and to keep the codebase succinct. Furthermore, this allows users to easily apply our algorithm to new tasks. We use Tensorflow [1] to obtain automatic gradient computations (often referred to as auto-diff), which simplifies the policy improvement step[4]. GPflow [18] is a Python package for Gaussian Process modelling built on Tensorflow. GPflow provides a full set of GP functionalities, and gives access to many specialised models. Having a Tensorflow back-end, gradients in all the GPflow models are also calculated automatically. Additionally, GPflow allows the user to readily define priors and to employ different optimisers or alternative implementations of sparse approximations for GPs. Finally, our implementation is interfaced with the Open-AI Gym [3], a suite of RL tasks widely used in the community. Gym tasks have consistent interfaces and detailed visualisation capabilities for a wide range of tasks varying over different sorts of complexity: dimensionality, smoothness of dynamics, length of episodes, and so on. Users can prototype their algorithms using easier tasks and move to more complex, time-consuming experiments, as the project matures.

## 3   Case Studies

To evaluate the performance of the package we run a set of experiments on different tasks. Visualisations of the environments used for the case studies are

---

[4] By way of comparison, all gradient calculations in the PILCO Matlab implementation are hand-coded, thus extensions are laborious as any additional user-defined controller or reward function has to include these gradient calculations too.

shown in Fig. 2. Details of the OpenAI gym [3] tasks used are on the gym website (https://gym.openai.com/). Experiments are presented in order of increasing complexity. As mentioned in Sect. 2, SafePILCO assumes a predetermined, closed-form reward function. Most of the tasks we apply our method on come with reward functions that are not in analytical form, as assumed. Thus we make the following distinction: the algorithm is evaluated on the original reward function coming with the environments, but is trained with a closed-from reward function of our design. Designing or "shaping" a reward function that results in a desired behaviour (in our case, a behaviour that maximises accumulated return measured with a different reward function), is broadly studied and shown to be in general challenging [17,20], however in our experience and for the environments used, it did not require extensive hyperparameter searches.

### 3.1  Plain PILCO

We give specific information for each environment and the associated reward functions. The *Mountain Car* experiment uses the `MountainCarContinuous-v0` gym environment. Small negative rewards are given at those states where the car is not at the top of the hill (goal state), whereas a large reward is given exclusively as the agent gets to the top. The goal state is the terminal state for the environment and no further reward is obtained. This is captured with a negative exponential reward, centered at the goal state.

For the *Inverted Pendulum*, the OpenAI gym [3] `InvertedPendulum-v2` environment is used. It is a variant of the cart-pole stabilisation task, where a pendulum is attached to a cart on a rail, and the controller applies a force to the cart. The pendulum starts close to the upright position and the controller stabilises it by moving the cart to the left or right on the rail. For this task, and for the double pendulum task below, the reward function provides a +1 reward when the pendulum angle is less than some given threshold, whereas for training an exponential reward centered at the upright position is used. Once out of this area the episode terminates, as the controller cannot exert a stabilising input.

For the *Pendulum Swing-Up* task, we modify part of the default behaviour of the gym `Pendulum-v0` environment: the initial starting state distribution of the pendulum is too wide for unimodal planning. We restrict the initial distribution to the pendulum starting close to the downward position. The environment penalises the agent with negative rewards correlated to the distance from the goal position, where the pendulum is upright. An exponential reward is again used.

The inverted *Double Pendulum* task uses `InvertedDoublePendulum-v2`. It is similar to `InvertedPendulum-v2`, except the pendulum now consists of two links. We apply force to the cart and have to stabilise the system to the upright position. We add a wrapper to the default environment that only slightly changes its interface, replacing a state variable that is an angle with its sine and cosine values. This corresponds to an existing functionality from PILCO [7,11].

In the *Swimmer* setup, from the `Swimmer-v2` OpenAI gym, a robot with two joints navigates a 2-d plane by "swimming" in a viscous fluid. Its joints are

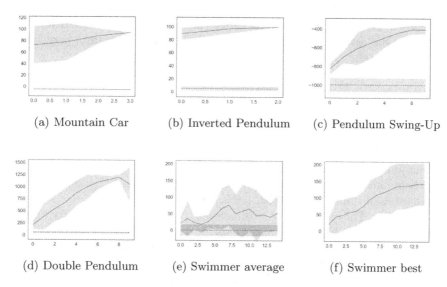

(a) Mountain Car      (b) Inverted Pendulum      (c) Pendulum Swing-Up

(d) Double Pendulum      (e) Swimmer average      (f) Swimmer best

**Fig. 3.** Results for different OpenAI gym tasks. Episode rewards on the y-axis and iteration number on the x-axis (blue, mean and two standard deviations around it). The performance of a random policy (red dashed line) is shown for comparison. For the *Swimmer* setup we report both the average performance of 10 random seeds at each iteration, and the best performance from all previous iterations of each random seed. (Color figure online)

controlled by an actuator, and the system is rewarded for moving in the direction of the x-axis. This is a more challenging task, with an 8D state space, 2D control space, and nonlinear dynamics. Furthermore, it requires coordination between the two controllers for the robot to start moving towards the right direction: this makes the acquisition of a reward signal at the early stages of training hard [16]. The rewards are given for distance travelled in the positive direction of the x-axis, based on the position of the root link. This position variable however is not one of the observed state variables and the task does not have a specific goal position. Thus we define a reward linear to the x-axis velocity of the agent, and we lightly penalise extreme angles at the joints, which leads to a smoother gait and improved generalisation outside of the planning horizon.

## 3.2 SafePILCO

In this section we add constraints over the state space of the environments. The experimental structure is similar to that in Sect. 3.1, however we report metrics differently (cf. Table 1): we list the number of constraint violations incurred during training (Const. viol.) and the best performance achieved in episodes where the system has respected the constraints. We also report the average number of episodes when the safety check has prohibited interaction with the system (Stopped epis.) and the maximum number of episodes the algorithm is allowed to interact with the system (Max epis.). The risk threshold for each case study (the estimated probability of violating the constraints over which interaction with the system is blocked) is reported as $\epsilon$.

The *Linear Cars* scenario is similar to the one in [21], where two cars are approaching a junction, and the algorithm controls one of them by either braking or accelerating. The goal is for the controlled car to cross the junction as soon as possible, without causing a collision. The state space is 4-dimensional, with

**Table 1.** SafePILCO results on constrained environments.

|  | LinearCars | BAS | SafeSwimmer |
|---|---|---|---|
| Const. Viol. | $0.0 \pm 0.0$ | $0.0 \pm 0.0$ | $0.4 \pm 0.49$ |
| Best return | $-10.7 \pm 2.7$ | $1.2 \pm 0.9$ | $11.6 \pm 8.2$ |
| Max Epis. | 8 | 4 | 12 |
| Stopped Epis. | $1.4 \pm 1.5$ | $0 \pm 0$ | $1.1 \pm 1.3$ |
| $\epsilon$ | 0.05 | 0.05 | 0.2 |

linear dynamics. The input $u$ has one dimension, proportional to the force applied to the first car. To avoid a collision, the cars must not be simultaneously adjacent to the junction (set at the origin $(0,0)$). This can be formulated as a constraint $|x^1| > a$ OR $|x^3| > a$ over the position of the two cars. We want to encourage the first car to cross the junction as soon as possible: a simple reward could be $-1$ for every time-step where first car has not crossed the junction, and $+1$ otherwise. However, this is discontinuous and cannot be used by SafePILCO directly, so we use instead a linear reward, proportional to the position of the first car.

In the *Building Automation Systems* (BAS) we consider a problem in the domain of automation systems for buildings. The environment comes from [4] (Case Study 2), which has developed a simulator[5] based on real measurement data. The task is to control the temperature in two adjacent rooms from a common heated air supply. The original cost function is the quadratic error between the temperatures in each of the two rooms and corresponding reference temperatures. For SafePILCO we use the standard exponential reward function.

The *Safe Swimmer* case study is based on the Swimmer-v2 environment, but we add the following constraints: we require that the angles at the two joints remain below a certain threshold ($95°$). Constraints of this sort are common in robotics, since pushing the joints to the edge of their functional ranges can lead to accumulated damage to the joints, the motors, or the robot links.

### 3.3 Results

The results in Fig. 3 (for Plain PILCO)and Table 1 (for SafePILCO) are averaged over 10 random seeds. To evaluate the controller at each iteration more accurately, for each random seed, we report the mean of 5 runs. The results show that SafePILCO is flexible enough to tackle a wide selection of RL problems, with good performance and data efficiency, and high rate of constraint compliance when constraints are imposed. For reference, in [27] model-based methods are evaluated on gym tasks, with a low-data (200k points) and a high-data ($2 \times 10^6$ points) regime, while for the Swimmer SafePILCO uses only 625 points from $\sim$3000 interactions steps.

---

[5] Code for the BAS simulator: https://gitlab.com/natchi92/BASBenchmarks.

# References

1. Abadi, M., et al.: TensorFlow: Large-scale machine learning on heterogeneous systems (2015). https://www.tensorflow.org/. software available from tensorflow.org
2. Berkenkamp, F., Krause, A., Schoellig, A.P.: Bayesian optimization with safety constraints: safe and automatic parameter tuning in robotics. CoRR abs/1602.04450 (2016). http://arxiv.org/abs/1602.04450
3. Brockman, G., et al.: OpenAI Gym. arXiv preprint arXiv:1606.01540 (2016)
4. Cauchi, N., Abate, A.: Benchmarks for cyber-physical systems: a modular model library for building automation systems. In: Proceedings of ADHS, pp. 49–54 (2018)
5. Chatzilygeroudis, K., Rama, R., Kaushik, R., Goepp, D., Vassiliades, V., Mouret, J.B.: Black-box data-efficient policy search for robotics. In: 2017 IEEE/RSJ International Conference on Intelligent Robots and Systems (IROS), pp. 51–58. IEEE (2017)
6. Chua, K., Calandra, R., McAllister, R., Levine, S.: Deep reinforcement learning in a handful of trials using probabilistic dynamics models. In: Advances in Neural Information Processing Systems, pp. 4754–4765 (2018)
7. Deisenroth, M.P.: Efficient reinforcement learning using Gaussian processes. Ph.D. thesis, Karlsruhe Institute of Technology (2010)
8. Deisenroth, M.P., Englert, P., Peters, J., Fox, D.: Multi-task policy search for robotics. In: 2014 IEEE International Conference on Robotics and Automation (ICRA), pp. 3876–3881. IEEE (2014)
9. Deisenroth, M.P., Neumann, G., Peters, J., et al.: A survey on policy search for robotics. Found. Trends® Robot. 2(1–2), 1–142 (2013)
10. Deisenroth, M.P., Rasmussen, C.E., Fox, D.: Learning to control a low-cost manipulator using data-efficient reinforcement learning. In: Robotics: Science and Systems (2011)
11. Deisenroth, M.P., Rasmussen, C.E.: PILCO: a model-based and data-efficient approach to policy search. In: In Proceedings of the International Conference on Machine Learning (2011)
12. Duan, Y., Chen, X., Houthooft, R., Schulman, J., Abbeel, P.: Benchmarking deep reinforcement learning for continuous control. In: International Conference on Machine Learning (ICML), pp. 1329–1338 (2016)
13. Duivenvoorden, R.R., Berkenkamp, F., Carion, N., Krause, A., Schoellig, A.P.: Constrained Bayesian optimization with particle swarms for safe adaptive controller tuning. In: Proceedings of the IFAC (International Federation of Automatic Control) World Congress, pp. 12306–12313 (2017)
14. Haarnoja, T., Zhou, A., Abbeel, P., Levine, S.: Soft actor-critic: off-policy maximum entropy deep reinforcement learning with a stochastic actor. arXiv preprint arXiv:1801.01290 (2018)
15. Koller, T., Berkenkamp, F., Turchetta, M., Krause, A.: Learning-based model predictive control for safe exploration. In: 2018 IEEE Conference on Decision and Control (CDC), pp. 6059–6066. IEEE (2018)
16. Levine, S., Finn, C., Darrell, T., Abbeel, P.: End-to-end training of deep visuomotor policies. J. Mach. Learn. Res. 17(1), 1334–1373 (2016)
17. Mataric, M.J.: Reward functions for accelerated learning. In: Machine Learning Proceedings 1994, pp. 181–189. Elsevier (1994)
18. Matthews, A.G.d.G., et al.: GPflow: a Gaussian process library using TensorFlow. J. Mach. Learn. Res. 18(40), 1–6 (2017). http://jmlr.org/papers/v18/16-537.html

19. Mnih, V., et al.: Human-level control through deep reinforcement learning. Nature **518**(7540), 529–533 (2015)
20. Ng, A.Y., Jordan, M.I.: Shaping and policy search in reinforcement learning. Ph.D. thesis, University of California, Berkeley Berkeley (2003)
21. Polymenakos, K., Abate, A., Roberts, S.: Safe policy search using Gaussian process models. In: Proceedings of the 18th International Conference on Autonomous Agents and MultiAgent Systems, pp. 1565–1573. International Foundation for Autonomous Agents and Multiagent Systems (2019)
22. Rasmussen, C.E., Williams, C.K.I.: Gaussian Processes for Machine Learning. MIT Press, Cambridge (2006)
23. Sui, Y., Gotovos, A., Burdick, J., Krause, A.: Safe exploration for optimization with Gaussian processes. In: Proceedings of The 32nd International Conference on Machine Learning, pp. 997–1005 (2015)
24. Vinogradska, J., Bischoff, B., Achterhold, J., Koller, T., Peters, J.: Numerical quadrature for probabilistic policy search. IEEE Trans. Pattern Anal. Mach. Intell. **42**, 164–175 (2018)
25. Vinogradska, J., Bischoff, B., Nguyen-Tuong, D., Romer, A., Schmidt, H., Peters, J.: Stability of controllers for gaussian process forward models. In: Proceedings of The 33rd International Conference on Machine Learning, pp. 545–554 (2016)
26. Vuong, T.L., Tran, K.: Uncertainty-aware model-based policy optimization. arXiv preprint arXiv:1906.10717 (2019)
27. Wang, T., et al.: Benchmarking model-based reinforcement learning (2019)

# StochNetV2: A Tool for Automated Deep Abstractions for Stochastic Reaction Networks

Denis Repin, Nhat-Huy Phung, and Tatjana Petrov$^{(\boxtimes)}$

Department of Computer and Information Sciences, University of Konstanz,
Konstanz, Germany
den.ne.repin@gmail.com, tatjana.petrov@uni-konstanz.de

**Abstract.** We present a toolbox for stochastic simulations with CRN models and their (automated) deep abstractions: a mixture density deep neural network trained on time-series data produced by the CRN. The optimal neural network architecture is learnt along with learning the transition kernel of the abstract process. Automated search of the architecture makes the method applicable directly to any given CRN, which is time-saving for deep learning experts and crucial for non-specialists. The tool was primarily designed to efficiently reproduce simulation traces of given complex stochastic reaction networks arising in systems biology research, possibly with multi-modal emergent phenotypes. It is at the same time applicable to any other application domain, where time-series measurements of a Markovian stochastic process are available by experiment or synthesised with simulation (e.g. are obtained from a rule-based description of the CRN).

## 1 Introduction

Predicting stochastic cellular dynamics as emerging from the mechanistic models of molecular interactions is a long-standing challenge in systems biology: low-level chemical reaction network (CRN) models give rise to a highly-dimensional continuous-time Markov chain (CTMC) which is computationally demanding and often prohibitive to analyse in practice. Deep abstractions of CRN models, proposed in [2], use deep learning to replace this CTMC with a discrete-time continuous state-space process, by training a mixture density deep neural network with traces sampled at regular time intervals (which can be obtained either by simulating a given CRN or as time-series data from experiment). Deep abstractions are dramatically cheaper to execute, while preserving the

TP's research is supported by the Ministry of Science, Research and the Arts of the state of Baden-Württemberg, and the DFG Centre of Excellence 2117 'Centre for the Advanced Study of Collective Behaviour' (ID: 422037984), DR's research is supported by Young Scholar Fund (YSF), project no. $P83943018FP430\_/18$ and by the 'Centre for the Advanced Study of Collective Behaviour'. The authors would like to thank to Luca Bortolussi for inspiring discussions on the topic.

© Springer Nature Switzerland AG 2020
M. Gribaudo et al. (Eds.): QEST 2020, LNCS 12289, pp. 27–32, 2020.
https://doi.org/10.1007/978-3-030-59854-9_4

statistical features of the training data. The abstraction accuracy improves with the amount of training data. However, the overall quality of the method will also depend on the choice of neural network architecture. In practice, the modeller has to find the suitable architecture manually, through a trial-and-error cycle. In [8], we proposed to learn the optimal neural network architecture along with learning the transition kernel of the abstract process [3,7]. A similar idea has been recently employed for emulating epidemiological spread [4]; However, this work has focused on a single, uni-modal model of epidemics and only stationary regime, while our method is generic - applicable to any given CRN.

In this paper, we present StochNetV2Toolbox[1]- a tool for MDN-based deep abstractions of CRNs. Deep abstractions provide time-series trajectories which abstract the trajectories of the original CRN. Abstract models are implemented with neural networks, which predict a distribution for sampling the next system state. Moreover, StochNetV2Toolbox allows to, in addition to the initial state, parametrise the neural network with the kinetic rates *(as a part of the input)*. The method is described in [8]. For illustration purposes, the tool includes a functionality for simulating multiple CRN instances on a spatial grid, where CRNs communicate via a subset of shared species which are diffused across neighbouring grid nodes.

## 2   Tool Architecture and Functionality

StochNetV2is implemented with four entities: `CRN_model`, `Dataset`, `StochNet`, and `Trainer` (see Fig. 1 for an overview). Two latter classes each have two different implementations: (i) a static implementation, used for standard deep abstractions as suggested in [2], and (ii) a dynamic implementation, used for *automated* deep abstractions, where the architecture of the neural network is learn along with the kernel of the process [8].

The general workflow proceeds in the following steps: (1) define a `CRN_model`, (2) produce trajectories, (3) create dataset from trajectories, (4) configure `StochNet`, (5) train it with `Trainer`, (6) produce trajectories. Finally, the user has the option to simulate multiple CRN instances on a spatial grid with class `Grid runner`.

### 2.1   CRN Models

The module contains base and example classes defining CRN models. These models can be simulated with Gillespie algorithm provided by `gillespy2` package. CRN models are used as a source of synthetic data to train and evaluate abstract models. An instance of `CRN_model` class can

– generate randomized initial concentrations (populations),
– generate randomized reaction rates,

---

[1] The tool name makes it transparent that the tool was inspired by [2] called 'StochNet'.

– set initial concentrations and reaction rates,
– produce trajectories.

A new CRN model should be inherited from `BaseCRNModel` class and implement all abstract methods. Several example models are provided (e.g. `SIR`, `Bees`, `Gene`, `X16`). In general, SBML models (Systems Biology Markup Language) can be imported, but it should be noted that the variability of the SBML format makes automated imports practically tedious, and for most models some pre-processing is required, e.g. editing reaction rates formulas, rewriting reversible reactions as two separate reactions, etc. see `BaseSBMLModel` and `EGFR` classes for examples.

## 2.2 Dataset

The `dataset` module implements functions and classes for creation and operations over trajectories data. It supports shuffling and applying pre-processing functions (such as adding noise) on-the-fly.

## 2.3 StochNet (Static)

`StochNet` class implements an interface for an abstract model. It is wrapped around a neural network (Mixture Density Network) which can be trained on simulation datasets and then used to produce trajectories. MDN consists of two parts: body (neural network extracting features of input state) and mixture (probability distribution with parameters depending on the extracted features). StochNet is initialized with body and mixture configuration files (config-file examples in `stochnet_v2/examples/configs`). A set of pre-defined building blocks for the body-part can be found in `stochnet_v2/static_classes/nn_bodies` file, which provides flexibility in the sense that custom building blocks can be added by the user. The supported distributions (the 'components' we use) for the mixture part can be found in `stochnet_v2/static_classes/top_layers`.

## 2.4 Training (Static)

Once `StochNet` is initialized, it can be trained with `Trainer`. When training is finished, all necessary files are saved to the model folder. A saved model can be loaded to produce trajectories at any time.

## 2.5 NASStochNet (Dynamic)

`NASStochnet` is an extension of the StochNet class. Instead of designing the body-part of MDN, it takes only a few hyper-parameters, such as an overall *depth* (number of layers) and *width* (number of neurons in layers). It starts with an over-parameterized probabilistic meta-model, which (by sampling so-called architecture parameters) represents many architectures at once. During training, the set of preferred layers, their order, and inter-connections are optimized automatically for given data.

## 2.6    Training (Architecture Search)

For the Architecture Search, training consists of two stages: (I) search for optimal configuration. This stage is a two-level optimisation, i.e. we run two separate optimisation procedures in altering manner for several epochs each: (main) update network parameters - weights in layers, (arch) update architecture parameters - weights of candidate operations in mix-layer, (II) fine-tuning of the found architecture after all redundancies are pruned.

After the search and fine-tuning stages, all necessary files are saved to the model folder, and the model can be loaded for simulations. Either `StochNet` or `NASStochNet` can be used to load trained model and run simulations.

## 2.7    Grid Runner

`GridRunner` implements a simulation of multiple CRN instances on a (spatial) grid with communication via spreading a subset of species across neighboring grid nodes. `GridRunner` is initialized with a model and `GridSpec`, which specifies a grid. Then, `GridRunner` stores state values for every model instance which can be updated by either in-node (one forward step of the model in every node) reactions, or on-grid interactions (diffusion of shared species across the grid).

## 2.8    Luigi Workflow Manager

The workflow is wrapped with the `luigi` library designed for running complex pipelines of inter-dependent tasks. Alternatively to manually run the above commands, one can fill a luigi configuration file, and it will run the whole sequence of tasks taking care of the right order and pre-requisites for every task.

# 3    Implementation

StochNetV2is written in Python 3 and uses Python libraries, mainly `tensorflow`, `gillespy2`, `luigi` and dependencies thereof. Source code of the tool with previously published Jupyter notebooks can be downloaded from GitHub - https://github.com/dennerepin/StochNetV2.

# 4    Evaluation and Applications

To evaluate the quality of abstract models, we compare distributions (histograms) of species of interest (e.g. Fig. 2a). For this, we simulate many trajectories of the original model starting from a set of random initial settings. Then an evaluation script runs from the same initial settings (example runtime comparison given in Fig. 2b). The evaluation script saves: (1) overall average value of histogram distance, (2) plots of species histograms after different number of steps, (3) plots of average (over different settings) distance between histograms produced by original and abstract model after different number of time-steps.

The case studies in the toolbox include applications in systems biology and collective behavior, such as the model of signaling pathway (EGFR [5], challenging multi-modal gene regulatory models (e.g. from [9]), and a reaction-based model of collective defence in honeybees (see [6] and GitHub page for details - https://github.com/dennerepin/StochNetV2²).

## 5   Related Tools

The original idea of using Mixture Density Networks was proposed in [2], which is followed by a theoretical work [1]. We are not aware of other tools using deep learning to abstract stochastic CRNs.

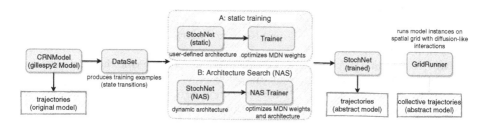

**Fig. 1.** Main components and workflow.

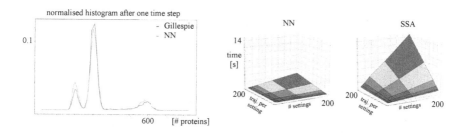

**Fig. 2.** (left) X40 case study from [9]: histograms of protein $P_2$ concentration after 1 time step and (right) the comparison of simulation run-time with NN and SSA, wrt. the number of initial settings and trajectories per setting.

## References

1. Bortolussi, L., Cairoli, F.: Bayesian abstraction of Markov population models. In: Parker, D., Wolf, V. (eds.) QEST 2019. LNCS, vol. 11785, pp. 259–276. Springer, Cham (2019). https://doi.org/10.1007/978-3-030-30281-8_15

---

² While we performed specific performance evaluation, e.g. in Fig. 2 and [8], a systematic scalability analysis is beyond the scope of this tool presentation.

2. Bortolussi, L., Palmieri, L.: Deep abstractions of chemical reaction networks. In: Češka, M., Šafránek, D. (eds.) CMSB 2018. LNCS, vol. 11095, pp. 21–38. Springer, Cham (2018). https://doi.org/10.1007/978-3-319-99429-1_2
3. Cai, H., Zhu, L., Han, S.: ProxylessNAS: direct neural architecture search on target task and hardware. CoRR abs/1812.00332 (2018). http://arxiv.org/abs/1812.00332
4. Davis, C.N., Hollingsworth, T.D., Caudron, Q., Irvine, M.A.: The use of mixture density networks in the emulation of complex epidemiological individual-based models. PLoS Comput. Biol. **16**(3), 1–16 (2020). https://doi.org/10.1371/journal.pcbi.1006869
5. Feret, J., Henzinger, T., Koeppl, H., Petrov, T.: Lumpability abstractions of rule-based systems. Theoret. Comput. Sci. **431**, 137–164 (2012)
6. Hajnal, M., Nouvian, M., Šafránek, D., Petrov, T.: Data-informed parameter synthesis for population Markov chains. In: Češka, M., Paoletti, N. (eds.) HSB 2019. LNCS, vol. 11705, pp. 147–164. Springer, Cham (2019). https://doi.org/10.1007/978-3-030-28042-0_10
7. Liu, H., Simonyan, K., Yang, Y.: DARTS: differentiable architecture search. In: International Conference on Learning Representations (2019). https://openreview.net/forum?id=S1eYHoC5FX
8. Petrov, T., Repin, D.: Automated deep abstractions for stochastic chemical reaction networks. arXiv preprint arXiv:2002.01889 (2020)
9. Plesa, T., Erban, R., Othmer, H.G.: Noise-induced mixing and multimodality in reaction networks. Eur. J. Appl. Math. **30**(5), 887–911 (2019)

# Model Checking and Verification

# Alternative Characterizations
# of Probabilistic Trace Equivalences
# on Coherent Resolutions
# of Nondeterminism

Marco Bernardo$^{(\boxtimes)}$

Dipartimento di Scienze Pure e Applicate, Università di Urbino, Urbino, Italy
marco.bernardo@uniurb.it

**Abstract.** For nondeterministic and probabilistic processes, the validity of some desirable properties of probabilistic trace semantics depends both on the class of schedulers used to resolve nondeterminism and on the capability of suitably limiting the power of the considered schedulers. Inclusion of probabilistic bisimilarity, compositionality with respect to typical process operators, and backward compatibility with trace semantics over fully nondeterministic or fully probabilistic processes, can all be achieved by restricting to coherent resolutions of nondeterminism. Here we provide alternative characterizations of probabilistic trace post-equivalence and pre-equivalence in the case of coherent resolutions. The characterization of the former is based on fully coherent trace distributions, whereas the characterization of the latter relies on coherent weighted trace sets.

## 1 Introduction

Quantitative models of computer, communication, and software systems combine, among others, functional and extra-functional aspects of system behavior. On the one hand, these models describe system activities and their execution order, possibly admitting nondeterminism in case of concurrency phenomena or to support implementation freedom. On the other hand, they include some information about the probabilities or the timing of activities and events in which the system is involved.

In the probabilistic setting, a particularly expressive model is given by *probabilistic automata* [24], because they encompass as special cases fully nondeterministic models like labeled transition systems [21], fully probabilistic models like action-labeled variants of discrete-time Markov chains [22], and reactive probabilistic models like Markov decision processes [13]. In a probabilistic automaton, the choice among the transitions departing from the current state is nondeterministic and can be influenced by the external environment, while the choice of the next state reached by the selected transition is probabilistic and made internally by the process.

Behavioral relations [1,4,16,20,28] play a fundamental role in the analysis of probabilistic models. They formalize observational mechanisms that permit

© Springer Nature Switzerland AG 2020
M. Gribaudo et al. (Eds.): QEST 2020, LNCS 12289, pp. 35–53, 2020.
https://doi.org/10.1007/978-3-030-59854-9_5

relating models that, despite their different representations in the same mathematical domain, cannot be distinguished by external entities when abstracting from certain internal details. Moreover, they support system modeling and verification by providing a means to relate system descriptions expressed at different levels of abstraction, as well as to reduce the size of a system representation while preserving specific properties to be assessed later.

In this paper, we focus on *trace semantics* for nondeterministic and probabilistic processes represented through a variant of simple probabilistic automata. A trace is a sequence of activities labeling a sequence of transitions performed by a process, thus abstracting from branching points in the process behavior. Several execution probabilities may be associated with the same trace, each corresponding to a different resolution of nondeterminism. The discriminating power of probabilistic trace semantics thus depends on the class of schedulers used to resolve nondeterminism, but in general it turns out to be excessive. This may hamper the achievement of a number of desirable properties.

The problem with almighty schedulers yielding a demonic view of nondeterminism is well known, both for trace semantics and for testing semantics. In the case of a process given by the parallel composition of several subprocesses, or in a testing scenario where a process is composed in parallel with a test, schedulers come into play after assembling the various components. As a consequence, schedulers can solve both choices local to the individual subprocesses and choices arising from their interleaving execution. This centralized approach thus gives the possibility to make decisions in one component on the basis of those made in other components, especially in the case of history-dependent schedulers [30].

To cope with the aforementioned information leakage, the idea of distributed scheduling was proposed in [10]. Given a number of modules, i.e., of variable-based versions of automata, that interact synchronously by updating all variables during every round, for each module there are several schedulers. One of them chooses the initial and updated values for the module external variables; for each atom, intended as a cluster of variables of the module, a further scheduler chooses the initial and updated values for the private and interface variables controlled by that atom. Compose-and-schedule is thus replaced by schedule-and-compose.

Distributed scheduling was then applied in [9] to the asynchronous model of switched probabilistic input/output automata. Following the terminology of [29], given a reactive interpretation to input actions and a generative interpretation to output actions, an input scheduler and an output scheduler are considered for each automaton occurring in a system. A token passing mechanism among the automata eliminates global choices by ensuring that a single automaton at a time can select a generative output action, to which the other automata can respond with reactive input actions having the same name.

Both [10] and [9] guarantee the compositionality of the probabilistic trace-distribution equivalence of [25]. This is not a congruence with respect to parallel composition under centralized scheduling; as shown in [23], the coarsest congruence contained in that linear-time equivalence turns out to be a variant of the simulation equivalence of [27], which is a branching-time equivalence. Distributed scheduling was further investigated in [15] for interleaved probabilistic input/output automata, a variant of switched ones in which an interleaving scheduler replaces the token passing mechanism. The examined problem was the attainment of the extremal probabilities of satisfying reachability properties under different classes of distributed schedulers (memoryless vs. history-dependent, deterministic vs. randomized), knowing that in the centralized case those probabilities are obtained when using memoryless deterministic schedulers [6].

Indeed, the overwhelming power of schedulers already shows up in the memoryless case, i.e., when neglecting the path followed to reach the current state. Under centralized scheduling, in [14] additional labels were used so that the same decisions are made by schedulers in distinct copies of the same state of a testing system, thus weakening the discriminating power of the probabilistic testing equivalences of [18,26,31] that, as shown in [11,19], can be characterized in terms of branching-time, simulation-like relations. An analogous weakening result under the same class of schedulers was obtained in [3] by means of a different definition of probabilistic testing equivalence, in which success probabilities are compared in a trace-by-trace fashion rather than cumulatively.

Likewise, under memoryless schedulers, a different definition of probabilistic trace equivalence allows compositionality to be recovered without resorting to distributed scheduling. In the probabilistic trace-distribution equivalence of [25], for each resolution of either process there must exist a resolution of the other process such that the two resolutions are fully matching, in the sense that, for every trace, both resolutions feature the same probability of executing that trace. We call it *probabilistic trace post-equivalence* as the quantification over traces occurs after the quantifications over resolutions. In [3] it was proposed to exchange the order of those quantifications, which avoids hardly justifiable process distinctions and regains compositionality. Given an arbitrary trace, for each resolution of either process there must exist a resolution of the other process such that both of them exhibit the same probability of executing that trace. In this case, we speak of partially matching resolutions, as a resolution of either process can be matched by different resolutions of the other process with respect to different traces. We call the resulting relation *probabilistic trace pre-equivalence*, because the quantification over traces occurs before the quantifications over resolutions.

Congruence with respect to parallel composition, which is ensured by distributed scheduling, is not the only desirable property of probabilistic trace equivalences. In addition to compositionality with respect to other typical process operators, it is necessary to address inclusion of probabilistic bisimilarity [27] as well as backward compatibility with trace equivalences over less expressive processes such as fully nondeterministic ones [8] and fully probabilistic ones [20]. As recently shown in [2], the validity of all these properties critically depends on the capability of limiting the freedom of schedulers and can be achieved if we restrict ourselves to *coherent resolutions* of nondeterminism. Similar to [14], the basic idea is that schedulers cannot select different continuations in states of a process that are equivalent to each other, so that also the states to which they correspond in any resolution of the process have equivalent continuations.

The focus of this paper is on alternative characterizations of trace semantics. In a fully nondeterministic setting, two processes are trace equivalent iff, for each trace $\alpha$, both processes can perform $\alpha$ or neither can. An immediate alternative characterization is that two trace equivalent processes possess the same trace set [8], where this set can be viewed as the language accepted by the automata underlying those processes. Likewise, two fully probabilistic processes are trace equivalent iff, for each trace $\alpha$, both processes can perform $\alpha$ with the same probability, which amounts to possessing the same set of traces each weighted with its execution probability [20], i.e., the same probabilistic language. In either case, process equivalence reduces to (possibly weighted) trace set equality.

Straightforward characterizations of that form are not possible in the case of nondeterministic and probabilistic processes, because (i) traces can have different execution probabilities in different coherent resolutions and (ii) trace semantics can be defined according to different approaches leading to probabilistic trace post-/pre-equivalences. This motivates the investigation of alternative characterizations for the two aforementioned equivalences under coherent resolutions arising from centralized, memoryless schedulers – i.e., as they were defined in [2] – which is the subject of this paper.

The construction developed in [2] to formalize the coherency constraints relies on *coherent trace distributions*, i.e., suitable families of sets of traces weighted with their execution probabilities in a given resolution. Therefore, one may expect that the coherency-based variant of the probabilistic trace-distribution equivalence of [25], i.e., probabilistic trace post-equivalence, can be characterized in terms of coherent trace distribution equality. We will show by means of an example that this is not the case. The characterization of the coherency-based variant of probabilistic trace post-equivalence relies on the equality of something stronger, which we will call *fully coherent trace distributions* and could also replace coherent trace distributions in the coherency constraints.

The coherency-based variant of the probabilistic trace pre-equivalence of [3] is less discriminating because it treats traces individually without keeping track of the resolutions in which they can be executed. We will show that it can thus be characterized through the equality of something weaker than coherent trace distributions, which we will call *coherent weighted trace sets* and is constituted by suitable sets of traces weighted with their execution probabilities. We will also illustrate by means of an example that we cannot use them to set up adequate coherency constraints. In conclusion, fully coherent trace distributions, coherent trace distributions, and coherent weighted trace sets form a hierarchy in which every layer serves a different purpose.

This paper is organized as follows. In Sect. 2 we recall simple probabilistic automata and resolutions of nondeterminism, while in Sect. 3 we recall the two probabilistic trace equivalences together with three anomalies that can be avoided by resorting to coherent resolutions. In Sect. 4 we show some properties of coherent trace distributions and coherent resolutions, which are then exploited in Sects. 5 and 6 to develop the alternative characterizations of the coherency-based variants of the two equivalences, respectively relying on the equality of fully coherent trace distributions and on the equality of coherent weighted trace sets. Finally, in Sect. 7 we provide some concluding remarks.

# 2    Nondeterministic and Probabilistic Models

We formalize systems featuring nondeterminism and probabilities through a variant of simple probabilistic automata [24], in which we do not distinguish between external and internal actions.

**Definition 1.** *A nondeterministic and probabilistic labeled transition system, NPLTS for short, is a triple $(S, A, \longrightarrow)$ where $S \neq \emptyset$ is an at most countable set of states, $A \neq \emptyset$ is a countable set of transition-labeling actions, and $\longrightarrow \subseteq S \times A \times Distr(S)$ is a transition relation, with $Distr(S)$ being the set of discrete probability distributions over $S$.* ∎

A transition $(s, a, \Delta)$ is written $s \xrightarrow{a} \Delta$. We say that $s' \in S$ is not reachable from $s$ via that $a$-transition if $\Delta(s') = 0$, otherwise we say that it is reachable with probability $p = \Delta(s')$. The reachable states form the support of the target distribution $\Delta$, i.e., $supp(\Delta) = \{s' \in S \mid \Delta(s') > 0\}$. An NPLTS can be depicted as a directed graph in which vertices represent states and action-labeled edges represent transitions, with states in the support of the same target distribution being linked by a dashed line and decorated with the respective probabilities when these are different from 1 (see the forthcoming Figs. 1, 2, 3, 4 and 5).

An NPLTS represents (i) a *fully nondeterministic* process when every transition has a target distribution with a singleton support, (ii) a *fully probabilistic* process when every state has at most one outgoing transition, or (iii) a *Markov decision process* when for each action any state has at most one outgoing transition labeled with that action implying the absence of *internal nondeterminism*.

**Definition 2.** *Let $\mathcal{L} = (S, A, \longrightarrow)$ be an NPLTS and $s, s' \in S$. We say that the finite sequence of steps:*

$$c \equiv s_0 \xrightarrow{a_1} s_1 \xrightarrow{a_2} s_2 \ldots s_{n-1} \xrightarrow{a_n} s_n$$

*is a computation of $\mathcal{L}$ of length $n \in \mathbb{N}$ from $s = s_0$ to $s' = s_n$ compatible with trace $\alpha = a_1 a_2 \ldots a_n \in A^*$, written $c \in \mathcal{CC}(s, \alpha)$, iff for each step $s_{i-1} \xrightarrow{a_i} s_i$ in $c$ there is a transition $s_{i-1} \xrightarrow{a_i} \Delta_i$ in $\mathcal{L}$ such that $s_i \in supp(\Delta_i)$, $1 \leq i \leq n$, where:*

- $\Delta_i(s_i)$ *is the execution probability of step $s_{i-1} \xrightarrow{a_i} s_i$ conditioned on the selection of transition $s_{i-1} \xrightarrow{a_i} \Delta_i$ at state $s_{i-1}$, or simply the execution probability of that step if $\mathcal{L}$ is fully probabilistic.*
- $prob(c) = \prod_{1 \leq i \leq n} \Delta_i(s_i)$ *is the execution probability of $c$ if $\mathcal{L}$ is fully probabilistic, assuming that $prob(c) = 1$ when $n = 0$.*
- *For $C \subseteq \mathcal{CC}(s, \alpha)$, we let $prob(C) = \sum_{c \in C} prob(c)$ if $\mathcal{L}$ is fully probabilistic, provided that no computation in $C$ is a proper prefix of one of the others.* ∎

When several transitions depart from the same state $s$ of an NPLTS $\mathcal{L}$, they describe a nondeterministic choice among different behaviors. A *resolution* of $s$ is the result of a possible way of resolving nondeterministic choices starting from $s$, as if a *scheduler* were applied that decides which activity has to be performed next. A resolution of nondeterminism can thus be formalized as a fully probabilistic NPLTS $\mathcal{Z}$ with a tree-like structure, whose branching points correspond to target distributions of transitions deriving from those of $\mathcal{L}$.

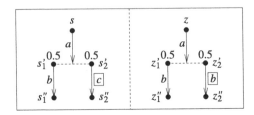

**Fig. 1.** Lack of injectivity breaks structure preservation

In [2] we examined two ways of resolving nondeterminism. The structure-preserving approach constructs a resolution by importing states and transitions from the original NPLTS via a *deterministic scheduler*. In a resolution of the structure-modifying approach (i) a transition can be produced by probabilistically combining transitions of the original model via a *randomized scheduler* [24], or (ii) a state can be obtained by probabilistically splitting states of the original model via an *interpolating scheduler* [12], or (iii) a combination thereof [7].

As in [2], we focus on structure-preserving resolutions arising from centralized, memoryless, deterministic schedulers. At each step, a scheduler of this kind selects one of the transitions departing from the current state, or no transitions at all thus stopping the execution. As a consequence, the resulting resolution is isomorphic to a submodel of the original model (or of its unfolding, should cycles be present), thereby preserving the structure of the original model (or of its unfolding). If the model is fully nondeterministic, each of its resolutions coincides with a computation of the model; if the model is fully probabilistic, its maximal resolution coincides with (the unfolding of) the entire model.

Following [5,17] we introduce a *correspondence function* $corr_{\mathcal{Z}} : Z \to S$ from the acyclic state space of the resolution $\mathcal{Z} = (Z, A, \longrightarrow_{\mathcal{Z}})$ being built, to the possibly cyclic state space of the considered model $\mathcal{L} = (S, A, \longrightarrow_{\mathcal{L}})$. For each transition $z \xrightarrow{a}_{\mathcal{Z}} \Delta$, the function $corr_{\mathcal{Z}}$ must preserve the probabilities of all the states corresponding to those in $supp(\Delta)$ and must be injective over $supp(\Delta)$. In the absence of injectivity, the original structure may not be preserved in the case that the target distribution of a transition assigns the same probability to several inequivalent states. This is exemplified in Fig. 1. The correspondence function that maps $z$ to $s$, $z'_1$ and $z'_2$ to $s'_1$, and $z''_1$ and $z''_2$ to $s''_1$ would cause the rightmost NPLTS to be considered a legal resolution of the leftmost NPLTS, which is not correct as the former is not isomorphic to any submodel of the latter.

**Definition 3.** *Let* $\mathcal{L} = (S, A, \longrightarrow_{\mathcal{L}})$ *be an NPLTS and* $s \in S$. *An acyclic NPLTS* $\mathcal{Z} = (Z, A, \longrightarrow_{\mathcal{Z}})$ *is a* structure-preserving resolution *of* $s$, *written* $\mathcal{Z} \in Res_{\mathrm{sp}}(s)$, *iff there exists a correspondence function* $corr_{\mathcal{Z}} : Z \to S$ *such that* $s = corr_{\mathcal{Z}}(z_s)$, *for some* $z_s \in Z$ *acting as the initial state of* $\mathcal{Z}$, *and for all* $z \in Z$ *it holds that:*

– If $z \xrightarrow{a}_{\mathcal{Z}} \Delta$ then $corr_{\mathcal{Z}}(z) \xrightarrow{a}_{\mathcal{L}} \Gamma$, with $corr_{\mathcal{Z}}$ being injective over $supp(\Delta)$ and satisfying $\Delta(z') = \Gamma(corr_{\mathcal{Z}}(z'))$ for all $z' \in supp(\Delta)$.
– At most one transition departs from $z$.  ∎

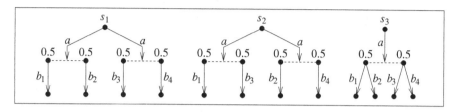

**Fig. 2.** $\sim_{\mathrm{PTr}}^{\mathrm{post}}$ is strictly finer than $\sim_{\mathrm{PTr}}^{\mathrm{pre}}$: $s_1 \not\sim_{\mathrm{PTr}}^{\mathrm{post}} s_2 \not\sim_{\mathrm{PTr}}^{\mathrm{post}} s_3$, $s_1 \sim_{\mathrm{PTr}}^{\mathrm{pre}} s_2 \sim_{\mathrm{PTr}}^{\mathrm{pre}} s_3$

## 3    Probabilistic Trace Equivalences and Their Anomalies

There is only one way of defining trace semantics for fully nondeterministic processes [8] and for fully probabilistic processes [20]. In contrast, this is not the case with processes featuring both nondeterminism and probabilities, as shown in the spectrum of behavioral equivalences for NPLTS models studied in [4].

The first probabilistic trace equivalence that we consider is the one of [25]. Two states are deemed equivalent when every resolution of either state is matched by a resolution of the other, in the sense that for each trace both resolutions execute that trace with the same probability. We call it probabilistic trace *post*-equivalence because the quantification over traces occurs *after* selecting the two fully matching resolutions as underlined in the definition below, where $z_{s_i}$ denotes both the initial state of $\mathcal{Z}_i$ and the state to which $s_i$ corresponds.

**Definition 4.** *Let $(S, A, \longrightarrow)$ be an NPLTS and $s_1, s_2 \in S$. We let $s_1 \sim_{\mathrm{PTr}}^{\mathrm{post}} s_2$ iff for each $\mathcal{Z}_1 \in Res_{\mathrm{sp}}(s_1)$ there exists $\mathcal{Z}_2 \in Res_{\mathrm{sp}}(s_2)$ such that <u>for all $\alpha \in A^*$</u>:*

$$prob(\mathcal{CC}(z_{s_1}, \alpha)) = prob(\mathcal{CC}(z_{s_2}, \alpha))$$

*and also the condition obtained by exchanging $\mathcal{Z}_1$ with $\mathcal{Z}_2$ is satisfied.*  ∎

The second probabilistic trace equivalence is the one of [3], which is a congruence with respect to parallel composition. It is less restrictive than the previous equivalence because, given two states, a resolution of either state can be matched by different resolutions of the other with respect to different traces. We call it probabilistic trace *pre*-equivalence because traces are fixed *before* selecting the two partially matching resolutions.

**Definition 5.** *Let $(S, A, \longrightarrow)$ be an NPLTS and $s_1, s_2 \in S$. We let $s_1 \sim_{\mathrm{PTr}}^{\mathrm{pre}} s_2$ iff, <u>for all $\alpha \in A^*$</u>, for each $\mathcal{Z}_1 \in Res_{\mathrm{sp}}(s_1)$ there is $\mathcal{Z}_2 \in Res_{\mathrm{sp}}(s_2)$ such that:*

$$prob(\mathcal{CC}(z_{s_1}, \alpha)) = prob(\mathcal{CC}(z_{s_2}, \alpha))$$

*and also the condition obtained by exchanging $\mathcal{Z}_1$ with $\mathcal{Z}_2$ is satisfied.*  ∎

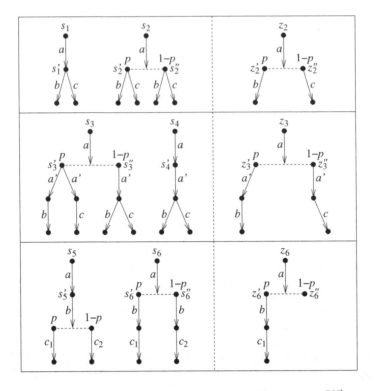

**Fig. 3.** Three anomalies of the probabilistic trace equivalences $\sim_{\mathrm{PTr}}^{\mathrm{post}}$ and $\sim_{\mathrm{PTr}}^{\mathrm{pre}}$

In Fig. 2 we show three NPLTS models whose initial states $s_1$, $s_2$, $s_3$ are pairwise distinguished by $\sim_{\mathrm{PTr}}^{\mathrm{post}}$ but identified by $\sim_{\mathrm{PTr}}^{\mathrm{pre}}$, because for all $i = 1, \ldots, 4$ the probability of executing trace $a\, b_i$ is the same in the three models.

Although deterministic schedulers are very intuitive, the rigid preservation they ensure about the structure of the original model, together with their free-dom of performing choices inconsistent with each other in states with equivalent continuations, causes the two considered probabilistic trace equivalences to be overdiscriminating. This results in the violation of a number of desirable prop-erties (a fact that also happens with structure-modifying schedulers, but to a much lesser extent). More precisely, in [2] we showed that $\sim_{\mathrm{PTr}}^{\mathrm{post}}$ and $\sim_{\mathrm{PTr}}^{\mathrm{pre}}$:

- are not coarser than probabilistic bisimilarity under deterministic schedulers;
- are not congruences w.r.t. action prefix under deterministic schedulers;
- are not compatible with their version for fully probabilistic processes.

The first anomaly is illustrated by the two NPLTS models in Fig. 3 whose initial states are $s_1$ and $s_2$. They are probabilistic bisimilar in the sense of [27] but $s_1 \not\sim_{\mathrm{PTr}}^{\mathrm{post}} s_2$ and $s_1 \not\sim_{\mathrm{PTr}}^{\mathrm{pre}} s_2$ because of the resolution whose initial state is $z_2$, where trace $a\, b$ is executable with probability $p$ instead of 1. This resolution belongs to $Res_{\mathrm{sp}}(s_2) \backslash Res_{\mathrm{sp}}(s_1)$ as it does not preserve the structure of the NPLTS whose initial state is $s_1$. Therefore, the two probabilistic trace equivalences do not include probabilistic bisimilarity.

The second anomaly is illustrated by the two NPLTS models in Fig. 3 whose initial states are $s_3$ and $s_4$. After the two $a$-transitions, two distributions are reached that are probabilistic trace equivalent, in the sense that for each class of equivalent states they both assign the same probability to that class. However, it holds that $s_3 \not\sim_{\mathrm{PTr}}^{\mathrm{post}} s_4$ and $s_3 \not\sim_{\mathrm{PTr}}^{\mathrm{pre}} s_4$ due to the resolution whose initial state is $z_3$, where trace $a\,a'\,b$ is executable with probability $p$ instead of 1. This resolution belongs to $Res_{\mathrm{sp}}(s_3) \setminus Res_{\mathrm{sp}}(s_4)$ as it does not preserve the structure of the NPLTS whose initial state is $s_4$. Therefore, the two probabilistic trace equivalences are not congruences with respect to the action prefix operator, which concatenates the execution of an action with a process distribution.

The third anomaly is illustrated by the two NPLTS models in Fig. 3 whose initial states are $s_5$ and $s_6$. They are identified by the trace equivalence for fully probabilistic processes of [20], which does not use schedulers at all as in those processes there are no nondeterministic choices to be solved. However, it turns out that $s_5 \not\sim_{\mathrm{PTr}}^{\mathrm{post}} s_6$ and $s_5 \not\sim_{\mathrm{PTr}}^{\mathrm{pre}} s_6$ because $\sim_{\mathrm{PTr}}^{\mathrm{post}}$ and $\sim_{\mathrm{PTr}}^{\mathrm{pre}}$ do make use of schedulers, and schedulers may decide of stopping the execution. This is witnessed by the resolution whose initial state is $z_6$ – notice that the scheduler has decided to stop the execution at $z_6''$ – where not only trace $a\,b\,c_1$ but also trace $a\,b$ is executable with probability $p$. This resolution belongs to $Res_{\mathrm{sp}}(s_6) \setminus Res_{\mathrm{sp}}(s_5)$ as it does not preserve the structure of the NPLTS whose initial state is $s_5$. Therefore, the two probabilistic trace equivalences are not backward compatible with the one for fully probabilistic processes.

## 4   Properties of Coherency

The anomalies shown in Fig. 3 are due to the freedom of schedulers of making different decisions in states enabling the same actions. In [2] we proposed to limit the excessive power of schedulers by restricting them to yield *coherent resolutions*. Intuitively, this means that, if several states in the support of the target distribution of a transition are equivalent, then the decisions made by the scheduler in those states have to be coherent with each other, so that the states to which they correspond in any resolution are equivalent as well.

The *coherency constraints* implementing this idea have been expressed in [2] by reasoning on *coherent trace distributions*, i.e., families of sets of traces weighted with their execution probabilities in a given resolution, built through the following operations.

**Definition 6.** *Let $A \neq \emptyset$ be a countable set. For $a \in A$, $p \in \mathbb{R}$, $TD \subseteq 2^{A^* \times \mathbb{R}}$, and $T \subseteq A^* \times \mathbb{R}$ we define:*

$$a \cdot TD = \{a \cdot T \mid T \in TD\} \qquad a \cdot T = \{(a\,\alpha, p') \mid (\alpha, p') \in T\}$$
$$p \cdot TD = \{p \cdot T \mid T \in TD\} \qquad p \cdot T = \{(\alpha, p \cdot p') \mid (\alpha, p') \in T\}$$
$$tr(TD) = \{tr(T) \mid T \in TD\} \qquad tr(T) = \{\alpha \in A^* \mid \exists p' \in \mathbb{R}. (\alpha, p') \in T\}$$

while for $TD_1, TD_2 \subseteq 2^{A^* \times \mathbb{R}}$ we define:

$$TD_1 + TD_2 = \begin{cases} \{T_1 + T_2 \mid T_1 \in TD_1 \wedge T_2 \in TD_2 \wedge tr(T_1) = tr(T_2)\} \\ \qquad\qquad\qquad\qquad\qquad\qquad\qquad \textit{if } tr(TD_1) = tr(TD_2) \\ \{T_1 + T_2 \mid T_1 \in TD_1 \wedge T_2 \in TD_2\} \\ \qquad\qquad\qquad\qquad\qquad\qquad\qquad \textit{otherwise} \end{cases}$$

where for $T_1, T_2 \subseteq A^* \times \mathbb{R}$ we define:

$$T_1 + T_2 = \{(\alpha, p_1 + p_2) \mid (\alpha, p_1) \in T_1 \wedge (\alpha, p_2) \in T_2\} \cup$$
$$\{(\alpha, p) \in T_1 \cup T_2 \mid \alpha \notin tr(T_1) \cap tr(T_2)\}$$

∎

Weighted trace set addition $T_1 + T_2$ is commutative and associative, with probabilities of identical traces in the two summands being always added up for coherency purposes. In constrast, trace distribution addition is only commutative. Essentially, the two summands in $TD_1 + TD_2$ represent two families of sets of weighted traces executable in the resolutions of two states in the support of a target distribution. Every weighted trace set $T_1 \in TD_1$ is summed with every weighted trace set $T_2 \in TD_2$ – so to characterize an overall resolution – unless $TD_1$ and $TD_2$ have the same family of trace sets, in which case summation is restricted to weighted trace sets featuring the same traces for the sake of coherency. Due to the lack of associativity, in the definition below all trace distributions $\Delta(s') \cdot TD^c_{n-1}(s')$ exhibiting the same family $\Theta$ of trace sets have to be summed up first, which is ensured by the presence of a double summation.

**Definition 7.** Let $(S, A, \longrightarrow)$ be an NPLTS and $s \in S$. The coherent trace distribution of $s$ is the subset of $2^{A^* \times \mathbb{R}_{]0,1]}}$ defined as follows:

$$TD^c(s) = \bigcup_{n \in \mathbb{N}} TD^c_n(s)$$

with the coherent trace distribution of $s$ whose traces have length at most $n$ being defined as:

$$TD^c_n(s) = \begin{cases} (\varepsilon, 1) \dagger \bigcup_{s \xrightarrow{a} \Delta} a \cdot \left( \sum_{\Theta \in tr(\Delta, n-1)} \overset{tr(TD^c_{n-1}(s'))=\Theta}{\sum_{s' \in supp(\Delta)}} \Delta(s') \cdot TD^c_{n-1}(s') \right) \\ \qquad\qquad\qquad\qquad \textit{if } n > 0 \textit{ and } s \textit{ has outgoing transitions} \\ \{\{(\varepsilon, 1)\}\} \\ \qquad\qquad\qquad\qquad \textit{otherwise} \end{cases}$$

where $tr(\Delta, n-1) = \{tr(TD^c_{n-1}(s')) \mid s' \in supp(\Delta)\}$ and the operator $(\varepsilon, 1) \dagger \_$ is such that $(\varepsilon, 1) \dagger TD = \{\{(\varepsilon, 1)\} \cup T \mid T \in TD\}$. ∎

In the case of a fully probabilistic NPLTS, due to the absence of nondeterminism any coherent trace distribution $TD^c_n(s)$ contains a single weighted trace set. This holds in particular for resolutions.

**Proposition 1.** Let $(S, A, \longrightarrow)$ be a fully probabilistic NPLTS, $s \in S$, $n \in \mathbb{N}$. Let $A^{\leq n} = \{\alpha \in A^* \mid |\alpha| \leq n\}$. Then $TD^c_n(s) = \{\{(\alpha, p) \in A^{\leq n} \times \mathbb{R}_{]0,1]} \mid prob(\mathcal{CC}(s, \alpha)) = p\}\}$. ∎

As for the relationship between $TD_n^c(s)$ and $TD_{n-1}^c(s)$, it turns out that every element of the former contains the same traces as an element of the latter. As we will see in the next section, their probabilities may differ.

**Proposition 2.** *Let $(S, A, \longrightarrow)$ be an NPLTS, $s \in S$, $n \in \mathbb{N}_{\geq 1}$. Then for all $T \in TD_n^c(s)$ there exists $T' \in TD_{n-1}^c(s)$ such that $tr(T') \subseteq tr(T)$.* ∎

For the NPLTS models in Fig. 3 we have that:

- $TD^c(s_2') = \{\{(\varepsilon, 1)\}, \{(\varepsilon, 1), (b, 1)\}, \{(\varepsilon, 1), (c, 1)\}\} = TD^c(s_2'')$ while in the related resolution states it holds that $TD^c(z_2') = \{\{(\varepsilon, 1)\}, \{(\varepsilon, 1), (b, 1)\}\} \neq \{\{(\varepsilon, 1)\}, \{(\varepsilon, 1), (c, 1)\}\} = TD^c(z_2'')$.
- $TD^c(s_3') = \{\{(\varepsilon, 1)\}, \{(\varepsilon, 1), (a', 1)\}, \{(\varepsilon, 1), (a', 1), (a'b, 1)\}, \{(\varepsilon, 1), (a', 1), (a'c, 1)\}\} = TD^c(s_3'')$ but $TD^c(z_3') = \{\{(\varepsilon, 1)\}, \{(\varepsilon, 1), (a', 1)\}, \{(\varepsilon, 1), (a', 1), (a'b, 1)\}\} \neq \{\{(\varepsilon, 1)\}, \{(\varepsilon, 1), (a', 1)\}, \{(\varepsilon, 1), (a', 1), (a'c, 1)\}\} = TD^c(z_3'')$.
- $TD_1^c(s_6') = \{\{(\varepsilon, 1), (b, 1)\}\} = TD_1^c(s_6'')$ but $TD_1^c(z_6') = \{\{(\varepsilon, 1), (b, 1)\}\} \neq \{\{(\varepsilon, 1)\}\} = TD_1^c(z_6'')$, which indicates that separate coherency constraints are needed relying on $TD_n^c$ sets for every $n \in \mathbb{N}$.

Further examples in [2] show that the coherency constraints should be based on $TD_n^c$ sets up to the probabilities they contain, i.e., the constraints should rely on $tr(TD_n^c)$ sets. Moreover, for every $n \in \mathbb{N}$, those examples call for a complete presence in each resolution of computations of length $n$ if any, including possible shorter maximal computations. Note that trace completeness up to length $n$ is looser than requiring resolution maximality.

**Definition 8.** *Let $\mathcal{L} = (S, A, \longrightarrow_{\mathcal{L}})$ be an NPLTS, $s \in S$, and $\mathcal{Z} = (Z, A, \longrightarrow_{\mathcal{Z}}) \in Res_{sp}(s)$ with correspondence function $corr_{\mathcal{Z}} : Z \to S$. We say that $\mathcal{Z}$ is a coherent resolution of $s$, written $\mathcal{Z} \in Res_{sp}^c(s)$, iff for all $z \in Z$, whenever $z \xrightarrow{a}_{\mathcal{Z}} \Delta$, then for all $n \in \mathbb{N}$:*

1. $tr(TD_n^c(corr_{\mathcal{Z}}(z'))) = tr(TD_n^c(corr_{\mathcal{Z}}(z''))) \implies tr(TD_n^c(z')) = tr(TD_n^c(z''))$ for all $z', z'' \in supp(\Delta)$.
2. For all $z' \in supp(\Delta)$, the only $T \in TD_n^c(z')$ admits $\bar{T} \in TD_n^c(corr_{\mathcal{Z}}(z'))$ such that $tr(T) = tr(\bar{T})$. ∎

Any complete submodel rooted at a state $z$ of a coherent resolution turns out to be coherent too, where complete means that no state reachable from $z$ in the resolution is cut off in the resolution submodel. Completeness is important for satisfying in particular the second coherency constraint of Definition 8.

**Proposition 3.** *Let $\mathcal{L} = (S, A, \longrightarrow_{\mathcal{L}})$ be an NPLTS, $s \in S$, $\mathcal{Z} = (Z, A, \longrightarrow_{\mathcal{Z}}) \in Res_{sp}^c(s)$ with correspondence function $corr_{\mathcal{Z}} : Z \to S$. Let $\mathcal{Z}_z' = (Z', A, \longrightarrow_{\mathcal{Z}'})$ be the complete submodel of $\mathcal{Z}$ rooted at $z \in Z$. Then $\mathcal{Z}_z' \in Res_{sp}^c(corr_{\mathcal{Z}}(z))$.* ∎

The resolutions in Fig. 3 do *not* respectively belong to $Res_{sp}^c(s_2)$, $Res_{sp}^c(s_3)$, $Res_{sp}^c(s_6)$. We proved in Thm. 1 of [2] that the examined anomalies disappear by substituting $Res_{sp}^c$ for $Res_{sp}$ in Definitions 4 and 5. This replacement yields the two coherency-based probabilistic trace equivalences $\sim_{\text{PTr}}^{\text{post,c}}$ and $\sim_{\text{PTr}}^{\text{pre,c}}$ for which we will investigate alternative characterizations in the next two sections by exploiting the properties shown in Propositions 1, 2, and 3.

# 5   Alternative Characterization of $\sim_{\mathrm{PTr}}^{\mathrm{post,c}}$

The definition of $\sim_{\mathrm{PTr}}^{\mathrm{post,c}}$ essentially requires that two states have the same trace distributions. Therefore, it is natural to expect an alternative characterization of $\sim_{\mathrm{PTr}}^{\mathrm{post,c}}$ based on the construction of Definition 7. Incidentally, this would fully justify the construction itself, given that the probabilities contained in the $TD_n^c$ sets have not been exploited in the coherency constraints of Definition 8. However, for an NPLTS $(S, A, \longrightarrow)$ and $s \in S$, the set $TD^c(s)$ may contain weighted traces that break coherency, hence that set cannot be used for characterization purposes.

For example, consider the NPLTS in Fig. 4. We have that:

$$TD_1^c(s_1) = \{\{(\varepsilon, 1), (b, 1)\}\} = TD_1^c(s_2)$$

and also:

$$TD_2^c(s_1) = \{\{(\varepsilon, 1), (b, 1), (b\,c, 1)\}, \{(\varepsilon, 1), (b, 1), (b\,d, 1)\}\} = TD_2^c(s_2)$$

because in the complete submodel rooted at $s_1$ it holds that:

$$TD_1^c(s_1') = \{\{(\varepsilon, 1), (c, 1)\}, \{(\varepsilon, 1), (d, 1)\}\} = TD_1^c(s_1'')$$

and hence, when applying Definition 7 to compute $TD_2^c(s_1)$, according to Definition 6 the summation is restricted to weighted trace sets featuring the same traces as:

$$tr(TD_1^c(s_1')) = \{\{\varepsilon, c\}, \{\varepsilon, d\}\} = tr(TD_1^c(s_1''))$$

Nevertheless, since:

$$TD_2^c(s_1') = \{\{(\varepsilon, 1), (c, 1), (c\,e_1, 1)\}, \{(\varepsilon, 1), (d, 1), (d\,e_2, 1)\}\}$$
$$TD_2^c(s_1'') = \{\{(\varepsilon, 1), (c, 1), (c\,e_3, 1)\}, \{(\varepsilon, 1), (d, 1), (d\,e_4, 1)\}\}$$

where:

$$tr(TD_2^c(s_1')) = \{\{\varepsilon, c, c\,e_1\}, \{\varepsilon, d, d\,e_2\}\} \neq \{\{\varepsilon, c, c\,e_3\}, \{\varepsilon, d, d\,e_4\}\} = tr(TD_2^c(s_1''))$$

we subsequently derive that:

$$
\begin{aligned}
TD_3^c(s_1) = (\varepsilon, 1) \, &\dagger \\
(\{&\{(b, p), (b\,c, p), (b\,c\,e_1, p)\}, \\
&\{(b, p), (b\,d, p), (b\,d\,e_2, p)\}\} + \\
&\{(b, 1-p), (b\,c, 1-p), (b\,c\,e_3, 1-p)\}, \\
&\{(b, 1-p), (b\,d, 1-p), (b\,d\,e_4, 1-p)\}\}) \\
= \{&\{(\varepsilon, 1), (b, 1), (b\,c, 1), (b\,c\,e_1, p), (b\,c\,e_3, 1-p)\}, \\
&\{(\varepsilon, 1), (b, 1), (b\,c, p), (b\,d, 1-p), (b\,c\,e_1, p), (b\,d\,e_4, 1-p)\}, \\
&\{(\varepsilon, 1), (b, 1), (b\,d, p), (b\,c, 1-p), (b\,d\,e_2, p), (b\,c\,e_3, 1-p)\}, \\
&\{(\varepsilon, 1), (b, 1), (b\,d, 1), (b\,d\,e_2, p), (b\,d\,e_4, 1-p)\}\}
\end{aligned}
$$

whereas:

$$
\begin{aligned}
TD_3^c(s_2) = \{&\{(\varepsilon, 1), (b, 1), (b\,c, 1), (b\,c\,e_1, p), (b\,c\,e_3, 1-p)\}, \\
&\{(\varepsilon, 1), (b, 1), (b\,d, 1), (b\,d\,e_2, p), (b\,d\,e_4, 1-p)\}\}
\end{aligned}
$$

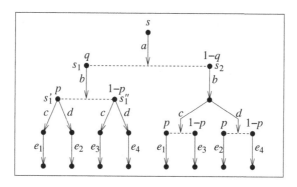

**Fig. 4.** Full coherency is necessary to reconcile $TD_3^c(s_1)$ and $TD_3^c(s_2)$

Therefore, in the calculation of $TD_4^c(s)$ we cannot simply sum up weighted trace sets in $TD_3^c(s_1)$ and in $TD_3^c(s_2)$ that exhibit the same traces. This is due to the presence in $TD_3^c(s_1)$ of the following two weighted trace sets:

$$\{(\varepsilon, 1), (b, 1), (b\,c, p), (b\,d, 1 - p), (b\,c\,e_1, p), (b\,d\,e_4, 1 - p)\}$$
$$\{(\varepsilon, 1), (b, 1), (b\,d, p), (b\,c, 1 - p), (b\,d\,e_2, p), (b\,c\,e_3, 1 - p)\}$$

which cannot be exposed by any coherent resolution. The key observation is that coherency constraints on traces like $b\,c$ and $b\,d$ are ignored, hence those two weighted trace sets in $TD_3^c(s_1)$ are *not extensions* of weighted trace sets in $TD_2^c(s_1)$. Indeed, neither of those weighted trace sets in $TD_3^c(s_1)$ includes as a subset a weighted trace set in $TD_2^c(s_1)$ because of the different probabilities of the aforementioned traces in the considered sets (see the sentence before Proposition 2).

This example reveals that the construction of Definition 7, together with weighted trace set addition and trace distribution addition as provided in Definition 6, are appropriate to set up the coherency constraints in Definition 8, but not to characterize the trace distributions of coherent resolutions. To achieve this, every set $TD_n^c(s)$, with $n > 0$ and $s$ having outgoing transitions, should *incrementally* build on $TD_{n-1}^c(s)$, in the sense that every weighted trace set in the former should include as a subset a weighted trace set in the latter (a *monotonicity*-like property stronger than the one of Proposition 2). We thus introduce a variant of coherent trace distribution, which we call *fully* coherent trace distribution.

**Definition 9.** *Let $(S, A, \longrightarrow)$ be an NPLTS and $s \in S$. The* fully coherent trace distribution *of $s$ is the subset of $2^{A^* \times \mathbb{R}_{]0,1]}}$ defined as follows:*
$$TD^{\mathrm{fc}}(s) = \bigcup_{n \in \mathbb{N}} TD_n^{\mathrm{fc}}(s)$$
*with the fully coherent trace distribution of $s$ whose traces have length at most $n$ being the subset of $TD_n^c(s)$ defined as:*

$$TD_n^{\mathrm{fc}}(s) = \begin{cases} \{T \in TD_n^{\mathrm{c}}(s) \mid \exists T' \in TD_{n-1}^{\mathrm{fc}}(s).\, T' \subseteq T\} \\ \qquad\qquad\qquad \textit{if } n > 0 \textit{ and } s \textit{ has outgoing transitions} \\ \{\{(\varepsilon, 1)\}\} \\ \qquad\qquad\qquad \textit{otherwise} \end{cases}$$

■

For the NPLTS in Fig. 4 we have that:

$$TD_3^{\mathrm{fc}}(s_1) = \{\{(\varepsilon, 1), (b, 1), (b\,c, 1), (b\,c\,e_1, p), (b\,c\,e_3, 1 - p)\},$$
$$\{(\varepsilon, 1), (b, 1), (b\,d, 1), (b\,d\,e_2, p), (b\,d\,e_4, 1 - p)\}\} = TD_3^{\mathrm{fc}}(s_2)$$

and overall $TD^{\mathrm{fc}}(s) = \bigcup_{0 \le n \le 4} TD_n^{\mathrm{fc}}(s)$ where:

$$TD_0^{\mathrm{fc}}(s) = \{\{(\varepsilon, 1)\}\}$$
$$TD_1^{\mathrm{fc}}(s) = \{\{(\varepsilon, 1), (a, 1)\}\}$$
$$TD_2^{\mathrm{fc}}(s) = \{\{(\varepsilon, 1), (a, 1), (a\,b, 1)\}\}$$
$$TD_3^{\mathrm{fc}}(s) = \{\{(\varepsilon, 1), (a, 1), (a\,b, 1), (a\,b\,c, 1)\},$$
$$\{(\varepsilon, 1), (a, 1), (a\,b, 1), (a\,b\,d, 1)\}\}$$
$$TD_4^{\mathrm{fc}}(s) = \{\{(\varepsilon, 1), (a, 1), (a\,b, 1), (a\,b\,c, 1), (a\,b\,c\,e_1, p), (a\,b\,c\,e_3, 1 - p)\},$$
$$\{(\varepsilon, 1), (a, 1), (a\,b, 1), (a\,b\,d, 1), (a\,b\,d\,e_2, p), (a\,b\,d\,e_4, 1 - p)\}\}$$

with the various sets $TD_n^{\mathrm{fc}}(s)$ precisely capturing the trace distributions of the coherent resolutions of $s$.

Fully coherent trace distributions $TD_n^{\mathrm{fc}}(s)$ coincide with coherent ones $TD_n^{\mathrm{c}}(s)$ when $n \le 2$ as a consequence of Definition 6. The example in Fig. 4 shows that, when $n \ge 3$, in general $TD_n^{\mathrm{fc}}(s)$ cannot be recursively characterized as $TD_n^{\mathrm{c}}(s)$ in Definition 7 even though each element of a fully coherent trace distribution can be expressed as a sum of elements of other fully coherent trace distributions.

**Proposition 4.** *Let* $(S, A, \longrightarrow)$ *be an NPLTS,* $s \in S$, $n \in \mathbb{N}$. *If* $n \le 2$ *or* $s$ *has no outgoing transitions, then* $TD_n^{\mathrm{fc}}(s) = TD_n^{\mathrm{c}}(s)$, *otherwise each element of* $TD_n^{\mathrm{fc}}(s)$ *is anyhow obtained by summing up a suitable element of* $TD_{n-1}^{\mathrm{fc}}(s')$ *for every* $s'$ *in the support of the target distribution of a transition of* $s$.     ■

By virtue of Proposition 1, the equality $TD_n^{\mathrm{fc}}(s) = TD_n^{\mathrm{c}}(s)$ extends to all $n \in \mathbb{N}$, i.e., fully coherent trace distributions boil down to coherent ones, in the case of a fully probabilistic NPLTS. This holds in particular for resolutions.

**Proposition 5.** *Let* $(S, A, \longrightarrow)$ *be a fully probabilistic NPLTS,* $s \in S$, $n \in \mathbb{N}$. *Then* $TD_n^{\mathrm{fc}}(s) = TD_n^{\mathrm{c}}(s)$.     ■

Before presenting the characterization result, we need to revisit the coherency constraints of Definition 8 in the light of the example of Fig. 4. Suppose that each of the two terminal states reached by an $e_1$-transition is replaced by a distribution with two states in its target, reached with probabilities $r$ and $1 - r$ and both featuring a nondeterministic choice between an $f$-transition to a terminal state and a $g$-transition to a terminal state. According to Definition 8, after the leftmost $e_1$-transition either both $f$-transitions are selected or both $g$-transitions are selected, and the same holds after the rightmost $e_1$-transition.

However, from $TD_3^c(s_1) \neq TD_3^c(s_2)$ it follows that $TD_4^c(s_1) \neq TD_4^c(s_2)$, so that on $s_1$ side both $f$-transitions may be selected while on $s_2$ side both $g$-transitions may be selected instead, or vice versa. If we consider an NPLTS whose initial state $s'$ has a single outgoing transition, which is labeled with $a$ and reaches a model isomorphic to the complete submodel rooted at $s_2$ in Fig. 4 (and extended in the aforementioned way after its only $e_1$-transition), $s'$ would be distinguished from $s$ instead of being identified with it. The coherency constraints of Definition 8 thus need to be strengthened by substituting $TD_n^{fc}$ for $TD_n^c$, in which case Thm. 1 of [2] is still valid and $\sim_{\mathrm{PTr}}^{\mathrm{post,c}}$ and $\sim_{\mathrm{PTr}}^{\mathrm{pre,c}}$ are modified accordingly.

The following lemma, where Proposition 1 is exploited again together with Propositions 3, 4, and 5, lays the basis for a characterization of $\sim_{\mathrm{PTr}}^{\mathrm{post,c}}$ in terms of fully coherent trace distribution equality. In the lemma, $z_s$ denotes both the initial state of $\mathcal{Z}$ and the state to which $s$ corresponds.

**Lemma 1.** *Let $(S, A, \longrightarrow)$ be an NPLTS, $s \in S$, $n \in \mathbb{N}$, and $T \subseteq A^* \times \mathbb{R}_{]0,1]}$. Then $T \in TD_n^{fc}(s)$ iff there exists $\mathcal{Z} \in Res_{\mathrm{sp}}^c(s)$ such that $TD_n^{fc}(z_s) = \{T\}$.* ∎

**Theorem 1.** *Let $(S, A, \longrightarrow)$ be an NPLTS and $s_1, s_2 \in S$. Then $s_1 \sim_{\mathrm{PTr}}^{\mathrm{post,c}} s_2$ iff $TD^{fc}(s_1) = TD^{fc}(s_2)$.* ∎

# 6    Alternative Characterization of $\sim_{\mathrm{PTr}}^{\mathrm{pre,c}}$

As far as $\sim_{\mathrm{PTr}}^{\mathrm{pre,c}}$ is concerned, similar to [3] we can provide an alternative characterization based on sets $T^c(s)$ built by considering all weighted traces executable from state $s$ at once, i.e., without keeping track of the resolutions of $s$ in which they are feasible. This is consistent with the focus of $\sim_{\mathrm{PTr}}^{\mathrm{pre,c}}$ on individual traces rather than on trace distributions. In the definition below, the double summation used in Definition 7 in the case that $n > 0$ and $s$ has outgoing transitions is not needed thanks to the commutativity and associativity of weighted trace set addition deriving from Definition 6.

**Definition 10.** *Let $(S, A, \longrightarrow)$ be an NPLTS and $s \in S$. The coherent weighted trace set of $s$ is the subset of $A^* \times \mathbb{R}_{]0,1]}$ defined as follows:*
$$T^c(s) = \bigcup_{n \in \mathbb{N}} T_n^c(s)$$
*with the coherent weighted trace set of $s$ whose traces have length at most $n$ being defined as:*

$$T_n^c(s) = \begin{cases} \{(\varepsilon, 1)\} \cup \displaystyle\bigcup_{s \xrightarrow{a} \Delta} a \cdot \left( \displaystyle\sum_{s' \in supp(\Delta)} \Delta(s') \cdot T_{n-1}^c(s') \right) \\ \qquad\qquad\qquad\qquad \textit{if } n > 0 \textit{ and } s \textit{ has outgoing transitions} \\ \{(\varepsilon, 1)\} \\ \qquad\qquad\qquad\qquad \textit{otherwise} \end{cases}$$
∎

For the NPLTS in Fig. 4 we have that $T^c(s) = \bigcup_{0 \leq n \leq 4} T_n^c(s)$ where:

$T_0^c(s) = \{(\varepsilon, 1)\}$
$T_1^c(s) = \{(\varepsilon, 1), (a, 1)\}$
$T_2^c(s) = \{(\varepsilon, 1), (a, 1), (a\,b, 1)\}$
$T_3^c(s) = \{(\varepsilon, 1), (a, 1), (a\,b, 1), (a\,b\,c, 1), (a\,b\,d, 1)\}$
$T_4^c(s) = \{(\varepsilon, 1), (a, 1), (a\,b, 1), (a\,b\,c, 1), (a\,b\,d, 1),$
$\qquad\qquad (a\,b\,c\,e_1, p), (a\,b\,c\,e_3, 1 - p), (a\,b\,d\,e_2, p), (a\,b\,d\,e_4, 1 - p)\}$

with the various sets $T_n^c(s)$ precisely capturing the weighted traces of the coherent resolutions of $s$.

It is easy to characterize $T_n^c(s)$ in the case of a fully probabilistic NPLTS. This holds in particular for resolutions.

**Proposition 6.** *Let $(S, A, \longrightarrow)$ be a fully probabilistic NPLTS, $s \in S$, $n \in \mathbb{N}$. Let $A^{\leq n} = \{\alpha \in A^* \mid |\alpha| \leq n\}$. Then $T_n^c(s) = \{(\alpha, p) \in A^{\leq n} \times \mathbb{R}_{]0,1]} \mid prob(\mathcal{CC}(s, \alpha)) = p\}$.* ∎

The construction in Definition 10 turns out to be monotonic, in the sense that $T_n^c(s)$ includes as a subset $T_{n-1}^c(s)$.

**Proposition 7.** *Let $(S, A, \longrightarrow)$ be an NPLTS, $s \in S$, $(\alpha, p) \in A^* \times \mathbb{R}_{]0,1]}$, and $n \in \mathbb{N}_{\geq |\alpha|}$. Then $(\alpha, p) \in T_n^c(s)$ implies $(\alpha, p) \in T_{n+1}^c(s)$.* ∎

The following lemma, which exploits Proposition 6 and 7, provides the basis for a characterization of $\sim_{\mathrm{PTr}}^{\mathrm{pre,c}}$ in terms of coherent weighted trace set equality. In the lemma, $z_s$ denotes both the initial state of $\mathcal{Z}$ and the state to which $s$ corresponds.

**Lemma 2.** *Let $(S, A, \longrightarrow)$ be an NPLTS, $s \in S$, and $(\alpha, p) \in A^* \times \mathbb{R}_{]0,1]}$. Then $(\alpha, p) \in T^c(s)$ iff there exists $\mathcal{Z} \in Res_{\mathrm{sp}}^c(s)$ such that $prob(\mathcal{CC}(z_s, \alpha)) = p$.* ∎

**Theorem 2.** *Let $(S, A, \longrightarrow)$ be an NPLTS and $s_1, s_2 \in S$. Then $s_1 \sim_{\mathrm{PTr}}^{\mathrm{pre,c}} s_2$ iff $T^c(s_1) = T^c(s_2)$.* ∎

We conclude with two remarks. The first one is that the construction in Definition 10 is identical to the one in Definition 3.5 of [3], but this should not be the case as coherency was neglected in [3]. Indeed, before Definition 3.5 of [3] the definition of $X + Y$ – i.e., $T_1 + T_2$ using the notation of this paper – should have included also $(\alpha, q_1) \in X$ and $(\alpha, q_2) \in Y$ without summing them up, otherwise the right-to-left implication in Lemma 3.7 of [3] does not hold as can be seen from trace $a\,b$ of the (incoherent) resolution in Fig. 3 of this paper whose initial state is $z_2$. That definition of $X + Y$ works here instead because the focus on coherency requires to always sum up the probabilities of weighted traces sharing the same trace.

The second remark is that looser coherency constraints, based on weighted trace sets rather than on trace distributions as in Definition 8, would not work. As anticipated in [2], if we used $T_n^c$ sets instead of $TD_n^c$ sets, then probabilistic trace equivalent NPLTS models like the ones in Fig. 5 would be told apart. Indeed, we would have $tr(T^c(s_1')) = \{\epsilon, b, b\,c_1, b\,c_2, b\,c\} = tr(T^c(s_2'))$ – whereas

$tr(TD^c(s_1')) \neq tr(TD^c(s_2'))$ – hence in any coherent resolution of $s'$ traces $a\,b\,c_1$, $a\,b\,c_2$, $a\,b\,c$ could only be executed with probability 0.5 if present, while $s''$ admits coherent resolutions in which those traces have execution probability 0.25.

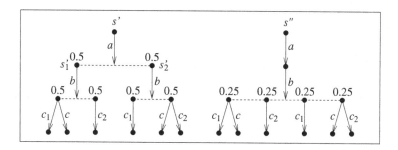

**Fig. 5.** Using weighted trace sets for coherency breaks probabilistic trace equivalence

## 7 Conclusions

Based on the notion of coherent resolution of nondeterminism, presented in [2] to avoid the anomalies of probabilistic trace semantics depicted in Fig. 3, in this paper we have provided alternative characterizations of $\sim_{\text{PTr}}^{\text{post,c}}$ [25] and $\sim_{\text{PTr}}^{\text{pre,c}}$ [3], respectively relying on the equality of fully coherent trace distributions and on the equality of coherent weighted trace sets. Both fully coherent trace distributions and coherent weighted trace sets are different from coherent trace distributions, introduced in [2] for defining coherency constraints on resolutions.

We plan to exploit the aforementioned alternative characterizations for studying properties and decision procedures of the two examined coherency-based probabilistic trace equivalences over nondeterministic and probabilistic processes.

## References

1. Baier, C., Katoen, J.P., Hermanns, H., Wolf, V.: Comparative branching-time semantics for Markov chains. Inf. Comput. **200**, 149–214 (2005)
2. Bernardo, M.: Coherent resolutions of nondeterminism. In: Gribaudo, M., Iacono, M., Phung-Duc, T., Razumchik, R. (eds.) EPEW 2019. LNCS, vol. 12039, pp. 16–32. Springer, Cham (2020). https://doi.org/10.1007/978-3-030-44411-2_2
3. Bernardo, M., De Nicola, R., Loreti, M.: Revisiting trace and testing equivalences for nondeterministic and probabilistic processes. Logical Methods Comput. Sci. **10**(116), 1–42 (2014)
4. Bernardo, M., De Nicola, R., Loreti, M.: Relating strong behavioral equivalences for processes with nondeterminism and probabilities. Theoret. Comput. Sci. **546**, 63–92 (2014)

5. Bernardo, M., Sangiorgi, D., Vignudelli, V.: On the discriminating power of testing equivalences for reactive probabilistic systems: results and open problems. In: Norman, G., Sanders, W. (eds.) QEST 2014. LNCS, vol. 8657, pp. 281–296. Springer, Cham (2014). https://doi.org/10.1007/978-3-319-10696-0_23

6. Bianco, A., de Alfaro, L.: Model checking of probabilistic and nondeterministic systems. In: Thiagarajan, P.S. (ed.) FSTTCS 1995. LNCS, vol. 1026, pp. 499–513. Springer, Heidelberg (1995). https://doi.org/10.1007/3-540-60692-0_70

7. Bonchi, F., Sokolova, A., Vignudelli, V.: The theory of traces for systems with nondeterminism and probability. In: Proceedings of the 34th ACM/IEEE Symposium on Logic in Computer Science (LICS 2019), pp. (19:62)1–14. IEEE-CS Press (2019)

8. Brookes, S.D., Hoare, C.A.R., Roscoe, A.W.: A theory of communicating sequential processes. J. ACM **31**, 560–599 (1984)

9. Cheung, L., Lynch, N.A., Segala, R., Vaandrager, F.: Switched PIOA: parallel composition via distributed scheduling. Theoret. Comput. Sci. **365**, 83–108 (2006)

10. de Alfaro, L., Henzinger, T.A., Jhala, R.: Compositional methods for probabilistic systems. In: Larsen, K.G., Nielsen, M. (eds.) CONCUR 2001. LNCS, vol. 2154, pp. 351–365. Springer, Heidelberg (2001). https://doi.org/10.1007/3-540-44685-0_24

11. Deng, Y., van Glabbeek, R.J., Hennessy, M., Morgan, C.: Characterising testing preorders for finite probabilistic processes. Logical Methods Comput. Sci. **4**(4:4), 1–33 (2008)

12. Deng, Y., van Glabbeek, R., Morgan, C., Zhang, C.: Scalar outcomes suffice for finitary probabilistic testing. In: De Nicola, R. (ed.) ESOP 2007. LNCS, vol. 4421, pp. 363–378. Springer, Heidelberg (2007). https://doi.org/10.1007/978-3-540-71316-6_25

13. Derman, C.: Finite State Markovian Decision Processes. Academic Press, Cambridge (1970)

14. Georgievska, S., Andova, S.: Probabilistic may/must testing: retaining probabilities by restricted schedulers. Formal Aspects Comput. **24**, 727–748 (2012). https://doi.org/10.1007/s00165-012-0236-5

15. Giro, S., D'Argenio, P.R.: On the expressive power of schedulers in distributed probabilistic systems. In: Proceedings of the 7th International Workshop on Quantitative Aspects of Programming Languages (QAPL 2009), ENTCS, vol. 253(3), pp. 45–71. Elsevier (2009)

16. Huynh, D.T., Tian, L.: On some equivalence relations for probabilistic processes. Fundamenta Informaticae **17**, 211–234 (1992)

17. Jonsson, B., Ho-Stuart, C., Yi, W.: Testing and refinement for nondeterministic and probabilistic processes. In: Langmaack, H., de Roever, W.-P., Vytopil, J. (eds.) FTRTFT 1994. LNCS, vol. 863, pp. 418–430. Springer, Heidelberg (1994). https://doi.org/10.1007/3-540-58468-4_176

18. Jonsson, B., Yi, W.: Compositional testing preorders for probabilistic processes. In: Proceedings of the 10th IEEE Symposium on Logic in Computer Science (LICS 1995), pp. 431–441. IEEE-CS Press (1995)

19. Jonsson, B., Yi, W.: Testing preorders for probabilistic processes can be characterized by simulations. Theoret. Comput. Sci. **282**, 33–51 (2002)

20. Jou, C.-C., Smolka, S.A.: Equivalences, congruences, and complete axiomatizations for probabilistic processes. In: Baeten, J.C.M., Klop, J.W. (eds.) CONCUR 1990. LNCS, vol. 458, pp. 367–383. Springer, Heidelberg (1990). https://doi.org/10.1007/BFb0039071

21. Keller, R.M.: Formal verification of parallel programs. Commun. ACM **19**, 371–384 (1976)

22. Kemeny, J.G., Snell, J.L.: Finite Markov Chains. Van Nostrand, London (1960)
23. Lynch, N., Segala, R., Vaandrager, F.: Compositionality for probabilistic automata. In: Amadio, R., Lugiez, D. (eds.) CONCUR 2003. LNCS, vol. 2761, pp. 208–221. Springer, Heidelberg (2003). https://doi.org/10.1007/978-3-540-45187-7_14
24. Segala, R.: Modeling and Verification of Randomized Distributed Real-Time Systems. Ph.D. thesis (1995)
25. Segala, R.: A compositional trace-based semantics for probabilistic automata. In: Lee, I., Smolka, S.A. (eds.) CONCUR 1995. LNCS, vol. 962, pp. 234–248. Springer, Heidelberg (1995). https://doi.org/10.1007/3-540-60218-6_17
26. Segala, R.: Testing probabilistic automata. In: Montanari, U., Sassone, V. (eds.) CONCUR 1996. LNCS, vol. 1119, pp. 299–314. Springer, Heidelberg (1996). https://doi.org/10.1007/3-540-61604-7_62
27. Segala, R., Lynch, N.: Probabilistic simulations for probabilistic processes. In: Jonsson, B., Parrow, J. (eds.) CONCUR 1994. LNCS, vol. 836, pp. 481–496. Springer, Heidelberg (1994). https://doi.org/10.1007/978-3-540-48654-1_35
28. van Glabbeek, R.J.: The linear time - branching time spectrum I. In: Handbook of Process Algebra, pp. 3–99. Elsevier (2001)
29. van Glabbeek, R.J., Smolka, S.A., Steffen, B.: Reactive, generative and stratified models of probabilistic processes. Inf. Comput. **121**, 59–80 (1995)
30. Vardi, M.Y.: Automatic verification of probabilistic concurrent finite-state programs. In: Proceedings of the 26th IEEE Symposium on Foundations of Computer Science (FOCS 1985), pp. 327–338. IEEE-CS Press (1985)
31. Yi, W., Larsen, K.G.: Testing probabilistic and nondeterministic processes. In: Proceedings of the 12th International Symposium on Protocol Specification, Testing and Verification (PSTV 1992), pp. 47–61. North-Holland (1992)

# Probabilistic Model Checking of AODV

Mojgan Kamali[✉] and Joost-Pieter Katoen

RWTH Aachen University, Aachen, Germany
{mojgan.kamali,katoen}@cs.rwth-aachen.de
https://moves.rwth-aachen.de

**Abstract.** This paper presents the formal modelling and verification of the Ad-hoc On-demand Distance Vector (AODV) routing protocol. Our study focuses on the quantitative aspects of AODV, in particular the influence of uncertainty (such as packet loss rates, collisions) on the probability to establish short routes. We present a compositional model of AODV's functionality using probabilistic timed automata. The strength of this model is that it combines hard real-time constraints with randomised protocol behaviour and can deal with non-determinism (due to e.g., queue behaviours at network nodes). An automated analysis by probabilistic model checking provides useful insights on the sensitivity of AODV's ability to establish shortest/longest routes and deliver data packets via such routes.

## 1 Introduction

*Wireless Networks.* Wireless technologies are on the rise: laptops and smart phones are ubiquitous, sensor networks monitor the environment generating vast amounts of data, and wireless technology is used in M2M (Machine-to-Machine) and V2V (Vehicle-to-Vehicle) communication and emergency response networks (Wireless Mesh Networks (WMNs)). WMN applications are diverse [22], including military networks, sensor networks like Body Area Networks (BAN) [33], and environmental data tracking services.

*Dependability.* Reliability and performance aspects play a crucial role for wireless protocols in particular for safety-critical applications such as M2M, V2V and WMN. For example, the quality of the wireless communication channel can greatly affect the resilience of M2M communication or platoon guidance using V2V. For safety-critical applications, an assessment of performance and reliability aspects prior to deployment is highly important. This in particular holds when network protocols become more complex.

*Routing Protocols.* They are key factors for the reliability and performance of wireless networks. They find appropriate paths through the network along which

This work is funded by the German Research Foundation DFG.

M. Gribaudo et al. (Eds.): QEST 2020, LNCS 12289, pp. 54–73, 2020.
https://doi.org/10.1007/978-3-030-59854-9_6

data packets are to be sent. The *Ad-hoc On-demand Distance Vector (AODV)* protocol [38] is a prominent routing protocol in wireless networks. It is one of the four protocols standardised by the IETF MANET Working Group and is used in Zigbee, a low-power, low data rate, and personal area wireless ad-hoc network. AODV finds routes on demand whenever needed. That is, it intends to establish a route between the source and destination node (route discovery) on injecting a data packet. This paper focuses on a *quantitative analysis* of AODV's functionality with respect to uncertainty factors such as message loss, message collision(s) and non-determinism due to concurrency (such as queueing behaviours at network nodes). The central question is how different message loss rates—how unreliable is a communication link?—and how the location of a message loss—which communication link is unreliable?—affect finding shortest/longest routes and packet delivery.

*Dependability Analysis by Simulation.* As the wireless channel is a random communication medium, system design methodologies need to account for stochastic metrics. This has mainly been addressed so far through simulations [2,30,32,36]. However, due to the extremely high reliability requirements—in M2M applications, reliability of 1E-6 are not uncommon—the simulation duration becomes prohibitive when evaluating protocols at a sufficient level of confidence. For safety-critical applications, statistical guarantees are insufficient, and hard guarantees are called for. In addition, while simulations require defining a given set of parameters, a bigger interest is in providing worst-case bounds over a set of design alternatives [42]. Certain alternatives are not quantifiable leading to non-determinism. For simulations this leads to an even higher computational burden as a much wider parameter set has to be evaluated.

*Our Approach.* In order to address these challenges, we apply *probabilistic model checking* [28] to the dependability analysis of routing protocols such as AODV. Given the importance of timing requirements as well as uncertainty aspects such as message loss rates and message collision(s), we employ probabilistic timed automata [31] (PTA) as modelling formalism. These extend finite-state automata with real-valued clock variables and discrete probabilistic branching. PTA are finite symbolic representations of uncountably large Markov decision processes. PTA model checking against a rich set of requirements—including dependability metrics and beyond—is decidable thanks to a finite quotienting technique [35].[1] This model allows for modelling hard time-out values, and unpredictable message delays (by only providing worst and best case delays but no probability distribution). We model the AODV protocol in a *compositional manner*, that is to say, as a network of communicating probabilistic timed automata. Dependability metrics are described as formal requirements using a probabilistic timed temporal logic. In contrast to the symbolic execution of protocols [40], this is a

---

[1] A digital clock semantics [35] where timer values are rounded to integral numbers gives rise to coarse finite abstractions.

model-based technique, i.e., it is applicable in early stages of the protocol development.

*Main Contributions.* The AODV protocol has been subject to different formal modelling and verification studies, e.g. [8,10,14,17,26]. Our work has two significant contributions: a new, quantitative model of the AODV protocol, and a rigorous assessment of key performance metrics of AODV. Our model differs from existing studies of the AODV protocol—or similar routing protocols—in the sense that in addition to formally model AODV's functionality, we consider (a) *different message loss rates*, (b) *possible message collisions*[2], (c) *nondeterminism* (due to e.g., concurrent treatment of message queues at network nodes) as well as (d) hard *network delays*. These aspects are captured by our *compositional, probabilistic timed automata model of AODV*. Related AODV models, analysed using Uppaal-SMC, do e.g., not model medium access control.[3] In contrast to simulations or a statistical model-checking analysis, we exploit *probabilistic model checking* [20], in particular mcsta in the MODEST toolset [21], to quantitatively assess AODV. This approach explores all possible protocol behaviour, thus providing *hard*—rather than statistical—*guarantees* on the quantitative metrics of AODV. By verifying varying message loss rates, our analysis reveals the sensitivity of the AODV's ability to establish short/long routes and deliver data packets via such routes for varying message loss rates.

## 2    Ad-Hoc On-Demand Distance Vector Protocol

AODV [38], a widely used reactive routing protocol applied in WMNs, tries to find a route from a source to a destination on an on-demand basis. We describe the functionality of the AODV with an injected packet to node s destined for node d via intermediate nodes l and m (Fig. 1). The figure shows a linear topology consisting of 4 nodes and their communication via transmission of control messages (broadcast and unicast). Dashed arrows depict broadcast communication and continuous arrows show unicast communication.

Presume node s has a data packet destined for node d. In order to send the data packet, s searches for a valid path to d in its routing table. If such a path exists, the data packet is routed along the path to d. Otherwise, s initiates the *route discovery process* to find a path to d by broadcasting a rreq message. The rreq is rebroadcast and forwarded by node l and m until it reaches d (or an intermediate node that has a valid route to d). On receiving the rreq message, nodes l and m create a routing table entry to establish a *reverse route* back to s. When d (or an intermediate node that has a valid route to d) receives the

---

[2] In our model, collisions can still take place, despite the presence of collision avoidance protocols, and we deliberately decided to consider both collisions and message losses due to other causes separately.

[3] When a message is broadcast or unicast, the message is immediately queued in the buffer of the recipient. This neither involves collisions nor MAC aspects.

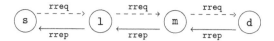

**Fig. 1.** The AODV routing protocol.

rreq, it sends (unicasts) a rrep back to s via the previously established reverse route. Upon receiving the rrep via s (at the end of the route discovery process), an end-to-end route between s and d is established which is used to transfer the data packet. If a rrep message is not received by s, it rebroadcasts another rreq up to RREQ_TRIES times.

When a node (e.g., here node d) unicasts a rrep, it waits to get a rrep-ack message from the rrep receiver (here node m). If such a message is not received, the rrep sender (here node d), broadcasts a rerr message, reporting an unsuccessful unicast of rrep (due to either broken link or lossy link).

## 3    Formal Modelling

Considering different aspects of the AODV's core functionality yields a rather complex protocol state machine. This considerably complicates reliability evaluation and analysis of the protocol. A significant question is to what extent an uncertain environment, e.g., message loss, worsens the protocol's reliability. The following features are crucial when modelling AODV:

– Stochastic impact, e.g., lossy communication and message collision(s);
– Non-determinism, e.g., node's queues and distributed processes interleaving;
– Real-time behaviour, e.g., message delay;
– Very rigorous requirements to be enforced, e.g., route establishment.

A key step in evaluating the reliability and analysis of AODV is a model that reflects the protocol's behaviour and detaches the ambiguities from the informal protocol specification. Traditional evaluation techniques for protocols, including test-bed experiments and simulation, are not suitable due to enormous simulation efforts. Neither analytical protocol evaluation approaches nor Markov-chain models are appropriate due to respectively AODV's complex behaviour and *hard timing* constraints. Therefore, we exploit Probabilistic Timed Automata (PTA) [31], a model that allows dealing with probabilistic behaviour, non-determinism and real-time characteristics [35]. In order to cope with the complexity of the AODV protocol, we follow a compositional modelling approach of PTA. To this end, PTAs are described by the MODEST language [5]. This language supports the composition of PTAs, is conceptually close to programming languages, and yields comprehensive and easily extensible models. The MODEST tool [19,21] allows a direct translation to PTAs [20] (with variables). Using the digital clock semantics [35], such variable PTAs are mapped onto finite (but typically huge) MDPs, by encoding the values of clocks in a symbolic manner. The models are then amenable to model checking by software tools such as PRISM and Storm.

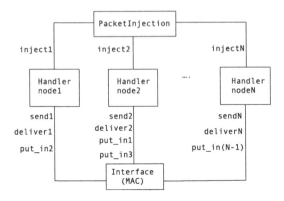

**Fig. 2.** A stacked model view on the modelling.

## 4   A PTA Model of AODV

The PTA model of AODV consists of 3 processes: the injection of data packets by the user (`PacketInjection` process), the AODV node's core-functionality (a `Handler` process for each network node), and the medium access controller (`Interface` process). These processes run in parallel and synchronise on the progression of time and on common actions. A high-level view on our model is given in Fig. 2. Process `PacketInjection` initiates the route discovery by synchronising with the `Handler` of the packet's source. Communication between nodes is modelled via the `Interface` process, receiving messages from the sender's `Handler` and sending them to receiver's `Handler`(s). The MODEST model of AODV (available online[4]) contains nearly thousand lines of codes formalising the protocol and its requirements for our analysis and verification.

In order to define the routing table of nodes, we introduce five matrices (two-dimensional arrays), keeping information about the destination address as `rtDestination`, the last sequence number received from the destination as `rtSeqNum`, the distance from the destination as `rtHops`, the next node along the destination as `rtNextHop` and validity of the route along the destination as `rtflag`, e.g., `rtflag[1][2]` indicates whether node1 has a valid route to node2.

Buffers of nodes are modelled as FIFO queues represented by eight arrays of integers[5] with constant size `QLength = 20`. When a node buffers a message, these arrays are updated accordingly. Variables `buffer` and `head_pointer` show where the next receiving message should be stored and which message is the next to be processed, respectively. For instance, `msgLocalOriginator` keeps track of the receiving message's originator and if a new message is subject to be stored, its originator is entered as: `msgLocalOriginator [(buffer+head_pointer)%QLength]=msgOriginator`.

---

[4] https://git.rwth-aachen.de/mojgan.kamali/qest2020.
[5] Due to limitation of MODEST in defining *struct* type, we define one array for every message element in the nodes' buffer.

We have also introduced several global variables for the nodes communication purposes. When a message is aimed for a broadcast, the corresponding Handler copies the message information into the global variables to hand them to the Interface. For instance, msgOriginator, msgSender, msgDestination are respectively used to store the global message's originator, sender and destination. The routing table of nodes are defined globally to carry out verification since properties cannot access processes' local variables.

## 4.1   Packet Injector Model

Process PacketInjection models injecting the users data packets to a source node. The first data packet is injected to the source, say OIP, targeted for the destination, say DIP, in order for AODV to start the route discovery. The rest of the data packets can be injected subsequently. The total number of injected packets is defined by the constant Max_sent_packets (equals to 1 in our experiments) while the variable counter is increased by 1 each time a new data packet is injected. It can be increased up to Max_sent_packets, and also indicates the number of times the process is executed recursively.

## 4.2   Node Model

The main functionality of AODV at a given node, e.g., generating control messages, broadcasting, receiving, discarding and rebroadcasting them, updating routing information, delays, etc., is modelled by the recursive Handler process. It includes a process main to avoid initialising the Handler's local variables on each recursive call. We call the main process recursively inside the Handler. In this section, we present a key fragment of the Handler model, conditions and updates (Listing 1.1 and Fig. 3). Numbers on the states of Fig. 3 represent the corresponding line numbers in Listing 1.1.

As described in Sect. 2, on receiving a packet, the source of the packet looks for the information about the packet destination in its routing table. If an entry exists, the packet is routed towards the destination. Otherwise, the source initiates a route discovery, broadcasting rreq messages through the network (line 7 ff.), where the condition of keyword when asserts that the received message is a packet (msgLocalType[head_pointer]==PACKET), valid information about destination is absent (line 8) and the handling node is not the destination of the packet (line 9). If all conditions are met, the Handler updates its unique sequence number (sn) as well as its rreqID which will be inserted into the soon-to-be broadcast rreq message, resets its local clock c and waits for broadcasting rreq. The sequence number and rreqID in rreq messages are used for indicating the freshness of the messages. In the time interval [Delay-Spread, Delay+Spread] where Delay = 40 (ms) and Spread = 5 (ms), there are two probabilistic choices denoted by Keyword palt (line 13):

**Listing 1.1.** MODEST description of the `Handler` process snippet.

```
1  Handler(int(1..N)ip,int(0..100)loss){
2  //initialise local variables
3  ...
4  urgent {= buffer=0, head_pointer=0 =};main()
5  main(){alt{...
6  //broadcast rreq (initiate route discovery)
7  :: when(msgLocalType[head_pointer] == PACKET
8        && !rtflag[ip][msgLocalDestination[head_pointer]]
9        && msgLocalDestination[head_pointer]!=ip
10       && ...)
11    tau{= sn=(sn==0) ? 0 : sn+1, rreqID++, c=0 =};
12    invariant(c<=Delay+Spread)
13    when(c>=Delay-Spread) palt{
14      :Collision: collision()
15      :100-Collision: send
16      {= rreqs[ip][rreqID]=true, msgType=RREQ,  msgHops=0,
17      msgrreqID=rreqID, ... =}}
18  //queue/lose messages
19  :: put_in; urgent alt{
20    :: when(buffer<QLength) urgent palt{
21      :100-loss: {=
22      msgLocalOriginator[(buffer+head_pointer)%QLength]=
23            msgOriginator,... =}
24      :loss: {==}}
25    :: when(buffer>=QLength){==}}
26  //discard rreq
27  :: when(msgLocalType[head_pointer]==RREQ
28        && rreqs[msgLocalOriginator[head_pointer]]
29          [msgLocalrreqID[head_pointer]])
30    tau{= rtDestination[ip][msgLocalSender[head_pointer]]=
31          msgLocalSender[head_pointer],
32      rtNextHop[ip][msgLocalSender[head_pointer]]=
33          msgLocalSender[head_pointer],
34      msgLocalType[head_pointer] = NONE, ... =}
35  //unicast rrep
36  :: when(msgLocalType[head_pointer]==RREQ
37        && !rreqs[msgLocalOriginator[head_pointer]]
38          [msgLocalrreqID[head_pointer]]
39        && msgLocalDestination[head_pointer]==ip && ...)
40    tau{= c=0 =}
41    invariant(c<=Delay+Spread)
42    when(c>=Delay-Spread) palt{
43      :Collision: collision()
44      :100-Collision: deliver
45      {= msgType=RREP, msgHops=0,... =};
46      urgent tau {= NextHop=rtNextHop[ip][msgOriginator]=}}
47  ...
48  }; main()}}
```

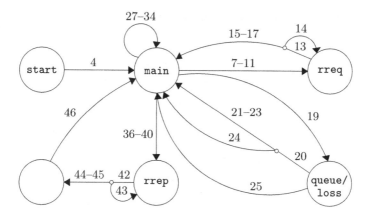

**Fig. 3.** PTA model of the `Handler` process snippet (Listing 1.1).

- A collision occurs. In this study, a collision happens when a node is broadcasting/unicasting a message while receiving messages from its neighbour(s) at the same time with weight `Collision`. In such a situation, the receiving node loses the received message(s) and does not broadcast/unicast its message to the node(s) it received message(s) from (while it still sends its message to other nodes in transmission range). The recursive process `collision()` handles collision(s) by marking nodes that cannot receive messages because of the collision(s) (line 14). In our study, the value of `Collision` is 1%, modelling the probabilistic collision given a low amount of network traffic.
- A broadcast of `rreq` happens without collision (with weight `100-Collision`): `Handler` and `Interface` synchronise on action `send`, broadcasting `rreq` (lines 15–17). This happens first by transferring the `rreq` to the `Interface`, and later by transmission of the `rreq` to the neighbouring nodes.

We model resending of `rreq` up to `RREQ_tries` by assuming: the first `rreq` is broadcast by the source, a `rrep` is not yet received by the source, there are no more messages in the network to be broadcast/unicast (no `rrep` is on the way). In this case, the next `rreq` is broadcast.

The `Interface` process is discussed in Sect. 4.3. For the sake of understandability, we assume that synchronisation with the `Interface` is first done via the `send` action and subsequently by the `put_in` action, transferring the `rreq` from the `Interface` to the neighbouring node (line 19). Action `put_in` is used for transferring all types of messages between `Handler` and `Interface`, e.g., `rreq`, `rrep`, `rerr` and `packet`. After synchronising on `put_in`, there are two non-deterministic choices labeled by `urgent alt` followed by conditions: (1) the node's buffer is not full (`buffer<QLength`) and (2) otherwise (`buffer>=QLength`). The `urgent` keyword forces immediate execution of the following process behaviour.

In case (1), the message will be either stored in the first free positions of the node's buffer with weight `100-loss` (lines 21–23) or be lost with weight `loss` (line 24). Keyword `palt` denotes the probabilistic choice. The statement means that a message is lost with probability `loss/100`. In case (2), the message is dropped as the buffer is full.

The internal actions indicated by `tau` actions cope with the internal behaviour of node (i.e., behaviours that are independent of other nodes), e.g., discarding queued messages. For instance in line 27, the `when` condition states that the node does not rebroadcast/forward the message, if the queued message is a `rreq` (`msgLocalType[head_pointer]==RREQ`) and the `Handler` has already originated/received the `rreq` with `rreqID` (lines 28–29). In such a situation, the node's routing table is updated only for the sender of the `rreq` (lines 30–34) and subsequently the `rreq` is deleted from the queue.

Handling the unicast of `rrep` by either the destination or the intermediate nodes is also modelled in this process. Line 36 ff. presents the case when a `rreq` is received by the destination. The `when` condition asserts that the message to be handled is a `rreq` message (`msgLocalType[head_pointer] == RREQ`), the `rreq` is not received before (lines 37–38) and the destination of the messages is the receiving node (line 39). If this condition holds, `c` is reset, and the `Handler` waits for unicasting a `rrep` to the originator of the `rreq`. In the interval `[Delay-Spread, Delay+Spread]`, there are two probabilistic choices: either a collision happens as described earlier, or a `rrep` is sent to the next node towards the `rreq` originator (`NextHop` in line 46) using the `deliver` action.

### 4.3   Medium Access Control (MAC) Model

The recursive `Interface` process models the communication between nodes, how messages/packets are flooded through the network (broadcast/unicast), and message collision. It synchronises with network nodes via non-deterministic actions, e.g., `send1,send2, deliver1`, etc.

When a node, say `Handler node1`, broadcasts a `rreq` or a `rerr` message, it synchronises with the `Interface` on the `send1` action, transferring the message from the node to the medium. The same is the case for `Handler node2`, etc. When `Handler node1` or `Handler node2` unicasts a `rrep` or a `packet`, they respectively synchronise with the `Interface` on `deliver1` or `deliver2` actions.

Based on the taken action, e.g., `send1` or`deliver1`, the interface realises that the message is either subject to broadcast or unicast.

– If broadcast, the `Interface` checks (1) which nodes are the receivers using the connectivity matrix `isconnected` (showing the direct connectivity between nodes) and (2) whether the receivers are allowed to receive the message (if they are not marked by the `Handler` as the effect of collision(s)).
– If unicast, the `Interface` investigates (1) who is the receiving node, and (2) whether the receiving node is allowed to receive the message (if it is not marked by the `Handler` as the effect of collision).

Then, the message is broadcast or unicast using the `put_in` action of the receiving node(s), or assumed collided (unmarking nodes from collision happens here).

# 5    Analysis Results

Developing a PTA model of AODV enabled deep insight into different aspects of the AODV protocol. Formal verification of the PTA model helps to check functional correctness of the protocol as well as QoS properties. We focus in particular on the following questions: (a) what is the influence of loss probabilities and message collision on AODV discovering short (or long) routes for different topologies?, (b) to what extent does it matter which link(s) in the network are faulty?, (c) how is route discovery affected if several links are unreliable?, (d) how likely are packets to be delivered once route discovery has completed?, and (e) how does broadcasting of `rreq` messages up to `RREQ_tries` times help the protocol to find routes (short or long) and to deliver data packets?

## 5.1    Evaluation Metrics

The focus of our formal verification is on route establishment (short or long) and packet delivery properties of AODV in presence of varying lossy communication channels, possible message collision(s) as well as broadcast retries by the packet source. We therefore consider the probability of messages getting lost in the reception of other nodes in the interval $[0, 100]$ percent and broadcasting of `rreq` messages by the packet source up to `RREQ_tries` times. In our experiments, we fix the message collision probability to 1%.

In order to investigate the correctness of our PTA model, we investigated a.o. (a) absence of a deadlock, (b) the probability of all nodes besides the packet source and destination, respectively, to broadcast `rreq` and `rrep` messages, and (c) the probability that irrelevant nodes update their routing tables (correct routes were found).

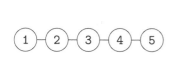

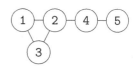

**Fig. 4.** Chain topology.        **Fig. 5.** Triangle topology.        **Fig. 6.** Grip topology.

## 5.2    Verified Properties

We verified the following three properties. They are all formulated using probabilistic (timed) CTL and are of the form `Pmax(<> predicate)` where `<>` stands for *eventually*. Intuitively, this formula denotes the maximal probability that a protocol state is eventually reached for which `predicate` holds. `Pmax` refers to considering all possible resolutions of the non-determinism in the AODV's PTA model such as the interleaving of all distributed MODEST processes such as sending and receiving messages.[6] Considering `Pmax` takes an angelic viewpoint on the protocol's behaviour. The three properties considered are:

---

[6] We do not verify our properties for `Pmin` as this probability is always 0 in our analysis.

**Table 1.** Evaluation scenarios.

|  | loss1 | loss2 | loss3 | loss4 | loss5 |
|---|---|---|---|---|---|
| **Scenario1** | 0 | $[0, 100]$ | $[0, 100]$ | $[0, 100]$ | 0 |
| **Scenario2** | 0 | 0 | 0 | 0 | $[0, 100]$ |

$$\textbf{P1:}\texttt{Pmax}(<> (\texttt{Hops} == \texttt{Distance}))$$

referring to the fact that the distance (`Hops`) of the route discovered by the source equals to the shortest distance (`Distance`) in the network between the source and destination. The second property

$$\textbf{P2:}\texttt{Pmax}(<> (\texttt{Hops} > \texttt{Distance}))$$

refers to the fact that the route discovered by the source is longer than the shortest possible route in the network. Finally,

$$\textbf{P3:}\texttt{Pmax}(<> (\texttt{deliver}[\text{DIP}] == \texttt{Max\_sent\_packet}))$$

refers to the fact that all packets are eventually delivered at the destination. We verified these properties for different message losses (Table 1), 1% probability of messages collision(s) as well as for different `RREQ_tries` between [0,2] times.

### 5.3 Formal Verification Results

*Verification Set-up.* We carried out the formal verification of our PTA model of AODV for several network topologies up to 5 nodes using the model checker `mcsta` in the MODEST toolset [21]. We show the verification results for the network topologies depicted in Fig. 4, 5 and 6. The results for these topologies are illustrative for the results obtained for other topologies. Our analysis is carried out on a CPU core of an Intel Skylake Platinum 8160 with 2.1 GHz clock speed and 192 GB memory.

For each verification experiment, we inject a packet to the source OIP $= 1$ to be delivered at destination DIP $= 5$. This initiates a route discovery process. We note that our formal analysis of AODV confirms that it does not always find a shortest path as also shown in [14]. This behaviour occurs if nodes on the short path are slower than the nodes on the long path when processing the messages in their queues. Our work extends [14] by adding the probability of losing messages and the probability of message collision(s) as well as considering the broadcast of `rreq` messages up to `RREQ_tries` times, and studies the effects on the ability of the protocol to find a shortest path.

We verify the properties **P1–P3** under varying channel conditions, 1% message collision and different values of `RREQ_tries`. We consider the loss probability for each node as a parameter with values in the interval $[0, 100]$ (as percentage) shown in Table 1. In **scenario1**, the loss probabilities for nodes 1 and 5 are zero (reliable communication) while the loss probability for nodes 2, 3, 4 varies

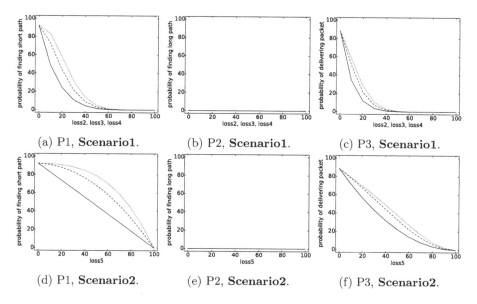

(a) P1, **Scenario1**.          (b) P2, **Scenario1**.          (c) P3, **Scenario1**.

(d) P1, **Scenario2**.          (e) P2, **Scenario2**.          (f) P3, **Scenario2**.

**Fig. 7.** Chain topology. **P1**: route discovery via short path, **P2**: route discovery via long path, **P3**: packet delivery. **Scenario1**: unreliable at node 2, 3 and 4 and **Scenario2**: unreliable at node 5. The solid curve corresponds to RREQ_tries = 0, the dashed curve to RREQ_tries = 1, and the dotted curve to RREQ_tries = 2.

(unreliable communication). In **scenario2**, the loss probability for nodes 1, 2, 3 and 4 are zero while the loss probability for node 5 is varying. The probabilities regarding the properties are multiplied by 100 (Y axis); 0% and 100% correspond to probability 0 and 1, respectively.

Chain *Topology:* **Scenario1:** As shown in Fig. 7a, the probability of finding the short path is drastically decreasing when all loss probabilities are increasing (this probability is 92% when all loss probabilities are 0% because of the message collision(s) probability of 1%)[7]. Figure 7b shows that the probability of finding a long path is 0% as expected. These figures together indicate that when all loss probabilities are about 70%, no route (neither short nor long) towards DIP is found even with the increase of RREQ_tries (dotted curves in both figures). Therefore, the probability of delivering packets to DIP (Fig. 7c) is decreasing with an even steeper slope than in Fig. 7a. For instance, when all loss probabilities are around 50%, no packet is delivered even when increasing RREQ_tries.

**Scenario2:** Fig. 7d and 7f show a decrease in the probability of finding the short path and packet delivery, respectively, when loss5 increases, whereas the

---

[7] In all of the topologies when all loss probabilities are 0%, the 1% probability of message collision(s) yields to a decrease in the probabilities of finding the short/long path (if any found) as well as the probability of packet delivery. This means that these probabilities are never 100% even in a fully reliable situation with all loss probabilities at 0% due to possible message collision(s) probability.

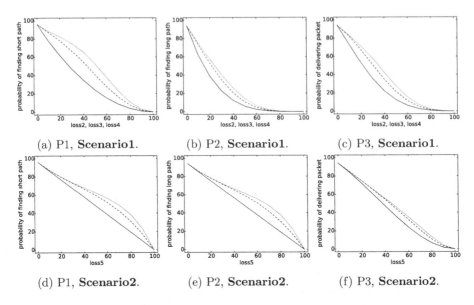

(a) P1, **Scenario1**.　　(b) P2, **Scenario1**.　　(c) P3, **Scenario1**.

(d) P1, **Scenario2**.　　(e) P2, **Scenario2**.　　(f) P3, **Scenario2**.

**Fig. 8.** `Triangle` topology. **P1**: route discovery via short path, **P2**: route discovery via long path, **P3**: packet delivery. **Scenario1**: unreliable at node 2, 3 and 4 and **Scenario2**: unreliable at node 5. The solid curve corresponds to `RREQ_tries` = 0, the dashed curve to `RREQ_tries` = 1, and the dotted curve to `RREQ_tries` = 2.

probability of finding the long path stays constant (always 0%). Increasing `loss5` does not have any effect on finding the long path even if `RREQ_tries` increases, since this long route is never found (as shown in Fig. 7e). Figure 7d and 7f show that increasing `RREQ_tries` can help finding the short path as well as delivering packets. This is in contrast to Fig. 7e.

`Triangle` *Topology:* **Scenario1:** Fig. 8a shows that the probability of finding the short path from `OIP` to `DIP` (**P1**) is monotonically decreasing for all curves on increasing the loss probabilities. The same occurs for the probability of finding the long path from `OIP` to `DIP` (**P2**): it decreases when all loss probabilities increase (Fig. 8b). For instance, if all loss probabilities are 20%, the probability of finding the short path to `DIP` for `RREQ_tries` = 0 is 61% (Fig. 8a, solid curve) while the probability of finding the long path for `RREQ_tries` = 0 is 38% (Fig. 8b, solid curve). This indicates that—as expected—the probability of discovering the short route increases when the long route becomes more unreliable. Otherwise when the loss probability on both short and long path is equal, both the short and long path are discoverable by AODV. The descending curve in Fig. 8c depicts a decrease in packet delivery probability when the loss probability on both routes increases.

Our analysis shows that finding the longer path by AODV is independent of the increase/decrease of the loss probability on the short path. For instance, in a fully reliable situation with all loss probabilities at 0%, a short path as well as a long path are found with the probabilities of 96% and 94%, respectively. So the probability of finding routes only depends on how lossy that specific route is. The reason for finding the long path is that when AODV receives a `rreq` with a unique ID it processes the message. After that, any `rreq` with the same ID will be discarded no matter if it has arrived via a shorter path.[8]

Our analysis also quantifies the impact of the number of retransmissions of `rreqs`. As shown in Fig. 8a–8c, when `RREQ_tries` increases from 0 to 2 (solid, dashed and dotted curves), the probabilities of finding a short and long path as well as packet delivery increase. For example, when all probabilities are 40%, the short path to `DIP` for `RREQ_tries` = 0 is discovered with probability about 34%. This value increases to around 54% for `RREQ_tries` = 1 and 65% for `RREQ_tries` = 2.

**Scenario2:** Fig. 8d, 8e and 8f present our results when destination `DIP` loses messages. Figure 8d presents that larger values for `loss5` yield a decrease in the probability of finding the short path to `DIP` (**P1**) as well as the probability of finding the long path to `DIP` (**P2**). Therefore when `loss5` increases, the probability of finding the short and long path to `DIP` decreases from 96% (`loss5` = 0) to 0% (`loss5` = 100). Figure 8f depicts that the probability of packets being delivered at `DIP` (**P3**) is also decreasing when `loss5` increases. This is because `loss5` is the loss probability of `DIP`. These figures show when `loss5` is less than 34%, increasing the value of `RREQ_tries` from 1 to 2 does not have any effect on the probability of finding the short/long path or packet delivery (dashed and dotted curves overlap). However when `loss5` increases, larger values of `RREQ_tries` increase the probabilities of short and long path as well packet delivery. For instance, considering `loss5` = 40, the probability of finding the short path to `DIP` for `RREQ_tries` = 0 is about 57%. This probability increases to around 66% for `RREQ_tries` = 1 and 67% for `RREQ_tries` = 2. This quantifies the effect of increasing `RREQ_tries`. As shown in Fig.8f, `RREQ_tries` does not play a significant role for packet delivery. For example, when `loss5` = 40%, the probability of delivering a packet is about 46% for `RREQ_tries` = 0. This probability increases to only around 54% for `RREQ_tries` = 1 and 55% for `RREQ_tries` = 2.

`Grip` *Topology:* **Scenario1:** Fig. 9a and 9c, respectively, show the decrease in the probability of finding the short path and the packet delivery probability when all loss probabilities increase. These two figures reflect a similar behaviour as in Fig. 8a and 8c for the `Triangle` topology. However, Fig. 9b does not have a monotonic trend (increases first and then decreases). The figure shows that when the loss probability is 0%, AODV finds the long path for such a topology

---

[8] We also analysed an alternative variant of the protocol which not only checks for the `rreq` ID, but also if the message has arrived via a short path. Then, the node processes the message and rebroadcasts it. The results of our analysis show a seemingly contradictory results in this situation: when the loss probability increases on the short path, then the probability of finding the long path is monotonically increasing (in contrast to Fig. 8b). This shows that finding the longer path by AODV is somehow dependent on the increase/decrease of the loss probability on the short path in this setting.

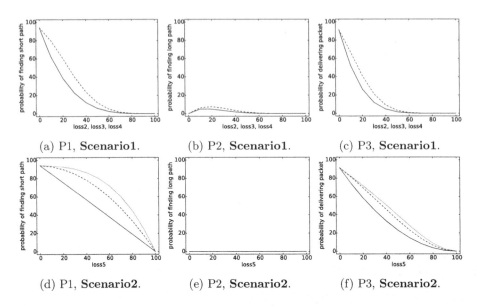

(a) P1, **Scenario1**.     (b) P2, **Scenario1**.     (c) P3, **Scenario1**.

(d) P1, **Scenario2**.     (e) P2, **Scenario2**.     (f) P3, **Scenario2**.

**Fig. 9.** `Grip` topology. **P1**: route discovery via short path, **P2**: route discovery via long path, **P3**: packet delivery. **Scenario1**: unreliable at node 2, 3 and 4 and **Scenario2**: unreliable at node 5. The solid curve corresponds to `RREQ_tries` = 0, the dashed curve to `RREQ_tries` = 1, and the dotted curve to `RREQ_tries` = 2.

with the probability of 0%. When all loss probabilities increase to 20%, the probability of finding the long path increases to about 4% for `RREQ_tries` = 0 and to 7% for `RREQ_tries` = 1. After that point, it decreases until it reaches 0%. The value of `RREQ_tries` has a considerable effect on the number of states and time for verification. For `RREQ_tries` = 2 we encounter a time out after 432 000 s. The figure depicts that all loss probabilities equal to 20% provide the peak for finding the long path. This unusual behaviour can be due to the fact that when `rreq` is broadcast via node 1, node 3 receives the message while node 2 loses it. Later when node 2 gets the message from node 3, it assumes that the messages is a new `rreq`. Node 2 processes it and rebroadcasts it. Then the longer path is discovered.

**Scenario2:** Fig. 9d and 9f depict a decrease in the probability of finding the short path and delivering packet, respectively, when `loss5` increases, whereas the probability of finding the long path stays constant (always 0%). This shows that increasing `loss5` does not have any effect on finding the long path even if `RREQ_tries` increases, since this route is never found. Figure 9d and 9f show that increasing `RREQ_tries` improves finding the short path as well as delivering packets, in contrast to Fig. 9e.

**Table 2.** Number of explored states and time for verification

| RREQ_tries | Chain topology | | | | Triangle topology | | | | Grip topology | | | |
|---|---|---|---|---|---|---|---|---|---|---|---|---|
| | Scenario1 | | Scenario2 | | Scenario1 | | Scenario2 | | Scenario1 | | Scenario2 | |
| | States | Time | States | Time | States | Time | States | Time | States | Time | States | Time |
| 0 | 5 646 | 1 | 2 698 | 1 | 75 092 | 13 | 60 304 | 9 | 254 958 | 78 | 62 486 | 11 |
| 1 | 58 984 | 9 | 4 288 | 1 | 1 225 652 | 1 314 | 155 988 | 34 | 13 278 354 | 150 907 | 220 618 | 67 |
| 2 | 545 200 | 244 | 5 878 | 1 | 17 594 464 | 213 646 | 347 088 | 146 | time out | time out | 693 674 | 626 |

## 5.4    Verification Statistics

Table 2 summarises the number of explored states of the underlying MDP of our PTA model together with the time (in seconds) for exploring the MDP for each topology and scenario. The verification statistics show that when the network becomes more complex (number of links and/or node connectivity increase), the number of explored states and time for exploring those states increase accordingly. For instance in the Chain topology, each node is connected to at most 2 nodes, where in the Triangle topology, every node is connected to exactly 2 nodes. The connectivity increases to 3 nodes for the Grip topology (node 2 is connected to 3 nodes). The value of RREQ_tries has a substantial impact on the state space since the retransmission of rreqs by the source happens only when all previous messages are handled. Increasing the number of nodes that lose messages also affect on the number of explored states due to the branching for loss probabilities, e.g. **Scenario2** has fewer number of explored states compared to **Scenario1**.

## 6    Related Work

Wireless protocols, in particular AODV [4,7,9,16,29], have so far been primarily analysed using simulation techniques [2,6,30,36]. In the later case, Cavin *et al.* [6] modelled a flooding algorithm in different simulators (OPNET, NS-2, and GloMoSim). Flooding is a very simple broadcast algorithm and is elementary for several wireless network protocols, e.g., routing protocols. Their study shows significantly different and barely comparable simulation results despite carefully setting the same parameters and using the same scenarios. This can be for instance due to mismatching of the modellisation of each·simulator.

Applying statistical model checking [1], the technique that combines discrete-event simulation and model checking, has also received quite some attention. In contrast to classical model checking, this technique monitors many possible traces of the formal model, and (as simulation) provides statistical guarantees. The Uppaal-SMC [12] has been applied to the DYMO [39] and the AODVv2 [37] protocols (the refined versions of AODV) [25,27]. Statistical model checking has also been applied to analyse the first version of the AODV protocol [11,23,24]. Our analysis of AODV differs from the latter studies in the sense that we apply a model-checking technique that gives us a *hard guarantee on the achieved results*. Moreover, we consider the probability of losing messages as a parameter that can be varied in the interval $[0, 100]$, facilitating the protocol analysis for different message loss values, whereas this probability was fixed in the other studies. Varying loss probability gives us the insight on the *sensitivity* of the AODV protocol relative to message loss probability.

Studies using Uppaal [3] include the formal modelling and analysis of the LMAC protocol [41] for wireless sensor networks [15], the OLSR protocol [26] and the BATMAN [8] routing protocol for mesh networks. The latter study verified properties such as loop freedom, bidirectional link discovery, and route discovery as well as revealing several ambiguities in the RFC [34]. Formal analysis of the AODV protocol applying model checking has been studied in [10,14,17]. Our work differs from the other existing studies of the AODV protocol in the sense that in addition to formally model the AODV protocol, we also consider lossy communication as well as message collision. We carry out a rigorous analysis on how lossy communication can affect the network reliability.

There are several studies that have applied the MODEST modelling language in the context of wireless networks [13,18,20]. The paper [20] analyses the Bounded Retransmission Protocol (BRP), the IEEE 802.11 wireless networks and the IEEE 802.3 CSMA/CD Protocol. In [13], the wireless EchoRing protocol is modelled and analysed. A formal study of the ZigBee is carried out [18].

# 7   Conclusions

In this paper, we have presented the formal modelling and the resulting verification of the AODV routing protocol, an employed routing protocol in WMNs. We provided a compositional modelling of AODV, reflecting the protocol's core functionality. The formal modelling revealed AODV's malfunctioning regarding finding the shortest path, discussed also in [14] (the AODV protocol does not always find the shortest path). The PTA model of AODV composed of different processes interleaving with each other allows us to model the realistic behaviour of wireless networks, in particular message delay, message loss, message collision(s), etc. Our approach allows determining the correctness as well as studying the reliability and performance under different conditions. In particular, we analysed AODV in the presence of random message loss and message collision and showed how varying these parameters affects the protocol reliability and performance. Future work includes applying parameter synthesis to AODV analysis.

**Acknowledgment.** We thank Peter Höfner (ANU) for discussions on AODV, Arnd Hartmanns (Twente) for assistance using `mcsta` and the reviewers for their helpful comments.

# References

1. Agha, G., Palmskog, K.: A survey of statistical model checking. ACM Trans. Model. Comput. Simul. **28**(1), 6:1–6:39 (2018)
2. Alsheikh, M.A., Hoang, D.T., Niyato, D., Tan, H.P., Lin, S.: Markov decision processes with applications in wireless sensor networks: a survey. IEEE Commun. Surv. Tutor. **17**(3), 1239–1267 (2015)
3. Behrmann, G., David, A., Larsen, K.G.: A tutorial on UPPAAL. In: Bernardo, M., Corradini, F. (eds.) SFM-RT 2004. LNCS, vol. 3185, pp. 200–236. Springer, Heidelberg (2004). https://doi.org/10.1007/978-3-540-30080-9_7
4. Biswas, A., Saha, B., Guha, S.: Performance analysis of AODV and DSR routing protocols for Ad-Hoc networks. In: Thilagam, P.S., Pais, A.R., Chandrasekaran, K., Balakrishnan, N. (eds.) ADCONS 2011. LNCS, vol. 7135, pp. 297–305. Springer, Heidelberg (2012). https://doi.org/10.1007/978-3-642-29280-4_36
5. Bohnenkamp, H.C., D'Argenio, P.R., Hermanns, H., Katoen, J.: MODEST: a compositional modeling formalism for hard and softly timed systems. IEEE Trans. Software Eng. **32**(10), 812–830 (2006)
6. Cavin, D., Sasson, Y., Schiper, A.: On the accuracy of MANET simulators. In: POMC, pp. 38–43. ACM (2002)
7. Chakeres, I.D., Belding-Royer, E.M.: AODV routing protocol implementation design. In: Proceedings of the 24th International Conference on Distributed Computing Systems Workshops, 2004, pp. 698–703 (2004)
8. Chaudhary, K., Fehnker, A., Mehta, V.: Modelling, verification, and comparative performance analysis of the B.A.T.M.A.N. protocol. In: MARS, EPTCS, vol. 244, pp. 53–65 (2017)
9. Chavan, A., Kurule, D., Dere, P.: Performance analysis of AODV and DSDV routing protocol in MANET and modifications in AODV against black hole attack. Procedia Comput. Sci. **79**, 835–844 (2016)
10. Chiyangwa, S., Kwiatkowska, M.: A timing analysis of AODV. In: Steffen, M., Zavattaro, G. (eds.) FMOODS 2005. LNCS, vol. 3535, pp. 306–321. Springer, Heidelberg (2005). https://doi.org/10.1007/11494881_20
11. Dal Corso, A., Macedonio, D., Merro, M.: Statistical model checking of ad hoc routing protocols in lossy grid networks. In: Havelund, K., Holzmann, G., Joshi, R. (eds.) NFM 2015. LNCS, vol. 9058, pp. 112–126. Springer, Cham (2015). https://doi.org/10.1007/978-3-319-17524-9_9
12. David, A., Larsen, K.G., Legay, A., Mikučionis, M., Poulsen, D.B.: Uppaal SMC tutorial. Int. J. Softw. Tools Tech. Trans. **17**(4), 397–415 (2015)
13. Dombrowski, C., Junges, S., Katoen, J., Gross, J.: Model-checking assisted protocol design for ultra-reliable low-latency wireless networks. In: IEEE 35th Symposium on Reliable Distributed Systems (SRDS), pp. 307–316. IEEE (2016)
14. Fehnker, A., van Glabbeek, R., Höfner, P., McIver, A., Portmann, M., Tan, W.L.: Automated analysis of AODV using UPPAAL. In: Flanagan, C., König, B. (eds.) TACAS 2012. LNCS, vol. 7214, pp. 173–187. Springer, Heidelberg (2012). https://doi.org/10.1007/978-3-642-28756-5_13
15. Fehnker, A., van Hoesel, L., Mader, A.: Modelling and verification of the LMAC protocol for wireless sensor networks. In: Davies, J., Gibbons, J. (eds.) IFM 2007. LNCS, vol. 4591, pp. 253–272. Springer, Heidelberg (2007). https://doi.org/10.1007/978-3-540-73210-5_14

16. Garg, S., Verma, A.K.: Simulation and comparison of AODV variants under different mobility models in MANETs. In: Vishwakarma, H.R., Akashe, S. (eds.) Computing and Network Sustainability. LNNS, vol. 12, pp. 333–342. Springer, Singapore (2017). https://doi.org/10.1007/978-981-10-3935-5_34

17. van Glabbeek, R., Höfner, P., Portmann, M., Tan, W.L.: Modelling and verifying the AODV routing protocol. Distrib. Comput. **29**(4), 279–315 (2016). https://doi.org/10.1007/s00446-015-0262-7

18. Groß, C., Hermanns, H., Pulungan, R.: Does clock precision influence Zigbee's energy consumptions? In: Tovar, E., Tsigas, P., Fouchal, H. (eds.) OPODIS 2007. LNCS, vol. 4878, pp. 174–188. Springer, Heidelberg (2007). https://doi.org/10.1007/978-3-540-77096-1_13

19. Hahn, E.M., Hartmanns, A., Hermanns, H., Katoen, J.P.: A compositional modelling and analysis framework for stochastic hybrid systems. Formal Methods Syst. Design **43**(2), 191–232 (2013). https://doi.org/10.1007/s10703-012-0167-z

20. Hartmanns, A., Hermanns, H.: A Modest approach to checking probabilistic timed automata. In: QEST, pp. 187–196. IEEE (2009)

21. Hartmanns, A., Hermanns, H.: The modest toolset: an integrated environment for quantitative modelling and verification. In: Ábrahám, E., Havelund, K. (eds.) TACAS 2014. LNCS, vol. 8413, pp. 593–598. Springer, Heidelberg (2014). https://doi.org/10.1007/978-3-642-54862-8_51

22. Hoebeke, J., Moerman, I., Dhoedt, B., Demeester, P.: An overview of mobile ad hoc networks: applications and challenges. Commun. Netw. **3**, 60–66 (2004)

23. Höfner, P., McIver, A.: Statistical model checking of wireless mesh routing protocols. In: Brat, G., Rungta, N., Venet, A. (eds.) NFM 2013. LNCS, vol. 7871, pp. 322–336. Springer, Heidelberg (2013). https://doi.org/10.1007/978-3-642-38088-4_22

24. Höfner, P., Kamali, M.: Quantitative analysis of AODV and its variants on dynamic topologies using statistical model checking. In: Braberman, V., Fribourg, L. (eds.) FORMATS 2013. LNCS, vol. 8053, pp. 121–136. Springer, Heidelberg (2013). https://doi.org/10.1007/978-3-642-40229-6_9

25. Kamali, M., Fehnker, A.: Adaptive formal framework for WMN routing protocols. In: Bae, K., Ölveczky, P.C. (eds.) FACS 2018. LNCS, vol. 11222, pp. 175–195. Springer, Cham (2018). https://doi.org/10.1007/978-3-030-02146-7_9

26. Kamali, M., Höfner, P., Kamali, M., Petre, L.: Formal analysis of proactive, distributed routing. In: Calinescu, R., Rumpe, B. (eds.) SEFM 2015. LNCS, vol. 9276, pp. 175–189. Springer, Cham (2015). https://doi.org/10.1007/978-3-319-22969-0_13

27. Kamali, M., Merro, M., Dal Corso, A.: AODVv2: performance vs. loop freedom. In: Tjoa, A.M., Bellatreche, L., Biffl, S., van Leeuwen, J., Wiedermann, J. (eds.) SOFSEM 2018. LNCS, vol. 10706, pp. 337–350. Springer, Cham (2018). https://doi.org/10.1007/978-3-319-73117-9_24

28. Katoen, J.: The probabilistic model checking landscape. In: LICS, pp. 31–45. ACM (2016)

29. Kolipaka, S., Bhandari, B.N., Rajani, A.: Performance analysis of AODV with multi-radio in hybrid wireless mesh network. In: Eleventh International Conference on Wireless and Optical Communications Networks (WOCN), pp. 1–5. IEEE (2014)

30. Kwak, B.J., Song, N.O., Miller, L.E.: Performance analysis of exponential backoff. IEEE/ACM Trans. Network. **13**(2), 343–355 (2005)

31. Kwiatkowska, M.Z., Norman, G., Segala, R., Sproston, J.: Automatic verification of real-time systems with discrete probability distributions. Theor. Comput. Sci. **282**(1), 101–150 (2002)
32. Misic, J., Shafi, S., Misic, V.B.: Performance of a beacon enabled IEEE 802.15.4 cluster with downlink and uplink traffic. IEEE Trans. Parallel Distrib. Syst. **17**(4), 361–376 (2006)
33. Negra, R., Jemili, I., Belghith, A.: Wireless body area networks: applications and technologies. Procedia Comput. Sci. **83**, 1274–1281 (2016)
34. Neumann, A., Aichele, C., Lindner, M., Wunderlich, S.: Better approach to mobile ad-hoc networking (BATMAN). Internet draft00 (2008). https://tools.ietf.org/html/draft-wunderlich-openmesh-manet-routing-00
35. Norman, G., Parker, D., Sproston, J.: Model checking for probabilistic timed automata. Formal Methods Syst. Des. **43**(2), 164–190 (2013). https://doi.org/10.1007/s10703-012-0177-x
36. Obaidat, M.S., Green, D.B.: Simulation of wireless networks. In: Obaidat, M.S., Papadimitriou, G.I. (eds.) Applied System Simulation. Springer, Boston (2003). https://doi.org/10.1007/978-1-4419-9218-5_6
37. Perkins, C., Stan, R., Dowdell, J., Steenbrink, L., Mercieca, V.: DynamicMANET On-demand (AODVv2) Routing draft-ietf-manet-aodvv2. Internet Draft 2016 (2016)
38. Perkins, C., Belding-Royer, E., Das, S.: Ad hoc on-demand distance vector (AODV) routing. RFC 3561 (2003). https://www.ietf.org/rfc/rfc3561
39. Perkins, C., Stan, R., Dowdell, J.: Dynamic manet on-demand (AODVv2) routing draft-ietf-manet-dymo. Internat draft26 (2013). https://tools.ietf.org/html/draft-ietf-manet-dymo-26
40. Schemmel, D., Büning, J., Soria Dustmann, O., Noll, T., Wehrle, K.: Symbolic liveness analysis of real-world software. In: Chockler, H., Weissenbacher, G. (eds.) CAV 2018. LNCS, vol. 10982, pp. 447–466. Springer, Cham (2018). https://doi.org/10.1007/978-3-319-96142-2_27
41. Shao, C., Hui, D., Pazhyannur, R., Bari, F., Zhang, R., Matsushima, S.: IEEE 802.11 medium access control (MAC) profile for control and provisioning of wireless access points (CAPWAP). RFC 7494 (2015). https://tools.ietf.org/html/rfc7494
42. Suriyachai, P., Roedig, U., Scott, A.: A survey of MAC protocols for mission-critical applications in wireless sensor networks. IEEE Commun. Surv. Tutor. **14**(2), 240–264 (2012)

# Multi-player Equilibria Verification for Concurrent Stochastic Games

Marta Kwiatkowska[1], Gethin Norman[2]([⊠]), David Parker[3],
and Gabriel Santos[1]

[1] Department of Computing Science, University of Oxford, Oxford, UK
[2] School of Computing Science, University of Glasgow, Glasgow, UK
gethin.norman@glasgow.ac.uk
[3] School of Computer Science, University of Birmingham, Birmingham, UK

**Abstract.** Concurrent stochastic games (CSGs) are an ideal formalism
for modelling probabilistic systems that feature multiple players or components with distinct objectives making concurrent, rational decisions.
Examples include communication or security protocols and multi-robot
navigation. Verification methods for CSGs exist but are limited to scenarios where agents or players are grouped into two *coalitions*, with those
in the same coalition sharing an identical objective. In this paper, we propose *multi-coalitional* verification techniques for CSGs. We use subgame-
perfect social welfare (or social cost) optimal Nash equilibria, which are
strategies where there is no incentive for any coalition to unilaterally
change its strategy in any game state, and where the total combined
objectives are maximised (or minimised). We present an extension of the
temporal logic rPATL (probabilistic alternating-time temporal logic with
rewards) to specify equilibria-based properties for any number of distinct
coalitions, and a corresponding model checking algorithm for a variant
of stopping games. We implement our techniques in the PRISM-games
tool and apply them to several case studies, including a secret sharing
protocol and a public good game.

## 1 Introduction

Stochastic multi-player games are a modelling formalism that involves a number of players making sequences of rational decisions, each of which results in
a probabilistic change in state. They are well suited to modelling systems that
feature competitive or collaborative behaviour between multiple components or
agents, operating in uncertain or stochastic environments. Examples include
communication or security protocols, which may employ randomisation or send
messages over unreliable channels, and multi-robot or multi-vehicle navigation,
where sensors and actuators are subject to noise or prone to failure. A game-
theoretic approach to modelling also allows rewards, incentives or resource usage
to be incorporated. For example, mechanism design can be used to create protocols reliant on incentive schemes to improve robustness against selfish behaviour
by participants, as utilised in network routing protocols [34] and auctions [14].

© Springer Nature Switzerland AG 2020
M. Gribaudo et al. (Eds.): QEST 2020, LNCS 12289, pp. 74–95, 2020.
https://doi.org/10.1007/978-3-030-59854-9_7

Designing reliable systems that comprise multiple components with differing objectives is a challenge. This is further complicated by the need to consider stochastic behaviour. Formal verification techniques for stochastic multi-player games can be a valuable tool for tackling this problem. The probabilistic model checker PRISM-games [25] has been developed for modelling and analysis of stochastic games: both the turn-based variant, where one player makes a decision in each state, and the concurrent variant, where players make decisions concurrently and without knowledge of each other's actions. PRISM-games also supports strategy synthesis, which allows automated generation of strategies for one or more players in the game, which are guaranteed to satisfy quantitative correctness specifications written in temporal logic.

The temporal logics used in PRISM-games for stochastic games are based on rPATL (probabilistic alternating-time temporal logic with rewards) [12], which combines features of the game logic ATL [4] and probabilistic temporal logics. For example, in a 3-player game, the formula $\langle\!\langle \mathtt{rbt_1} \rangle\!\rangle \mathtt{P}_{\geqslant p}[\,\mathtt{F}\,\mathtt{g_1}\,]$ states "robot 1 has a strategy under which the probability of it successfully reaching its goal is at least $p$, regardless of the strategies of robots 2 and 3". Model checking and strategy synthesis algorithms for rPATL exist for both turn-based [12] and concurrent stochastic games [22].

rPATL uses ATL's coalition operator $\langle\!\langle \cdot \rangle\!\rangle$ to formulate properties. In the above example, there are two coalitions, one containing robot 1 and the other robots 2 and 3. The coalitions have distinctly opposing (zero-sum) objectives, aiming either to maximise or minimise the probability of robot 1 reaching its goal. A recent extension [23] allows the two coalitions to have distinct objectives, using Nash equilibria. More precisely, it uses subgame-perfect social welfare optimal Nash equilibria, which are strategies for all players where there is no incentive for either coalition to unilaterally change its strategy in any state, and where the total combined objectives are maximised. For example, $\langle\!\langle \mathtt{rbt_1}{:}\mathtt{rbt_2},\mathtt{rbt_3} \rangle\!\rangle_{\max=?}(\mathtt{P}[\,\mathtt{F}\,\mathtt{g_1}\,] + \mathtt{P}[\,\mathtt{F}\,(\mathtt{g_2} \wedge \mathtt{g_3})\,])$ asks for such an equilibrium, where the two coalitions' objectives are to maximise the probability of reaching their own (distinct) goals. Model checking rPATL for both the zero-sum [12,22] and equilibria-based [23] properties has the advantage that it essentially reduces to the analysis of 2-player stochastic games, for which various algorithms exist (e.g. [2,3,10]). However, a clear limitation is the assumption that agents can, or would be willing to, collaborate and form two distinct coalitions.

In this paper, we propose *multi-coalitional* verification techniques for concurrent stochastic games (CSGs). We extend the temporal logic rPATL to allow reasoning about any number of distinct coalitions with different quantitative objectives, expressed using a variety of temporal operators capturing either the probability of an event occurring or a reward measure. We then give a model checking algorithm for the logic against CSGs, restricting our attention to a variant of stopping games [13], which, with probability 1, eventually reach a point where the outcome of each player's objective does not change by continuing. Our algorithm uses a combination of backward induction (for finite-horizon operators) and value iteration (for infinite-horizon operators). A key ingredient of the computation is finding optimal Nash equilibria for $n$-player games, which we perform using support enumeration [33] and a mixture of SMT and non-linear optimisation solvers. We implement our techniques in the PRISM-games tool and apply them to several case studies, including a secret sharing protocol and a public good game. This allows us to verify multi-player scenarios that could not be analysed with existing techniques [23].

**Related Work.** As summarised above, there are various algorithms to solve CSGs, e.g., [2,3,10], and model checking techniques have been developed for both zero-sum [22] and equilibria-based [23] versions of rPATL on CSGs, implemented in PRISM-games [25]. However, all of this work assumes or reduces to the 2-player case. Equilibria for $n$-player CSGs are considered in [8], but only complexity results, not algorithms, are presented. Other tools exist to reason about equilibria, including PRALINE [7], EAGLE [37], EVE [18], MCMAS-SLK [9] (via strategy logic) and Gambit [26], but these are all for *non-stochastic* games.

## 2  Preliminaries

We let $Dist(X)$ denote the set of probability distributions over set $X$. For any vector $v$ we use $v(i)$ to denote the $i$th entry of the vector.

**Definition 1 (Normal form game).** *A (finite, n-person) normal form game (NFG) is a tuple* $\mathsf{N} = (N, A, u)$ *where:* $N = \{1, \ldots, n\}$ *is a finite set of players;* $A = A_1 \times \cdots \times A_n$ *and* $A_i$ *is a finite set of actions available to player* $i \in N$; $u = (u_1, \ldots, u_n)$ *and* $u_i : A \to \mathbb{R}$ *is a utility function for player* $i \in N$.

For an NFG $\mathsf{N}$, the players choose actions at the same time, where the choice for player $i \in N$ is over the action set $A_i$. When each player $i$ chooses $a_i$, the utility received by player $j$ equals $u_j(a_1, \ldots, a_n)$. A (mixed) strategy $\sigma_i$ for player $i$ is a distribution over its action set. Let $\eta_{a_i}$ denote the pure strategy that selects action $a_i$ with probability 1 and $\Sigma_\mathsf{N}^i$ the set of strategies for player $i$. A *strategy profile* $\sigma = (\sigma_1, \ldots, \sigma_n)$ is a tuple of strategies for each player and under $\sigma$ the expected utility of player $i$ equals:

$$u_i(\sigma) \stackrel{\text{def}}{=} \sum_{(a_1, \ldots, a_n) \in A} u_i(a_1, \ldots, a_n) \cdot \left( \prod_{j=1}^n \sigma_j(a_j) \right).$$

For strategy $\sigma_i$ of a player, the *support* of $\sigma_i$ is the set of actions it chooses with nonzero probability, i.e., $\{a_i \in A_i \mid \sigma_i(a_i) > 0\}$. Furthermore, the support of a profile is the product of the supports of the individual strategies and a profile is said to have full support if it includes all available action tuples.

We now fix an NFG $\mathsf{N} = (N, A, u)$ and introduce the notion of Nash equilibrium and the variants we require. For profile $\sigma = (\sigma_1, \ldots, \sigma_n)$ and player $i$ strategy $\sigma_i'$, we define the sequence $\sigma_{-i} \stackrel{\text{def}}{=} (\sigma_1, \ldots, \sigma_{i-1}, \sigma_{i+1}, \ldots, \sigma_n)$ and profile $\sigma_{-i}[\sigma_i'] \stackrel{\text{def}}{=} (\sigma_1, \ldots, \sigma_{i-1}, \sigma_i', \sigma_{i+1}, \ldots, \sigma_n)$.

**Definition 2 (Best and least response).** *For player $i$ and strategy sequence $\sigma_{-i}$, a* best response *for player $i$ to $\sigma_{-i}$ is a strategy $\sigma_i^\star$ for player $i$ such that $u_i(\sigma_{-i}[\sigma_i^\star]) \geqslant u_i(\sigma_{-i}[\sigma_i])$ for all $\sigma_i \in \Sigma_{\mathsf{N}}^i$ and a* least response *for player $i$ to $\sigma_{-i}$ is a strategy $\sigma_i^\star$ for player $i$ such that $u_i(\sigma_{-i}[\sigma_i^\star]) \leqslant u_i(\sigma_{-i}[\sigma_i])$ for all $\sigma_i \in \Sigma_{\mathsf{N}}^i$.*

**Definition 3 (Nash equilibrium).** *A strategy profile $\sigma^\star$ is a* Nash equilibrium *(NE) if $\sigma_i^\star$ is a best response to $\sigma_{-i}^\star$ for all $i \in N$.*

**Definition 4 (Social welfare NE).** *An NE $\sigma^\star$ is a* social welfare optimal NE *(SWNE) and $\langle u_i(\sigma^\star) \rangle_{i \in N}$ are SWNE values if $u_1(\sigma^\star) + \cdots + u_n(\sigma^\star) \geqslant u_1(\sigma) + \cdots + u_n(\sigma)$ for all NE $\sigma$ of N.*

**Definition 5 (Social cost NE).** *A profile $\sigma^\star$ of N is a* social cost optimal NE *(SCNE) and $\langle u_i(\sigma^\star) \rangle_{i \in N}$ are SCNE values if $\sigma^\star$ is an NE of $\mathsf{N}^- = (N, A, -u)$ and $u_1(\sigma^\star) + \cdots + u_n(\sigma^\star) \leqslant u_1(\sigma) + \cdots + u_n(\sigma)$ for all NE $\sigma$ of $\mathsf{N}^-$. Furthermore, $\sigma^\star$ is an SWNE of $\mathsf{N}^-$ if and only if $\sigma^\star$ is an SCNE of N.*

The notion of SWNE is standard [29] and applies when utility values represent profits or rewards. We use the dual notion of SCNE for utilities that represent losses or costs. Example objectives in this category include minimising the probability of a fault or the expected time to complete a task. We have chosen to represent SCNE directly since this is more natural than the alternative of simply negating utilities, particularly in the case of probabilities.

**Example 1.** Consider the NFG representing a variant of a *public good game* [20], in which three players each receive a fixed amount of capital (10€) and can choose to invest none, half or all of it in a common stock (represented by the actions $in_i^0$, $in_i^5$ and $in_i^{10}$ respectively). The total invested by the players is multiplied by a factor $f$ and distributed equally among the players, and the aim of the players is to maximise their profit. The utility function of player $i$ is therefore given by:

$$u_i(in_1^{k_1}, in_2^{k_2}, in_3^{k_3}) = (f/3) \cdot (k_1 + k_2 + k_3) - k_i \,.$$

for $k_i \in \{0, 5, 10\}$ and $1 \leqslant i \leqslant 3$. If $f = 2$, then the profile where each investor chooses not to invest is an NE and each player's utility equals 0. More precisely, if a single player was to deviate from this profile by investing half or all of their capital, then their utility would decrease to $(2/3) \cdot 5 - 5 = -5/3$ or $(2/3) \cdot 10 - 10 = -10/3$, respectively. Since this is the only NE it is also the only SWNE and $(0, 0, 0)$ are the only SWNE values. The profile where each player invests all of their capital is not an NE as, under this profile, a player's utility equals $(2/3) \cdot 30 - 10 = 10$ and any player can increase their utility to $(2/3) \cdot 25 - 5 = 35/3$ by deviating and investing half of their capital.

On the other hand, if $f = 3$, then there are two NE: when all players invest either none or invest all of their capital. The sum of utilities of the players under these profiles are $0 + 0 + 0 = 0$ and $20 + 20 + 20 = 60$ respectively, and therefore the second profile is the only SWNE.

**Definition 6 (Concurrent stochastic game).** *A concurrent stochastic multi-player game (CSG) is a tuple* $\mathsf{G} = (N, S, \bar{S}, A, \Delta, \delta, AP, L)$ *where:*

- $N = \{1, \ldots, n\}$ *is a finite set of players;*
- $S$ *is a finite set of states and* $\bar{S} \subseteq S$ *is a set of initial states;*
- $A = (A_1 \cup \{\bot\}) \times \cdots \times (A_n \cup \{\bot\})$ *where* $A_i$ *is a finite set of actions available to player* $i \in N$ *and* $\bot$ *is an idle action disjoint from the set* $\cup_{i=1}^{n} A_i$;
- $\Delta \colon S \to 2^{\cup_{i=1}^{n} A_i}$ *is an action assignment function;*
- $\delta \colon S \times A \to Dist(S)$ *is a (partial) probabilistic transition function;*
- $AP$ *is a set of atomic propositions and* $L \colon S \to 2^{AP}$ *is a labelling function.*

A CSG $\mathsf{G}$ starts in an initial state $\bar{s} \in \bar{S}$ and, when in state $s$, each player $i \in N$ selects an action from its available actions $A_i(s) \overset{\text{def}}{=} \Delta(s) \cap A_i$ if this set is non-empty, and from $\{\bot\}$ otherwise. For any state $s$ and action tuple $a = (a_1, \ldots, a_n)$, the partial probabilistic transition function $\delta$ is defined for $(s, a)$ if and only if $a_i \in A_i(s)$ for all $i \in N$. We augment CSGs with *reward structures*, which are tuples of the form $r = (r_A, r_S)$ where $r_A \colon S \times A \to \mathbb{R}$ and $r_S \colon S \to \mathbb{R}$ are action and state reward functions, respectively.

A *path* is a sequence $\pi = s_0 \xrightarrow{\alpha_0} s_1 \xrightarrow{\alpha_1} \cdots$ such that $s_i \in S$, $\alpha_i = (a_1^i, \ldots, a_n^i) \in A$, $a_j^i \in A_j(s_i)$ for $j \in N$ and $\delta(s_i, \alpha_i)(s_{i+1}) > 0$ for all $i \geqslant 0$. Given a path $\pi$, we denote by $\pi(i)$ the $(i+1)$th state, $\pi[i]$ the $(i+1)$th action, and if $\pi$ is finite, $last(\pi)$ the final state. The sets of finite and infinite paths (starting in state $s$) of $\mathsf{G}$ are given by $FPaths_\mathsf{G}$ and $IPaths_\mathsf{G}$ ($FPaths_{\mathsf{G},s}$ and $IPaths_{\mathsf{G},s}$).

*Strategies* are used to resolve the choices of the players. Formally, a strategy for player $i$ is a function $\sigma_i \colon FPaths_\mathsf{G} \to Dist(A_i \cup \{\bot\})$ such that, if $\sigma_i(\pi)(a_i) > 0$, then $a_i \in A_i(last(\pi))$. A *strategy profile* is a tuple $\sigma = (\sigma_1, \ldots, \sigma_n)$ of strategies for all players. The set of strategies for player $i$ and set of profiles are denoted $\Sigma_\mathsf{G}^i$ and $\Sigma_\mathsf{G}$. Given a profile $\sigma$ and state $s$, let $IPaths_{\mathsf{G},s}^\sigma$ denote the infinite paths with initial state $s$ corresponding to $\sigma$. We can then define, using standard techniques [21], a probability measure $Prob_{\mathsf{G},s}^\sigma$ over $IPaths_{\mathsf{G},s}^\sigma$ and, for a random variable $X \colon IPaths_\mathsf{G} \to \mathbb{R}$, the expected value $\mathbb{E}_{\mathsf{G},s}^\sigma(X)$ of $X$ in $s$ under $\sigma$.

In a CSG, a player's utility or *objective* is represented by a random variable $X_i \colon IPaths_\mathsf{G} \to \mathbb{R}$. Such variables can encode, for example, the probability of reaching a target or the expected cumulative reward before reaching a target. Given an objective for each player, social welfare and social cost NE can be defined as for NFGs. As in [23], we consider *subgame-perfect* NE [32], which are NE in *every state* of the CSG. In addition, for infinite-horizon objectives, the existence of NE is an open problem [6] so, for such objectives, we use $\varepsilon$-NE, which exist for any $\varepsilon > 0$. Formally, we have the following definition.

**Definition 7 (Subgame-perfect $\varepsilon$-NE).** *For CSG $\mathsf{G}$ and $\varepsilon > 0$, a profile $\sigma^\star$ is a subgame-perfect $\varepsilon$-NE for the objectives $\langle X_i \rangle_{i \in N}$ if and only if:* $\mathbb{E}_{\mathsf{G},s}^{\sigma^\star}(X_i) \geqslant \sup_{\sigma_i \in \Sigma_i} \mathbb{E}_{\mathsf{G},s}^{\sigma_{-i}^\star[\sigma_i]}(X_i) - \varepsilon$ *for all $i \in N$ and $s \in S$.*

**Example 2.** We now extend Example 1 to allow the players to invest their capital (and subsequent profits) over a number of months and assume that, at the end of each month, the parameter $f$ can either increase or decrease by 0.2 with probability 0.1. This can be modelled as a CSG G whose states are tuples of the form $(m, f, c_1, c_2, c_3)$, where $m$ is the current month, $f$ the parameter value and $c_i$ is the current capital of player $i$ (the initial capital plus or minus any profits or losses made in previous months). If $f$ has initial value 2 and the players start with a capital of 10€, then the initial state of G equals $(0, 2, 10, 10, 10)$. The actions of player $i$ are of the form $in_i^{k_i}$, which corresponds to $i$ investing $k_i$ in the current month. The probabilistic transition function of the game is such that:

$$\delta((m, f, c_1, c_2, c_3), (in_1^{k_1}, in_2^{k_2}, in_3^{k_3}))(m', f', c_1', c_2', c_3')$$
$$= \begin{cases} 0.8 & \text{if } m' = m + 1, f' = f \text{ and } c_i' = c_i + p_i \\ 0.1 & \text{if } m' = m + 1, f' = f + 0.2 \text{ and } c_i' = c_i + p_i \\ 0.1 & \text{if } m' = m + 1, f' = f - 0.2 \text{ and } c_i' = c_i + p_i \\ 0 & \text{otherwise} \end{cases}$$

where $p_i = (f/3) \cdot (k_1 + k_2 + k_3) - k_i$ for $k_i \in \{0, 5, 10\}$ and $1 \leqslant i \leqslant 3$.

If we are interested in the profits of the players after $k$ months, then we can consider a random variable for player $i$ which would return, for a path with $(k + 1)$th state $(k, f, c_1, c_2, c_3)$, the value $c_i - 10$.

# 3 Extended rPATL with Nash Formulae

We now consider the logic rPATL with Nash formulae [23] and enhance it with equilibria-based properties that can separate players into more than two coalitions.

**Definition 8 (Extended rPATL syntax).** *The syntax of our extended version of* rPATL *is given by the grammar:*

$$\phi ::= \texttt{true} \mid a \mid \neg\phi \mid \phi \wedge \phi \mid \langle\!\langle C \rangle\!\rangle \mathsf{P}_{\sim q}[\psi] \mid \langle\!\langle C \rangle\!\rangle \mathsf{R}_{\sim x}^r[\rho] \mid \langle\!\langle C_1{:}\cdots{:}C_m \rangle\!\rangle_{opt\sim x}(\theta)$$
$$\theta ::= \mathsf{P}[\psi] + \cdots + \mathsf{P}[\psi] \mid \mathsf{R}^r[\rho] + \cdots + \mathsf{R}^r[\rho]$$
$$\psi ::= \mathsf{X}\phi \mid \phi\,\mathsf{U}^{\leqslant k}\,\phi \mid \phi\,\mathsf{U}\,\phi$$
$$\rho ::= \mathsf{I}^{=k} \mid \mathsf{C}^{\leqslant k} \mid \mathsf{F}\,\phi$$

*where* a *is an atomic proposition,* $C$ *and* $C_1, \ldots, C_m$ *are coalitions of players such that* $C_i \cap C_j = \varnothing$ *for all* $1 \leqslant i \neq j \leqslant m$ *and* $\cup_{i=1}^m C_i = N$, $opt \in \{\min, \max\}$, $\sim \in \{<, \leqslant, \geqslant, >\}$, $q \in \mathbb{Q} \cap [0, 1]$, $x \in \mathbb{Q}$, $r$ *is a reward structure and* $k \in \mathbb{N}$.

Our addition to the logic is *Nash formulae* of the form $\langle\!\langle C_1{:}\cdots{:}C_m \rangle\!\rangle_{opt\sim x}(\theta)$, where the *nonzero sum* formulae $\theta$ comprises a sum of $m$ probability or reward objectives (for full details of the rest of the logic see [22,23]). The formula $\langle\!\langle C_1{:}\cdots{:}C_m \rangle\!\rangle_{\max\sim x}(\mathsf{P}[\psi_1] + \cdots + \mathsf{P}[\psi_m])$ holds in a state if, when the players form the coalitions $C_1, \ldots, C_m$, there is a subgame-perfect SWNE for which the *sum* of the values of the objectives $\mathsf{P}[\psi_1], \ldots, \mathsf{P}[\psi_m]$ for the coalitions $C_1, \ldots, C_m$ satisfies $\sim x$. The case for reward objectives is similar and, for formulae of the form $\langle\!\langle C_1{:}\cdots{:}C_m \rangle\!\rangle_{\min\sim x}(\theta)$, we require the existence of an SCNE rather than an

SWNE. We also allow *numerical* queries of the form $\langle\!\langle C_1:\cdots:C_m\rangle\!\rangle_{opt=?}(\theta)$, which return the sum of the SWNE or SCNE values.

In a probabilistic nonzero-sum formula $\theta = \mathsf{P}[\,\psi_1\,] + \cdots + \mathsf{P}[\,\psi_m\,]$, each objective $\psi_i$ can be a next ($\mathsf{X}\,\phi$), bounded until ($\phi_1\,\mathsf{U}^{\leqslant k}\,\phi_2$) or until ($\phi_1\,\mathsf{U}\,\phi_2$) formula, with the usual equivalences, e.g., $\mathsf{F}\,\phi \equiv \mathtt{true}\,\mathsf{U}\,\phi$. For the reward case $\theta = \mathsf{R}^{r_1}[\,\rho_1\,] + \cdots + \mathsf{R}^{r_m}[\,\rho_m\,]$, each $\rho_i$ refers to a reward formula with respect to reward structure $r_i$ and can be bounded instantaneous reward ($\mathtt{I}^{=k}$), bounded accumulated reward ($\mathtt{C}^{\leqslant k}$) or reachability reward ($\mathsf{F}\,\phi$).

**Example 3.** Recall the public good CSG from Example 2. Examples of nonzero-sum formulae in our logic include:

- $\langle\!\langle p_1:p_2:p_3\rangle\!\rangle_{\max\geqslant 3}(\mathsf{P}[\,\mathsf{F}\,c_1\,\geqslant\,20\,] + \mathsf{P}[\,\mathsf{F}\,c_2\,\geqslant\,20\,] + \mathsf{P}[\,\mathsf{F}\,c_3\,\geqslant\,20\,])$ states that the three players can collaborate such that they each eventually double their capital with probability 1;
- $\langle\!\langle p_1:p_2:p_3\rangle\!\rangle_{\max=?}(\mathsf{R}^{cap_1}[\,\mathtt{I}^{=4}\,] + \mathsf{R}^{cap_2}[\,\mathtt{I}^{=4}\,] + \mathsf{R}^{cap_3}[\,\mathtt{I}^{=4}\,])$ asks for the sum of the expected capital of the players at 4 months when they collaborate, where the state reward function of $cap_i$ returns the capital of player $i$.
- $\langle\!\langle p_1:p_2:p_3\rangle\!\rangle_{\max\geqslant 50}(\mathsf{R}^{pro_1}[\,\mathtt{C}^{\leqslant 6}\,] + \mathsf{R}^{pro_2}[\,\mathtt{C}^{\leqslant 6}\,] + \mathsf{R}^{pro_3}[\,\mathtt{C}^{\leqslant 6}\,])$ states that the sum of the expected cumulative profit of the players after 6 months when they collaborate is at least 50, where the action reward function of $pro_i$ returns the expected profit of player $i$ from a state for the given action tuple.

In order to give the semantics of the logic, we require an extension of the notion of *coalition games* [22] which, given a CSG $\mathsf{G}$ and partition $\mathcal{C}$ of the players into $m$ coalitions, reduces $\mathsf{G}$ to an $m$-player coalition game, where each player corresponds to one of the coalitions in $\mathcal{C}$. Without loss of generality, we assume $\mathcal{C}$ is of the form $\{\{1,\ldots,n_1\},\{n_1+1,\ldots n_2\},\ldots,\{n_{m-1}+1,\ldots n_m\}\}$ and let $j_{\mathcal{C}}$ denote player $j$'s position in its coalition.

**Definition 9 (Coalition game).** *For CSG* $\mathsf{G} = (N, S, \bar{s}, A, \Delta, \delta, AP, L)$ *and partition of the players into $m$ coalitions* $\mathcal{C} = \{C_1, \ldots, C_m\}$, *we define the coalition game* $\mathsf{G}^{\mathcal{C}} = (M, S, \bar{s}, A^{\mathcal{C}}, \Delta^{\mathcal{C}}, \delta^{\mathcal{C}}, AP, L)$ *as an $m$-player CSG where:*

- $M = \{1,\ldots,m\}$;
- $A^{\mathcal{C}} = (A_1^{\mathcal{C}} \cup \{\bot\}) \times \cdots \times (A_m^{\mathcal{C}} \cup \{\bot\})$;
- $A_i^{\mathcal{C}} = (\prod_{j\in C_i}(A_j \cup \{\bot\}) \setminus \{(\bot,\ldots,\bot)\})$ *for all* $i \in M$;
- *for any* $s \in S$ *and* $i \in M$: $a_i^{\mathcal{C}} \in \Delta^{\mathcal{C}}(s)$ *if and only if either* $\Delta(s) \cap A_j = \varnothing$ *and* $a_i^{\mathcal{C}}(j_{\mathcal{C}}) = \bot$ *or* $a_i^{\mathcal{C}}(j_{\mathcal{C}}) \in \Delta(s)$ *for all* $j \in C_i$;
- *for any* $s \in S$ *and* $(a_1^{\mathcal{C}}, \ldots, a_m^{\mathcal{C}}) \in A^{\mathcal{C}}$: $\delta^{\mathcal{C}}(s,(a_1^{\mathcal{C}},\ldots,a_m^{\mathcal{C}})) = \delta(s,(a_1,\ldots,a_n))$ *where for* $i \in M$ *and* $j \in C_i$ *if* $a_i^{\mathcal{C}} = \bot$, *then* $a_j = \bot$ *and otherwise* $a_j = a_i^{\mathcal{C}}(j_{\mathcal{C}})$.

*Furthermore, for a reward structure* $r = (r_A, r_S)$, *by abuse of notation we use* $r = (r_A^{\mathcal{C}}, r_S^{\mathcal{C}})$ *for the corresponding reward structure of* $\mathsf{G}^{\mathcal{C}}$ *where:*

- *for any* $s \in S$, $a_i^{\mathcal{C}} \in A^{\mathcal{C}}$: $r_{A^{\mathcal{C}}}^{\mathcal{C}}(s,(a_1^{\mathcal{C}},\ldots,a_m^{\mathcal{C}})) = r_A(s,(a_1,\ldots,a_n))$ *where for* $i \in M$ *and* $j \in C_i$, *if* $a_i^{\mathcal{C}} = \bot$, *then* $a_j = \bot$ *and otherwise* $a_j = a_i^{\mathcal{C}}(j_{\mathcal{C}})$;
- *for any* $s \in S$: $r_S^{\mathcal{C}}(s) = r_S(s)$.

The logic includes infinite-horizon objectives (U, F), for which the existence of SWNE and SCNE is open [6]. However, $\varepsilon$-SWNE and $\varepsilon$-SCNE *do* exist for any $\varepsilon > 0$.

**Definition 10 (Extended rPATL semantics).** *For a CSG* G, $\varepsilon > 0$ *and a formula* $\phi$, *the satisfaction relation* $\models$ *is defined inductively over the structure of* $\phi$. *The propositional logic fragment* (true, a, $\neg$, $\wedge$) *is defined in the usual way. The zero-sum formulae* $\langle\langle C \rangle\rangle P_{\sim q}[\,\psi\,]$ *and* $\langle\langle C \rangle\rangle R^r_{\sim x}[\rho]$ *are defined as in [22,23]. For a Nash formula and state* $s \in S$ *in CSG* G, *we have:*

$$s \models \langle\langle C_1 : \cdots : C_m \rangle\rangle_{opt \sim x}(\theta) \Leftrightarrow \exists \sigma^\star \in \Sigma_{\mathsf{G}^c} . \left( \mathbb{E}^{\sigma^\star}_{\mathsf{G}^c,s}(X_1^\theta) + \cdots + \mathbb{E}^{\sigma^\star}_{\mathsf{G}^c,s}(X_m^\theta) \right) \sim x$$

*and* $\sigma^\star = (\sigma_1^\star, \ldots, \sigma_m^\star)$ *is a subgame perfect* $\varepsilon$-SWNE *if opt* = max, *and a subgame perfect* $\varepsilon$-SCNE *if opt* = min, *for the objectives* $(X_1^\theta, \ldots, X_m^\theta)$ *in* $\mathsf{G}^c$ *where* $C = \{C_1, \ldots, C_m\}$ *and for* $1 \leqslant i \leqslant m$ *and* $\pi \in IPaths^{\sigma^\star}_{\mathsf{G}^c,s}$ :

$$X_i^{\mathsf{P}[\psi^1]+\cdots+\mathsf{P}[\psi^m]}(\pi) = 1 \text{ if } \pi \models \psi^i \text{ and } 0 \text{ otherwise}$$
$$X_i^{\mathsf{R}^{r_1}[\rho^1]+\cdots+\mathsf{R}^{r_m}[\rho^m]}(\pi) = rew(r_i, \rho^i)(\pi)$$
$$\pi \models \mathsf{X}\,\phi \Leftrightarrow \pi(1) \models \phi$$
$$\pi \models \phi_1 \, \mathsf{U}^{\leqslant k} \, \phi_2 \Leftrightarrow \exists i \leqslant k . (\pi(i) \models \phi_2 \wedge \forall j < i . \pi(j) \models \phi_1)$$
$$\pi \models \phi_1 \, \mathsf{U} \, \phi_2 \Leftrightarrow \exists i \in \mathbb{N} . (\pi(i) \models \phi_2 \wedge \forall j < i . \pi(j) \models \phi_1)$$
$$rew(r, \mathsf{I}^{=k})(\pi) = r_S(\pi(k))$$
$$rew(r, \mathsf{C}^{\leqslant k})(\pi) = \sum_{i=0}^{k-1} \left( r_A(\pi(i), \pi[i]) + r_S(\pi(i)) \right)$$
$$rew(r, \mathsf{F}\,\phi)(\pi) = \begin{cases} \infty & \text{if}\,\forall j \in \mathbb{N} . \pi(j) \not\models \phi \\ \sum_{i=0}^{k_\phi} \left( r_A(\pi(i), \pi[i]) + r_S(\pi(i)) \right) & \text{otherwise} \end{cases}$$

*and* $k_\phi = \min\{k - 1 \mid \pi(k) \models \phi\}$.

## 4   Model Checking CSGs Against Nash Formulae

rPATL is a branching-time logic and so the model checking algorithm works by recursively computing the set $Sat(\phi)$ of states satisfying formula $\phi$ over the structure of $\phi$. Therefore, to extend the existing algorithm of [22,23], we need only consider formulae of the form $\langle\langle C_1 : \cdots : C_m \rangle\rangle_{opt \sim x}(\theta)$. From Definition 10, this requires the computation of subgame-perfect SWNE or SCNE values of the objectives $(X_1^\theta, \ldots, X_m^\theta)$ and a comparison of their sum to the threshold $x$.

We first explain how we compute SWNE values in NFGs. Next we consider CSGs, and show how to compute subgame-perfect SWNE and SCNE values for finite-horizon objectives and approximate values for infinite-horizon objectives. For the remainder of this section we fix an NFG N and CSG G.

As in [23], to check nonzero-sum properties on CSGs, we have to work with a restricted class of games. This can be seen as a variant of *stopping games* [13], as used for multi-objective turn-based stochastic games. Compared to [13], we use a weaker, objective-dependent assumption, which ensures that, under all profiles, with probability 1, eventually the outcome of each player's objective does not change by continuing. This can be checked using graph algorithms [1].

**Assumption 1.** *For each subformula* $P[\phi_1^i \cup \phi_2^i]$, *set* $Sat(\neg\phi_1^i \vee \phi_2^i)$ *is reached with probability 1 from all states under all profiles. For each subformula* $R^r[F \phi^i]$, *the set* $Sat(\phi^i)$ *is reached with probability 1 from all states under all profiles.*

**Computing SWNE Values of NFGs.** Computing NE values for an $n$-player game is a complex task when $n > 2$, as it can no longer be reduced to a linear programming problem. The algorithm for the two-player case presented in [23], based on *labelled polytopes*, starts by considering all the regions of the strategy profile space and then iteratively reduces the search space as positive probability assignments are found and added as restrictions on this space. The efficiency of this approach deteriorates when analysing games with large numbers of actions and when one or more players are indifferent, as the possible assignments resulting from action permutations need to be exhausted.

Going in the opposite direction, support enumeration [33] is a method for computing NE that exhaustively examines all sub-regions, i.e., supports, of the strategy profile space, one at a time, checking whether that sub-region contains equilibria. The number of supports is exponential in the number of actions and equals $\prod_{i=1}^{n}(2^{|A_i|} - 1)$. Therefore computing SWNE values through support enumeration will only be efficient for games with a small number of actions.

We now show how, for a given support, using the following lemma, the computation of SWNE profiles can be encoded as a *nonlinear programming problem*. The lemma states that a profile is an NE if and only if any player switching to a single action in the support of the profile yields the same utility for the player and switching to an action outside the support can only decrease its utility.

**Lemma 1** ([33]). *The strategy profile* $\sigma = (\sigma_1, \ldots, \sigma_n)$ *of* N *is an NE if and only if the following conditions are satisfied:*

$$\forall i \in N. \forall a_i \in A_i. \ \sigma_i(a_i) > 0 \rightarrow u_i(\sigma_{-i}[\eta_{a_i}]) = u_i(\sigma) \tag{1}$$

$$\forall i \in N. \forall a_i \in A_i. \ \sigma_i(a_i) = 0 \rightarrow u_i(\sigma_{-i}[\eta_{a_i}]) \leqslant u_i(\sigma). \tag{2}$$

Given the support $B = B_1 \times \cdots \times B_n \subseteq A$, to construct the problem, we first choose *pivot* actions[1] $b_i^p \in B_i$ for $i \in N$, then the problem is to minimise:

$$\left(\sum_{i \in N} \max_{a \in A} u_i(a)\right) - \sum_{i \in N} \left(\sum_{b \in B} u_i(b) \cdot \left(\prod_{j \in N} p_{j,b_j}\right)\right) \tag{3}$$

---

[1] For each $i \in N$ this can be any action in $B_i$.

subject to:

$$\sum_{c \in B_{-i}(b_i^p)} u_i(c) \cdot \left( \prod_{j \in N_{-i}} p_{j,c_j} \right) - \sum_{c \in B_{-i}(b_i)} u_i(c) \cdot \left( \prod_{j \in N_{-i}} p_{j,c_j} \right) = 0 \quad (4)$$

$$\sum_{c \in B_{-i}(b_i^p)} u_i(c) \cdot \left( \prod_{j \in N_{-i}} p_{j,c_j} \right) - \sum_{c \in B_{-i}(a_i)} u_i(c) \cdot \left( \prod_{j \in N_{-i}} p_{j,c_j} \right) \geqslant 0 \quad (5)$$

$$\sum_{b_i \in B_i} p_{i,b_i} = 1 \quad \text{and} \quad p_{i,b_i} > 0 \quad (6)$$

for all $i \in N$, $b_i \in B_i \backslash \{b_i^p\}$ and $a_i \in A_i \backslash B_i$ where $B_{-i}(c_i) = B_1 \times \cdots \times B_{i-1} \times \{c_i\} \times B_{i+1} \times \cdots \times B_n$ and $N_{-i} = N \backslash \{i\}$. The variables in the above program represent the probabilities players choose different actions, i.e. $p_{i,b_i}$ is the probability $i$ selects $b_i$. The constraints (6) ensure the probabilities of each player sum to one and the support of the corresponding profile equals $B$. The constraints (4) and (5) require that the solution corresponds to an NE as these encode the constraints (1) and (2), respectively, of Lemma 1 when restricting to pivot actions. This restriction is sufficient as (1) requires all actions in the support to yield the same utility. The first term in (3) corresponds to the maximum possible sum of utilities for the players, i.e. it sums the maximum utility of each player, and the second sums the individual utilities of the players when they play according to the profile corresponding to the solution. By minimising the difference between these two terms, we require the solution to be social welfare optimal.

SMT solvers with nonlinear modules can be used to solve such problems, although they can be inefficient. Alternative approaches include *barrier* or *interior-point* methods [30].

**Table 1.** Utilities for an instance of a three-player prisoner's dilemma.

| $a$ | $u_1$ | $u_2$ | $u_3$ | $a$ | $u_1$ | $u_2$ | $u_3$ | $a$ | $u_1$ | $u_2$ | $u_3$ | $a$ | $u_1$ | $u_2$ | $u_3$ |
|---|---|---|---|---|---|---|---|---|---|---|---|---|---|---|---|
| $(c_1, c_2, c_3)$ | 7 | 7 | 7 | $(c_1, d_2, c_3)$ | 3 | 9 | 3 | $(d_1, c_2, c_3)$ | 9 | 3 | 3 | $(d_1, d_2, c_3)$ | 5 | 5 | 0 |
| $(c_1, c_2, d_3)$ | 3 | 3 | 9 | $(c_1, d_2, d_3)$ | 0 | 5 | 5 | $(d_1, c_2, d_3)$ | 5 | 0 | 5 | $(d_1, d_2, d_3)$ | 1 | 1 | 1 |

**Example 4.** Consider the instance of three prisoner's dilemma with utilities described in Table 1 where $A_i = \{c_i, d_i\}$ for $1 \leqslant i \leqslant 3$. For the full support $B^{fs}$ the utility of player $i$ equals:

$$u_i(B^{fs}) = p_{i,c_i} \cdot u_i(B^{fs}_{-i}(c_i)) + p_{i,d_i} \cdot u_i(B^{fs}_{-i}(d_i))$$

where $u_i(B^{fs}_{-i}(c_i))$ and $u_i(B^{fs}_{-i}(d_i))$ are the utilities of player $i$ when switching to choosing action $c_i$ and $d_i$ with probability 1 and are given by:

$$u_i(B^{fs}_{-i}(c_i)) = 7 \cdot p_{j,c_j} \cdot p_{k,c_k} + 3 \cdot p_{j,c_j} \cdot p_{k,d_k} + 3 \cdot p_{j,d_j} \cdot p_{k,c_k}$$
$$u_i(B^{fs}_{-i}(d_i)) = 9 \cdot p_{j,c_j} \cdot p_{k,c_k} + 5 \cdot p_{j,c_j} \cdot p_{k,d_k} + 5 \cdot p_{j,d_j} \cdot p_{k,c_k} + p_{j,d_j} \cdot p_{k,d_k}$$

for $1 \leqslant i \neq j \neq k \leqslant 3$. Now, choosing $c_i$ as the pivot action for $1 \leqslant i \leqslant 3$, we obtain the nonlinear program of minimising:

$$27 - (u_1(B^{fs}) + u_2(B^{fs}) + u_3(B^{fs}))$$

subject to: $u_i(B_{-i}^{fs}(c_i)) - u_i(B_{-1}^{fs}(d_i)) = 0$, $p_{i,c_i} + p_{i,d_i} = 1$, $p_{i,c_i} > 0$ and $p_{i,d_i} > 0$ for $1 \leqslant i \leqslant 3$. When trying to solving this problem, we find that there is no NE as the constraints reduce to $p_{3,c_3} \cdot (p_{2,d_2} + 1) = -1$, which cannot be satisfied.

For the partial support $B^{ps} = \{(d_1, d_2, d_3)\}$, $d_i$ is the only choice of pivot action for player $i$ and, after a reduction, we obtain the program of minimising:

$$27 - (p_{2,d_2} p_{3,d_3} + p_{1,d_1} p_{3,d_3} + p_{1,d_1} p_{2,d_2})$$

subject to: $p_{i,d_i} \cdot p_{j,d_j} \geqslant 0$, $p_{i,d_i} = 1$ and $p_{i,d_i} > 0$ for $1 \leqslant i \neq j \leqslant 3$. Solving this problem we see it is satisfied, and therefore the profile where each player $i$ chooses $d_i$ is an NE. This demonstrates that, as for the two-player prisoner's dilemma, defection dominates cooperation for all players, which leads to the only NE.

**Computing Values of Nash Formulae.** We now show how to compute the SWNE values of a Nash formula $\langle\!\langle C_1 : \cdots : C_m \rangle\!\rangle_{opt\sim x}(\theta)$. The case for SCNE values can be computed similarly, where to compute SCNE values of a NFG N, we use Definition 5, negate the utilities of N, find SWNE values of the resulting NFG and return the negation of these values as SCNE values of N.

If all the objectives in the nonzero sum formula $\theta$ are finite-horizon, *backward induction* [28,35] can be applied to compute (precise) subgame-perfect SWNE values. On the other hand, when all the objectives are infinite-horizon, we extend the techniques of [23] for two coalitions and use *value iteration* [11] to approximate subgame-perfect SWNE values. In cases when there is a combination of finite- and infinite-horizon objectives, we can extend the techniques of [23] and make all objectives infinite-horizon by modifying the game in a standard manner.

Value computation for each type of objective is described below. The extension of the two-player case [23] is non-trivial: in that case, when one player reaches their goal, we can apply MDP verification techniques by making the players form a coalition to reach the remaining goal. However, in the $n$-player case, if one player reaches their goal we cannot reduce the analysis to an $(n-1)$-player game, as the choices of the player that has reached its goal can still influence the outcomes of the remaining players, and making the player form a coalition with one of the other players will give the other player an advantage. Instead, we need to keep track of the set of players that have reached their goal (denoted $D$) and can no longer reach their goal in the case of until formulae (denoted $E$), and define the values at each iteration using these sets.

We use the notation $V_{G^C}(s, \theta)$ $(V_{G^C}(s, \theta, n))$ for the vector of computed values of the objectives $(X_1^\theta, X_2^\theta, \ldots, X_m^\theta)$ in state $s$ of $G^C$ (at iteration $n$). We also use $\mathbf{1}_m$ and $\mathbf{0}_m$ to denote a vector of size $m$ whose entries all equal to 1 or 0, respectively. For any set of states $S'$ and state $s$ we let $\eta_{S'}(s)$ equal 1 if $s \in S'$ and 0 otherwise. Furthermore, to simplify the presentation the step bounds appearing in path and reward formulae can take negative values.

*Bounded Probabilistic Until.* If $\theta = P[\phi_1^1 \cup^{\leqslant k_1} \phi_2^1] + \cdots + P[\phi_1^m \cup^{\leqslant k_m} \phi_2^m]$, we compute SWNE values of the objectives for the nonzero-sum formulae $\theta_n = P[\phi_1^1 \cup^{\leqslant k_1 - n} \phi_2^1] + \cdots + P[\phi_1^m \cup^{\leqslant k_m - n} \phi_2^m]$ for $0 \leqslant n \leqslant k$ recursively, where $k =$

$\max\{k_1,\ldots,k_l\}$ and $V_{G^c}(s,\theta) = V_{G^c}(s,\varnothing,\varnothing,\theta_0)$. For any state $s$ and $0 \leqslant n \leqslant k$, $D, E \subseteq M$ such that $D \cap E = \varnothing$:

$$V_{G^c}(s,D,E,\theta_n) = \begin{cases} (\eta_D(1),\ldots,\eta_D(m)) & \text{if } D \cup E = M \\ V_{G^c}(s,D \cup D',E,\theta_n) & \text{else if } D' \neq \varnothing \\ V_{G^c}(s,D,E \cup E',\theta_n) & \text{else if } E' \neq \varnothing \\ val(\mathsf{N}) & \text{otherwise} \end{cases}$$

where $D' = \{l \in M \backslash (D \cup E) \mid s \in Sat(\phi_2^l)\}$, $E' = \{l \in M \backslash (D \cup E) \mid s \in Sat(\neg\phi_1^l \wedge \neg\phi_2^l)\}$ and $val(\mathsf{N})$ equals SWNE values of the game $\mathsf{N} = (M, A^{\mathcal{C}}, u)$ in which for any $1 \leqslant l \leqslant m$ and $a \in A^{\mathcal{C}}$:

$$u_l(a) = \begin{cases} 1 & \text{if } l \in D \\ 0 & \text{else if } l \in E \\ 0 & \text{else if } n_l - n \leqslant 0 \\ \sum_{s' \in S} \delta^{\mathcal{C}}(s,a)(s') \cdot v_{n-1}^{s',l} & \text{otherwise} \end{cases}$$

and $(v_{n-1}^{s',1}, v_{n-1}^{s',2}, \ldots, v_{n-1}^{s',m}) = V_{G^c}(s',D,E,\theta_{n-1})$ for all $s' \in S$.

*Instantaneous Rewards.* If $\theta = \mathsf{R}^{r_1}[\,\mathsf{I}^{=k_1}\,] + \cdots + \mathsf{R}^{r_m}[\,\mathsf{I}^{=k_m}\,]$, we compute SWNE values of the objectives for the nonzero-sum formulae $\theta_n = \mathsf{R}^{r_1}[\,\mathsf{I}^{=n_1-n}\,] + \cdots + \mathsf{R}^{r_m}[\,\mathsf{I}^{=n_l-n}\,]$ for $0 \leqslant n \leqslant k$ recursively, where $k = \max\{k_1,\ldots,k_l\}$ and $V_{G^c}(s,\theta) = V_{G^c}(s,\theta_0)$. For any state $s$ and $0 \leqslant n \leqslant k$, $V_{G^c}(s,\theta_n)$ equals SWNE values of the game $\mathsf{N} = (M, A^{\mathcal{C}}, u)$ in which for any $1 \leqslant l \leqslant m$ and $a \in A^{\mathcal{C}}$:

$$u_l(a) = \begin{cases} 0 & \text{if } n_l - n < 0 \\ \sum_{s' \in S} \delta^{\mathcal{C}}(s,a)(s') \cdot r_S^l(s') & \text{else if } n_l - n = 0 \\ \sum_{s' \in S} \delta^{\mathcal{C}}(s,a)(s') \cdot v_{n+1}^{s',l} & \text{otherwise} \end{cases}$$

and $(v_{n+1}^{s',1},\ldots,v_{n+1}^{s',m}) = V_{G^c}(s',\theta_{n+1})$ for all $s' \in S$.

*Bounded Cumulative Rewards.* If $\theta = \mathsf{R}^{r_1}[\,\mathsf{C}^{\leqslant k_1}\,] + \cdots + \mathsf{R}^{r_m}[\,\mathsf{C}^{\leqslant k_m}\,]$, we compute SWNE values of the objectives for the nonzero-sum formulae $\theta_n = \mathsf{R}^{r_1}[\,\mathsf{C}^{\leqslant n_1-n}\,] + \cdots + \mathsf{R}^{r_l}[\,\mathsf{C}^{\leqslant n_m-n}\,]$ for $0 \leqslant n \leqslant k$ recursively, where $k = \max\{k_1,\ldots,k_l\}$ and $V_{G^c}(s,\theta) = V_{G^c}(s,\theta_0)$. For any state $s$ and $0 \leqslant n \leqslant k$, $V_{G^c}(s,\theta_n)$ equals SWNE values of the game $\mathsf{N} = (M, A^{\mathcal{C}}, u)$ in which for any $1 \leqslant l \leqslant m$ and $a \in A^{\mathcal{C}}$:

$$u_l(a) = \begin{cases} 0 & \text{if } n_l - n \leqslant 0 \\ r_S^l(s) + r_A^l(s,a) + \sum_{s' \in S} \delta^{\mathcal{C}}(s,a)(s') \cdot v_{n+1}^{s',l} & \text{otherwise} \end{cases}$$

and $(v_{n+1}^{s',1},\ldots,v_{n+1}^{s',m}) = V_{G^c}(s',\theta_{n+1})$ for all $s' \in S$.

*Probabilistic Until.* If $\theta = \mathsf{P}[\,\phi_1^1 \ \mathsf{U} \ \phi_2^1\,] + \cdots + \mathsf{P}[\,\phi_1^m \ \mathsf{U} \ \phi_2^m\,]$, values can be computed through value iteration as the limit $V_{G^c}(s,\theta) = \lim_{n \to \infty} V_{G^c}(s,\theta,n)$ where $V_{G^c}(s,\theta,n) = V_{G^c}(s,\varnothing,\varnothing,\theta,n)$ and for any $D, E \subseteq M$ such that $D \cap E = \varnothing$:

$$V_{G^c}(s,D,E,\theta,n) = \begin{cases} (\eta_D(1),\ldots,\eta_D(m)) & \text{if } D \cup E = M \\ (\eta_{Sat(\phi_2^1)}(s),\ldots,\eta_{Sat(\phi_2^m)}(s)) & \text{else if } n = 0 \\ V_{G^c}(s,D \cup D',E,\theta,n) & \text{else if } D' \neq \varnothing \\ V_{G^c}(s,D,E \cup E',\theta,n) & \text{else if } E' \neq \varnothing \\ val(\mathsf{N}) & \text{otherwise} \end{cases}$$

where $D' = \{l \in M \backslash (D \cup E) \mid s \in Sat(\phi_2^l)\}$, $E' = \{l \in M \backslash (D \cup E) \mid s \in Sat(\neg\phi_1^l \wedge \neg\phi_2^l)\}$ and $val(\mathsf{N})$ equals SWNE values of the game $\mathsf{N} = (M, A^{\mathcal{C}}, u)$ in which for any $1 \leqslant l \leqslant m$ and $a \in A^{\mathcal{C}}$:

$$u_l(a) = \begin{cases} 1 & \text{if } l \in D \\ 0 & \text{else if } l \in E \\ \sum_{s' \in S} \delta^{\mathcal{C}}(s,a)(s') \cdot v_{n-1}^{s',l} & \text{otherwise} \end{cases}$$

and $(v_{n-1}^{s',1}, v_{n-1}^{s',2}, \dots, v_{n-1}^{s',m}) = \mathsf{V}_{\mathsf{G}^{\mathcal{C}}}(s', D, E, \theta, n-1)$ for all $s' \in S$.

*Expected Reachability.* If $\theta = \mathsf{R}^{r_1}[\mathsf{F}\ \phi^1] + \cdots + \mathsf{R}^{r_m}[\mathsf{F}\ \phi^m]$, values can be computed through value iteration as the limit $\mathsf{V}_{\mathsf{G}^{\mathcal{C}}}(s, \theta) = \lim_{n \to \infty} \mathsf{V}_{\mathsf{G}^{\mathcal{C}}}(s, \theta, n)$ where $\mathsf{V}_{\mathsf{G}^{\mathcal{C}}}(s, \theta, n) = \mathsf{V}_{\mathsf{G}^{\mathcal{C}}}(s, \varnothing, \theta, n)$ and for any $D \subseteq M$:

$$\mathsf{V}_{\mathsf{G}^{\mathcal{C}}}(s, D, \theta, n) = \begin{cases} \mathbf{0}_m & \text{if } D = M \\ \mathbf{0}_m & \text{else if } n = 0 \\ \mathsf{V}_{\mathsf{G}^{\mathcal{C}}}(s, D \cup D', \theta, n) & \text{else if } D' \neq \varnothing \\ val(\mathsf{N}) & \text{otherwise} \end{cases} \tag{7}$$

$D' = \{l \in M \backslash D \mid s \in Sat(\phi^l)\}$ and $val(\mathsf{N})$ equals SWNE values of the game $\mathsf{N} = (M, A^{\mathcal{C}}, u)$ in which for any $1 \leqslant l \leqslant m$ and $a \in A^{\mathcal{C}}$:

$$u_l(a) = \begin{cases} 0 & \text{if } l \in D \\ r_S^l(s) + r_A^l(s,a) + \sum_{s' \in S} \delta^{\mathcal{C}}(s,a)(s') \cdot v_{n-1}^{s',l} & \text{otherwise} \end{cases}$$

and $(v_{n-1}^{s',1}, v_{n-1}^{s',2}, \dots, v_{n-1}^{s',m}) = \mathsf{V}_{\mathsf{G}^{\mathcal{C}}}(s', D, \theta, n-1)$ for all $s' \in S$.

**Strategy Synthesis.** When performing property verification, it is usually beneficial to include *strategy synthesis*, that is, construct a witness to the satisfaction of a property. When verifying a Nash formula $\langle\langle C_1{:}\cdots{:}C_m \rangle\rangle_{opt \sim x}(\theta)$, we can also return a subgame-perfect SWNE or SCNE for the objectives $(X_1^\theta, \dots, X_m^\theta)$. This is achieved by keeping track of an SWNE for the NFG solved in each state. The synthesised strategies require randomisation and memory. Randomisation is needed for NE of NFGs. Memory is required for finite-horizon properties and since choices change after a path formula becomes true or a target is reached. For infinite-horizon properties, only approximate $\varepsilon$-NE profiles are synthesised.

**Correctness and Complexity.** The proof of correctness of the algorithm can be found in an extended version of this paper [24]. In the case of finite-horizon nonzero-sum formulae the correctness of the model checking algorithm follows from the fact that we use backward induction [28,35]. For infinite-horizon nonzero-sum formulae the proof is based on showing that the values of the players computed during value iteration correspond to subgame-perfect SWNE or SCNE values of finite game trees, and the values of these game trees converge uniformly to the actual values of $\mathsf{G}^{\mathcal{C}}$. The complexity of the algorithm is linear in the formula size, and finding subgame-perfect NE for reachability objectives in $n$-player games is PSPACE [8]. Value iteration requires finding all NE for a NFG in each state of the model, and computing NE of an NFG with three (or more) players is PPAD-complete [15].

# 5   Case Studies and Experimental Results

We have implemented our approach on top of PRISM-games 3.0 [25], extending the implementation to support multi-coalitional equilibria-based properties. The files for the case studies and results in this section are available from [41].

**Table 2.** Finding SWNE in NFGs (timeout of 20 ms for Z3).

| Game | Players | Actions | Supports | Supports returned by Z3 | | | Time (s) |
|---|---|---|---|---|---|---|---|
| | | | | *unsat* | *sat* | *unknown* | IPOPT |
| Majority voting | 3 | 3,3,3 | 343 | 330 | 12 | 1 | 0.309 |
| | 3 | 4,4,4 | 3,375 | 3,236 | 110 | 29 | 18.89 |
| | 3 | 5,5,5 | 29,791 | 26,250 | 155 | 3,386 | 336.5 |
| | 4 | 2,2,2,2 | 81 | 59 | 22 | 0 | 0.184 |
| | 4 | 3,3,3,3 | 2,401 | 2,212 | 87 | 102 | 6.847 |
| | 4 | 4,4,4,4 | 50,625 | 41,146 | 518 | 8,961 | 1,158 |
| | 5 | 2,2,2,2,2 | 243 | 181 | 62 | 0 | 0.591 |
| | 5 | 3,3,3,3,3 | 16,807 | 14,950 | 266 | 1,591 | 253.3 |
| Covariant game | 3 | 3,3,3 | 343 | 304 | 6 | 33 | 7.645 |
| | 3 | 4,4,4 | 3,375 | 2,488 | 16 | 871 | 203.8 |
| | 3 | 5,5,5 | 29,791 | 14,271 | 8 | 15,512 | 5,801 |
| | 4 | 2,2,2,2 | 81 | 76 | 3 | 2 | 0.106 |
| | 4 | 3,3,3,3 | 2,401 | 1,831 | 0 | 570 | 183.0 |
| | 5 | 2,2,2,2,2 | 243 | 221 | 8 | 14 | 4.128 |
| | 5 | 3,3,3,3,3 | 16,807 | 6,600 | 7 | 10,200 | 5,002 |

**Implementation.** CSGs are specified using the PRISM-games 3.0 modelling language, as described in [23,25]. Models are built and stored using the tool's Java-based 'explicit' engine, which employs sparse matrices. Finding SWNE of NFGs, which can be reduced to solving a nonlinear programming problem (see Sect. 4), is performed using a combination of the SMT solver Z3 [16] and the nonlinear optimisation suite IPOPT [39]. Although SMT solvers are able to find solutions to nonlinear problems, they are not guaranteed to do so and are only efficient in certain cases. These cases include when there is a small number of actions per player or finding support assignments for which an equilibrium is not possible. To mitigate the inefficiencies of the SMT solver, we use Z3 for filtering out unsatisfiable support assignments with a timeout: given a support assignment, Z3 returns either *unsat*, *sat* or *unknown* (if the timeout is reached). If either *sat* or *unknown* are returned, then the assignment is passed to IPOPT, which checks for satisfiability (if required) and computes SWNE values using an interior-point filter line-search algorithm [40]. To speed up the overall computation the support assignments are analysed in parallel. We also search for and filter out *dominated strategies* as a precomputation step. The NFGs are built on the fly, as well as the gradient of the objective function (3) and the Jacobian of the constraints (4)–(6), which are required as an input to IPOPT.

**Table 3.** Statistics for a representative set of CSG verification instances.

| Case study & property [parameters] | Players | Param. values | CSG statistics | | | Constr. time(s) | Verif. time (s) |
|---|---|---|---|---|---|---|---|
| | | | States | Max. Act. | Trans. | | |
| Secret Sharing $R_{max=?}[F\ d\lor r=r_{max}]$ $model/[\alpha,r_{max},p_{fail}]$ | 3 | $raa/0.3,10,\_$ | 4,279 | 2,1,1 | 5,676 | 0.057 | 0.565 |
| | | $rba/0.3,10,0.2$ | 7,095 | 2,1,1 | 9,900 | 0.090 | 0.939 |
| | | $rra/0.3,10,\_$ | 8,525 | 2,2,1 | 11,330 | 0.250 | 25.79 |
| | | $rrr/0.3,10,\_$ | 17,017 | 2,2,2 | 22,638 | 0.250 | 96.07 |
| Public Good $R_{max=?}[I=k_{max}]$ $[f,k_{max}]$ | 3 | 2.9,2 | 758 | 3,3,3 | 1,486 | 0.098 | 7.782 |
| | | 2.9,3 | 16,337 | 3,3,3 | 36,019 | 0.799 | 110.1 |
| | | 2.9,4 | 279,182 | 3,3,3 | 703,918 | 6.295 | 1,459 |
| | 4 | 2.9,1 | 83 | 3,3,3,3 | 163 | 0.046 | 0.370 |
| | | 2.9,2 | 6,644 | 3,3,3,3 | 13,204 | 0.496 | 7.111 |
| | | 2.9,3 | 399,980 | 3,3,3,3 | 931,420 | 11.66 | 99.86 |
| | 5 | 2.9,1 | 245 | 3,3,3,3,3 | 487 | 0.081 | 2.427 |
| | | 2.9,2 | 59,294 | 3,3,3,3,3 | 118,342 | 2.572 | 2,291 |
| Aloha (deadline) $P_{max=?}[F\ s_i\land t\leqslant D]$ $[b_{max},D]$ | 3 | 1,8 | 3,519 | 2,2,2 | 5,839 | 0.168 | 11.23 |
| | | 2,8 | 14,230 | 2,2,2 | 28,895 | 0.430 | 14.05 |
| | | 3,8 | 72,566 | 2,2,2 | 181,438 | 1.466 | 18.41 |
| | | 4,8 | 413,035 | 2,2,2 | 1,389,128 | 7.505 | 43.23 |
| | 4 | 1,8 | 23,251 | 2,2,2,2 | 42,931 | 0.708 | 75.59 |
| | | 2,8 | 159,892 | 2,2,2,2 | 388,133 | 3.439 | 131.7 |
| | | 3,8 | 1,472,612 | 2,2,2,2 | 4,777,924 | 28.69 | 819.2 |
| | 5 | 1,8 | 176,777 | 2,2,2,2,2 | 355,209 | 3.683 | 466.3 |
| Aloha $R_{min=?}[F\ s_i]$ $[b_{max}]$ | 3 | 1 | 1,034 | 2,2,2 | 1,777 | 0.096 | 40.76 |
| | | 2 | 5,111 | 2,2,2 | 10,100 | 0.210 | 29.36 |
| | | 3 | 22,812 | 2,2,2 | 56,693 | 0.635 | 51.22 |
| | | 4 | 107,799 | 2,2,2 | 355,734 | 2.197 | 150.1 |
| Medium access $R_{max=?}[C^{\leqslant k}]$ $[e_{max},k]$ | 3 | 5,10 | 1,546 | 2,2,2 | 17,100 | 0.324 | 147.9 |
| | | 10,10 | 10,591 | 2,2,2 | 135,915 | 1.688 | 682.7 |
| | | 15,20 | 33,886 | 2,2,2 | 457,680 | 4.663 | 6,448 |
| | 4 | 5,5 | 15,936 | 2,2,2,2 | 333,314 | 4.932 | 3,581 |

Table 2 presents experimental results for solving various NFGs (generated with GAMUT [31]) using Z3 (with a timeout of 20 ms) and IPOPT. For each NFG, the table lists the numbers of players, actions of each player and support assignments. The table also includes the supports of each type returned by Z3 and the solution time of IPOPT. As can be seen, using Z3 significantly reduces the assignments IPOPT needs to analyse, by orders of magnitude in some cases. However, as the number of actions grows, the number of assignments that remain for IPOPT to solve increases rapidly, and therefore so does the solution time. Furthermore, increasing the number of players only magnifies this issue.

The results show that solving NFGs can be computationally very expensive. Note that just finding an NE is already a difficult problem, whereas we search for SWNE, and hence need to find *all* NE. For example, in [33], using a backtracking search algorithm or either of the Simplicial Subdivision [38] and the Govindan-Wilson [17] algorithms for finding a sample NE, there are instances of NFGs with 6 players and 5 actions that timeout after 30 min.

We also comment that care needs to be taken with numerical computations. The value iteration part of the model checking algorithm is (as usual) implemented using floating point arithmetic, and may therefore exhibit small rounding errors. However, the intermediate results are passed to solvers, which may expect inputs in terms of rational numbers (Z3 in this case). It could be beneficial to investigate the use of arbitrary precision arithmetic instead.

We now present case studies and experimental results to demonstrate the applicability and performance of our approach and implementation.

**Efficiency and Scalability.** Table 3 presents a selection of results demonstrating the performance of the implementation. The models in the table are discussed in more detail below. The results were carried out using a 2.10 GHz Intel Xeon Gold with 16GB of JVM memory. The table includes statistics for the models: number of players, states, (maximum) actions for each player in a state, transitions and the times to both build and verify the models. All models have been verified in under 2 h and in most cases much less than this. The largest model, verified in under 15 min, has 4 players, almost 1.5 million states and 5 million transitions. The majority of the time is spent solving NFG games and, as shown in Table 2, this varies depending on the number of choices and players.

**Secret Sharing.** The first case study is the *secret sharing* protocol of [19], which uses uncertainty to induce cooperation. The protocol is defined for 3 agents and can be extended to more agents by partitioning the agents into three groups. Since the 3 agents act independently, this protocol could not be analysed with the two-coalitional variant of rPATL [23]. Each agent has an unfair coin with the same bias ($\alpha$). In the first step of the protocol, agents flip their coins, and if their coins land on heads, they are supposed to send their share of the secret to the other agents. In the second step, everyone reveals the value of their coin to the other agents. The game ends if all agents obtain all shares and therefore can all reconstruct the secret, or an agent cheats, i.e., fails to send their share to another agent when they are supposed to. If neither of these conditions hold, new shares are issued to the agents and a new round starts. The protocol assumes that each agent prefers to learn the secret and that others do not learn. This is expressed by the utilities $u_3$, $u_2$, $u_1$ and $u_0$ that an agent $i$ gets if all the agents, two agents (including $i$), only $i$ and no agent is able to learn the secret, respectively.

A *rational* agent in this context is one that has the choice of cheating and ignoring the coin toss in order to maximise their utility. An *altruistic* agent is one who strictly follows the protocol and a *byzantine* agent has a probability ($p_{fail}$) of failing and subsequently sending or computing the wrong values. Figure 1 presents the expected utilities when there are two altruistic and one rational agent and when there is one altruistic, one byzantine and one rational agent as $\alpha$ varies. The results when there is one altruistic and two rational agents or three rational agents yield the same graph as Fig. 1(a), where the one or two additional rational agents utilities match those of the altruistic agents. According to the theoretical results of [19], for a model with one rational and two altruistic agents, the rational agent only has an incentive to cheat if:

$$(u_1 \cdot \alpha^2 + u_0 \cdot (1 - \alpha)^2)/(\alpha^2 + (1 - \alpha)^2) > u_3. \tag{8}$$

This result is validated by Fig. 1(a) for the given utility values; the rational agent only cheats when $\alpha \geqslant 0.5$ (for $\alpha < 0.5$ all agents receive a utility of 1 corresponding to all agents getting the secret), which corresponds to when (8) holds for our chosen utility values. Furthermore, Fig. 1 also shows that the closer $\alpha$ is to one then the greater the expected utility of a rational agent. Figure 1(b) also shows that, with a byzantine agent, the rational agent cheats when $\alpha \geqslant 0.4$.

Figure 2 plots the expected utilities of the agents when the protocol stops after a maximum number of rounds $(r_{max})$ when $\alpha = 0.3$ and $\alpha = 0.8$. The utilities converge more slowly for $\alpha = 0.3$, since, when $\alpha$ is small, there is a higher chance that an agent flips tails in a round, meaning not all agents will share their secret in this round and the protocol will move into another round. Again we see that there are more incentives for a rational agent to cheat as $\alpha$ gets closer to 1. However, when $\alpha = 0.3$ and there are altruistic agents, the incentive decreases and eventually disappears as the number of rounds increases.

**Public Good Game.** We consider a variant of the public good game presented in Example 2, in which the parameter $f$ is fixed, where each player receives an initial amount of capital $(e_{init})$ and, in each of $k$ months, can invest none, half or all of their current capital. A 2-player version of the game was modelled in [25].

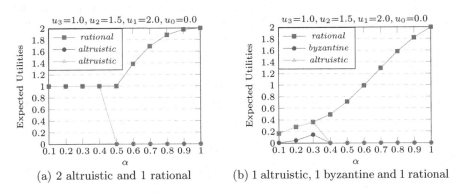

(a) 2 altruistic and 1 rational        (b) 1 altruistic, 1 byzantine and 1 rational

**Fig. 1.** $\langle\!\langle usr_1 : usr_2 : usr_3 \rangle\!\rangle_{\max=?} (\text{R}[\,\text{F done}\,] + \text{R}[\,\text{F done}\,] + \text{R}[\,\text{F done}\,])$ $(p_{fail} = 0.2)$.

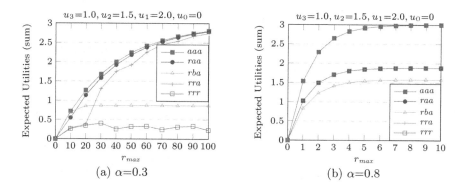

**Fig. 2.** Expected utilities over a bounded number of rounds ($p_{fail} = 0.2$ for $rba$).

Figure 3 presents results for the 3-player public good game as $f$ varies, plotting the expected utilities when the players act in isolation and, for comparison, when player 1 acts in isolation and players 2 and 3 form a coalition (indicated by $\langle\langle\rangle\rangle$), which would be required if the two-coalitional variant of rPATL [23] was used. When the players act in isolation, if $f \leqslant 2$, then there is no incentive for the players to invest. As $f$ increases, the players start to invest some of their capital in some of the months, and when $f = 3$ each player invests all their capital in each month. On the other hand, when players 2 and 3 act in a coalition, there is incentive to invest capital for smaller values of $f$, as players 2 and 3 can coordinate their investments to ensure they both profit; however, player 1 also gains from these investments, and therefore has no incentive to invest in the final month. As $f$ increases, there is a greater incentive for player 1 to invest and the final capital for all the players increases. The drop in the capital of player 1, as $f$ increases, is caused by players 2 and 3 coordinating against player 1 and decreasing their investments. This forces player 1 to invest to increase its investment which, as profits are shared, also increases the capital of players 2 and 3.

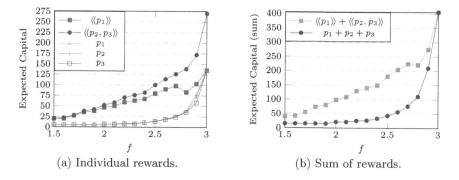

**Fig. 3.** $\langle\langle p_1 : p_2 : p_3 \rangle\rangle_{\max=?}(\mathtt{R}^{c_1}[\,\mathtt{I}^{=r_{max}}\,] + \mathtt{R}^{c_2}[\,\mathtt{I}^{=r_{max}}\,] + \mathtt{R}^{c_3}[\,\mathtt{I}^{=r_{max}}\,])$ ($e_{init} = 5$, $k = 3$).

**Aloha.** This case study concerns a number of users trying to send packets using the slotted ALOHA protocol introduced in [23]. In a time slot, if a single user tries to send a packet, there is a probability $q$ that the packet is sent; if $k$ users try and send, then the probability decreases to $q/k$. If sending a packet fails, the number of slots a user waits before resending is set according to an exponential backoff scheme. The analysis of the model in [23] consisted of considering three users with two acting in coalition. We extend the analysis by considering the case when the three act in isolation and extend the model with a fourth user. The objectives concern maximising the probability of sending a packet within a deadline, e.g. $\langle\!\langle usr_1 : \cdots : usr_m \rangle\!\rangle_{\max=?}(\mathrm{P}[\,\mathrm{F}\,(\mathsf{s}_1 \wedge t \leqslant D)\,] + \cdots + \mathrm{P}[\,\mathrm{F}\,(\mathsf{s}_m \wedge t \leqslant D)\,])$, and the expected time to send a packet. By allowing the users to act independently we find that the expected time required for all users to send their packets reduces compared to when two of the players act as a coalition.

**Medium Access Control.** This case study is based on a deterministic concurrent game model of medium access control [7]. The model consists of two users that have limited energy and share a wireless channel. The users repeatedly choose to transmit or wait and, if both transmit, the transmissions fail due to interference. We previously extended the model to three users and added the probability of transmissions failing (which is dependent on the number of users transmitting) [23]. However, the analysis was restricted to the scenario where two users were in coalition [23]. We can now remove this restriction and analyse the case when each user tries to maximise the expected number of messages they send over a bounded number of steps and extend this analysis to four users.

# 6   Conclusions

We have presented a logic and algorithm for model checking *multi-coalitional* equilibria-based properties of CSGs, focusing on a variant of stopping games. We have implemented the approach in PRISM-games and demonstrated its applicability on a range of case studies and properties. The main limitation of the approach is the time required for solving NFGs during value iteration as the number of players increases. Efficiency improvements that could be employed include filtering out conditionally dominated strategies [36]. Future work will also include investigating correlated equilibria [5] and mechanism design [27].

**Acknowledgements.** This project has received funding from the European Research Council (ERC) under the European Union's Horizon 2020 research and innovation programme (grant agreement No. 834115) and the EPSRC Programme Grant on Mobile Autonomy (EP/M019918/1).

# References

1. de Alfaro, L.: Formal verification of probabilistic systems. Ph.D. thesis, Stanford University (1997)
2. de Alfaro, L., Henzinger, T., Kupferman, O.: Concurrent reachability games. Theoret. Comput. Sci. **386**(3), 188–217 (2007)
3. de Alfaro, L., Majumdar, R.: Quantitative solution of omega-regular games. J. Comput. Syst. Sci. **68**(2), 374–397 (2004)
4. Alur, R., Henzinger, T.A., Kupferman, O.: Alternating-time temporal logic. J. ACM **49**(5), 672–713 (2002)
5. Aumann, R.: Subjectivity and correlation in randomized strategies. J. Math. Econ. **1**(1), 67–96 (1974)
6. Bouyer, P., Markey, N., Stan, D.: Mixed Nash equilibria in concurrent games. In: Proceedings of FSTTCS 2014, LIPICS, vol. 29, pp. 351–363. Leibniz-Zentrum für Informatik (2014)
7. Brenguier, R.: PRALINE: a tool for computing Nash equilibria in concurrent games. In: Sharygina, N., Veith, H. (eds.) CAV 2013. LNCS, vol. 8044, pp. 890–895. Springer, Heidelberg (2013). https://doi.org/10.1007/978-3-642-39799-8_63
8. Brihaye, T., Bruyère, V., Goeminne, A., Raskin, J.F., van den Bogaard, M.: The complexity of subgame perfect equilibria in quantitative reachability games. In: Proceedings of CONCUR 2019, LIPICS, vol. 140, pp. 13:1–13:16. Leibniz-Zentrum für Informatik (2019)
9. Čermák, P., Lomuscio, A., Mogavero, F., Murano, A.: MCMAS-SLK: a model checker for the verification of strategy logic specifications. In: Biere, A., Bloem, R. (eds.) CAV 2014. LNCS, vol. 8559, pp. 525–532. Springer, Cham (2014). https://doi.org/10.1007/978-3-319-08867-9_34
10. Chatterjee, K., de Alfaro, L., Henzinger, T.: Strategy improvement for concurrent reachability and turn-based stochastic safety games. J. Comput. Syst. Sci. **79**(5), 640–657 (2013)
11. Chatterjee, K., Henzinger, T.A.: Value iteration. In: Grumberg, O., Veith, H. (eds.) 25 Years of Model Checking. LNCS, vol. 5000, pp. 107–138. Springer, Heidelberg (2008). https://doi.org/10.1007/978-3-540-69850-0_7
12. Chen, T., Forejt, V., Kwiatkowska, M., Parker, D., Simaitis, A.: Automatic verification of competitive stochastic systems. Formal Methods Syst. Design **43**(1), 61–92 (2013)
13. Chen, T., Forejt, V., Kwiatkowska, M., Simaitis, A., Wiltsche, C.: On stochastic games with multiple objectives. In: Chatterjee, K., Sgall, J. (eds.) MFCS 2013. LNCS, vol. 8087, pp. 266–277. Springer, Heidelberg (2013). https://doi.org/10.1007/978-3-642-40313-2_25
14. Cramton, P., Shoham, Y., Steinberg, R.: An overview of combinatorial auctions. SIGecom Exchanges **7**, 3–14 (2007)
15. Daskalakis, C., Goldberg, P., Papadimitriou, C.: The complexity of computing a Nash equilibrium. Commun. ACM **52**(2), 89–97 (2009)
16. de Moura, L., Bjørner, N.: Z3: an efficient SMT solver. In: Ramakrishnan, C.R., Rehof, J. (eds.) TACAS 2008. LNCS, vol. 4963, pp. 337–340. Springer, Heidelberg (2008). https://doi.org/10.1007/978-3-540-78800-3_24
17. Govindan, S., Wilson, R.: A global newton method to compute Nash equilibria. J. Econ. Theory **110**(1), 65–86 (2003)

18. Gutierrez, J., Najib, M., Perelli, G., Wooldridge, M.: EVE: a tool for temporal equilibrium analysis. In: Lahiri, S.K., Wang, C. (eds.) ATVA 2018. LNCS, vol. 11138, pp. 551–557. Springer, Cham (2018). https://doi.org/10.1007/978-3-030-01090-4_35

19. Halpern, J., Teague, V.: Rational secret sharing and multiparty computation: extended abstract. In: Proceedings of STOC 2004, pp. 623–632. ACM (2004)

20. Hauser, O., Hilbe, C., Chatterjee, K., Nowak, M.: Social dilemmas among unequals. Nature **572**, 524–527 (2019)

21. Kemeny, J., Snell, J., Knapp, A.: Denumerable Markov Chains. Springer, New York (1976). https://doi.org/10.1007/978-1-4684-9455-6

22. Kwiatkowska, M., Norman, G., Parker, D., Santos, G.: Automated verification of concurrent stochastic games. In: McIver, A., Horvath, A. (eds.) QEST 2018. LNCS, vol. 11024, pp. 223–239. Springer, Cham (2018). https://doi.org/10.1007/978-3-319-99154-2_14

23. Kwiatkowska, M., Norman, G., Parker, D., Santos, G.: Equilibria-based probabilistic model checking for concurrent stochastic games. In: ter Beek, M.H., McIver, A., Oliveira, J.N. (eds.) FM 2019. LNCS, vol. 11800, pp. 298–315. Springer, Cham (2019). https://doi.org/10.1007/978-3-030-30942-8_19

24. Kwiatkowska, M., Norman, G., Parker, D., Santos, G.: Multi-player equilibria verification for concurrent stochastic games (2020). arXiv:2007.03365

25. Kwiatkowska, M., Norman, G., Parker, D., Santos, G.: PRISM-games 3.0: stochastic game verification with concurrency, equilibria and time. In: Proceedings of CAV 2020, LNCS. Springer (2020, to appear). http://www.prismmodelchecker.org/games

26. McKelvey, R., McLennan, A., Turocy, T.: Gambit: software tools for game theory, version 16.0.1 (2016). http://www.gambit-project.org

27. Narahari, Y., Narayanam, R., Garg, D., Prakash, H.: Foundations of mechanism design. In: Game Theoretic Problems in Network Economics and Mechanism Design Solutions, Advanced Information and Knowledge Processing, pp. 1–131. Springer, London (2009). https://doi.org/10.1007/978-1-84800-938-7_2

28. von Neumann, J., Morgenstern, O., Kuhn, H., Rubinstein, A.: Theory of Games and Economic Behavior. Princeton University Press, Princeton (1944)

29. Nisan, N., Roughgarden, T., Tardos, E., Vazirani, V.: Algorithmic Game Theory. CUP, Cambridge (2007)

30. Nocedal, J., Wächter, A., Waltz, R.: Adaptive barrier update strategies for nonlinear interior methods. SIAM J. Optim. **19**(4), 1674–1693 (2009)

31. Nudelman, E., Wortman, J., Shoham, Y., Leyton-Brown, K.: Run the GAMUT: a comprehensive approach to evaluating game-theoretic algorithms. In: Proceedings of AAMAS 2004, pp. 880–887. ACM (2004). http://www.gamut.stanford.edu

32. Osborne, M., Rubinstein, A.: An Introduction to Game Theory. OUP, Oxford (2004)

33. Porter, R., Nudelman, E., Shoham, Y.: Simple search methods for finding a Nash equilibrium. In: Proceedings of AAAI 2004, pp. 664–669. AAAI Press (2004)

34. Roughgarden, T., Tardos, E.: How bad is selfish routing? J. ACM **49**, 236–259 (2002)

35. Schwalbe, U., Walker, P.: Zermelo and the early history of game theory. Games Econ. Behav. **34**(1), 123–137 (2001)

36. Shimoji, M., Watson, J.: Conditional dominance, rationalizability, and game forms. J. Econ. Theory **83**, 161–195 (1998)

37. Toumi, A., Gutierrez, J., Wooldridge, M.: A tool for the automated verification of Nash equilibria in concurrent games. In: Leucker, M., Rueda, C., Valencia, F.D. (eds.) ICTAC 2015. LNCS, vol. 9399, pp. 583–594. Springer, Cham (2015). https://doi.org/10.1007/978-3-319-25150-9_34

38. Van Der Laan, G., Talman, A., Van Der Heyden, L.: Simplicial variable dimension algorithms for solving the nonlinear complementarity problem on a product of unit simplices using a general labelling. Math. Oper. Res. **12**(3), 377–397 (1987)

39. Wächter, A.: Short tutorial: getting started with IPOPT in 90 minutes. In: Combinatorial Scientific Computing, no. 09061 in Dagstuhl Seminar Proceedings. Leibniz-Zentrum für Informatik (2009). http://www.github.com/coin-or/Ipopt

40. Wächter, A., Biegler, L.: On the implementation of an interior-point filter line-search algorithm for large-scale nonlinear programming. Math. Program. **106**(1), 25–57 (2006)

41. Supporting material. http://www.prismmodelchecker.org/files/qest20

# Loss-Size and Reliability Trade-Offs Amongst Diverse Redundant Binary Classifiers

Kizito Salako[(✉)]

The Centre for Software Reliability, City, University of London, Northampton Sq.,
London EC1V 0HB, UK
k.o.salako@city.ac.uk

**Abstract.** Many applications involve the use of binary classifiers, including applications where safety and security are critical. The quantitative assessment of such classifiers typically involves *receiver operator characteristic* (ROC) methods and the estimation of sensitivity/specificity. But such techniques have their limitations. For safety/security critical applications, more relevant measures of reliability and risk should be estimated. Moreover, ROC techniques do not explicitly account for: 1) inherent uncertainties one faces during assessments, 2) reliability evidence other than the observed failure behaviour of the classifier, and 3) how this observed failure behaviour alters one's uncertainty about classifier reliability. We address these limitations using *conservative Bayesian inference* (CBI) methods, producing statistically principled, conservative values for risk/reliability measures of interest. Our analyses reveals trade-offs amongst all binary classifiers with the same expected loss – the most reliable classifiers are those most likely to experience high impact failures. This trade-off is harnessed by using diverse redundant binary classifiers.

**Keywords:** Reliability assessment · Binary classification · Diverse redundancy · Conservative Bayesian inference

## 1  Introduction

Numerous applications that society relies upon involve binary classification [21,22,26]. Examples include medical diagnosis, autonomous vehicle safety, crime detection/forensic science and IT network protection. The failure of classifiers in such applications can have a significant impact – affecting the well-being, safety and security of those reliant on these technologies. Classifiers *must* be "good enough" to be deployed. But, demonstrating this can be challenging. The primary challenge here is uncertainty: an assessor of such systems is uncertain about

**Electronic supplementary material** The online version of this chapter (https://doi.org/10.1007/978-3-030-59854-9_8) contains supplementary material, which is available to authorized users.

M. Gribaudo et al. (Eds.): QEST 2020, LNCS 12289, pp. 96–114, 2020.
https://doi.org/10.1007/978-3-030-59854-9_8

if/when the classifiers will fail during operation, the nature of failures should they occur, and the resulting impact failures will have on the wider system. Consequently, the *statistical* assessment of classifiers is necessary. And any serious attempts at quantifying classifier reliability – say, the probability of a classifier's correct functioning on a sequence of classification tasks – *must* account for these uncertainties. This is not easy to do, because the probability distributions that characterise these uncertainties are often unknown or unknowable.

As an approach to assessing classifiers, *receiver operator characteristic* (ROC) methods are well-suited for comparing certain statistical properties of classifiers [8,11]. But, these methods do not account for all of the aforementioned forms of uncertainty. In this paper, we offer a complementary approach to ROC methods, and the following contributions to the assessment of binary classifiers:

1. We critique the sole use of ROC approaches in the statistical assessment of binary classifiers, particularly for safety/security critical applications;
2. We formalise a statistical model of classifier failure and loss, in terms of loss distributions. We argue, it is loss distributions that classifier assessments should be concerned with – these subsume "point" measures such as sensitivity or specificity.
3. We highlight the statistical challenge an assessor faces – i.e. infinitely many loss distributions, all consistent with a classifier's observed failure behaviour;
4. We show, for a given expected loss, that a trade-off exists between classifier reliability and the size of losses when failures occur. This trade-off has not been reported in the literature before;
5. Using this trade-off we prove the range of those loss distributions – from the most reliable classifiers to the least reliable ones – that are consistent with a given expected loss. Our results allow an assessor to reason conservatively about a classifier's reliability, given it's observed failure behaviour;
6. We demonstrate that convex combinations of diverse classifiers can be used to harness this trade-off. In this way, infinitely many hybrid classifiers may be constructed from a few diverse ones. As a curious aside, we also show how such classifier combinations share striking similarities with the optimal allocation of assets in an investment portfolio – a famous problem in Finance;
7. "Optimal" ways of combining the outputs of diverse classifiers have been argued for in the literature – optimal here means smallest expected loss. These ignore the aforementioned trade-off, and we illustrate how such optimal adjudication schemes do not necessarily produce the most reliable systems;
8. A Bayesian formalisation of an assessor's uncertainty about a classifier's loss distribution. Furthermore, by way of mathematical proof, we show that our Bayesian assessments are guaranteed to be conservative – no other similarly constrained Bayesian prior gives more conservative conclusions than ours.

The outline of the rest of the paper is as follows. Critical context and related work is given in Sect. 2. In Sect. 3, we introduce our statistical model of binary classification. Section 4 then presents analyses of trade-offs between the size of losses (due to classifier failures) and classifier reliability. The consequences of such trade-offs, for optimally combining classifiers, are also explored in some detail. This is followed in Sect. 5 by a novel application of CBI methods, to explicitly account for uncertainties surrounding the assessment of classifiers. This also takes into account the trade-offs in previous sections. The paper concludes with final considerations in Sect. 6.

## 2  Critical Context and Related Work

During assessment, a classifier's failure propensity and associated risks are estimated by subjecting the classifier to a sequence of statistically representative classification tasks (i.e. operational testing), and averaging over the classifier's observed failure behaviour. Popular statistics computed in this way, and compared using *receiver operator characteristic* (ROC) methods, include estimates of a classifier's false-positive rate (FPR) and true-positive rate (TPR) [32,33].

A useful graphical tool for comparing classifier performance is ROC-*space* [8,25], shown in Fig. 1. Each point in the unit-square represents all those *discrete classifiers* with the specific FPR, TPR values at that point. All useful discrete classifiers can be made to lie above the 45° diagonal [10]. The diagonal, itself, represents classifiers that make random or blanket classifications. The best classifiers are located at $(0,1)$. *Probabilistic classifiers* can have their FPR, TPR values altered by changing the threshold at which they distinguish positive classifications from negative ones. Continuously altering such a threshold produces, in essence, a range of discrete classifiers – a unique ROC curve (e.g. dashed curve) represents the set of discrete classifiers obtained in this way. Given a suitable stochastic process that generates the classification tasks, and given the losses when classifiers fail, an *isocost line* (e.g. line $l_1$) represents classifiers with the same expected loss. Any parallel line that is closer to the (0,1) point (e.g. line $l_2$) represents classifiers with smaller expected loss [24]. Consequently, under appropriate invariance assumptions [9,34], regions **A** and **B** contain classifiers that, respectively, are guaranteed to have expected loss no larger, or smaller, than the expected loss for those classifiers at the point where **A** and **B** meet. For a given ROC curve, the *area under the curve* (AUC) measures the"size" of the set of all classifiers guaranteed to have expected loss at least as small as some classifier on the curve (i.e. the "size" of the union of all **B** regions with north-west corners on the curve). The AUC is also a measure of how likely it is that a probabilistic classifier will rank a randomly chosen positive classification task more highly than a randomly chosen negative task [15].

But, by themselves, these statistics and ROC methods do not explicitly account for an assessor's uncertainty about the accuracy of FPR and TPR estimates, nor the uncertainty about the stochastic process generating the classification tasks. And, they do not explicitly incorporate reliability evidence obtained before the classifier is subjected to operational testing. Further still, they do not provide a means of updating an assessor's uncertainty (about classifier reliability, future failures or losses) upon observing the classifier during operational testing. Moreover, for those safety or security critical applications where any future failure is unacceptable, there are arguably more relevant quantitative measures of

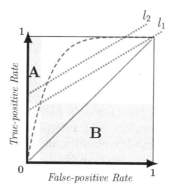

**Fig. 1.** ROC-space

reliability to consider, other than FPR and TPR. For instance, the probability that a classifier succeeds on the next $n$ classification tasks, for very large $n$.

To complement ROC methods, we choose a Bayesian approach [4,13,14]. Classifiers either succeed or fail according to some *unknown* loss distribution – only one of infinitely many plausible distributions that could characterise the classifier's failure behaviour. *Which one* of these is the true loss distribution is an assessor's best guess. In principle, based on evidence gathered prior to operational testing, an assessor's ignorance about a classifier's loss distribution is formalised as a *prior distribution* over the set $\mathcal{L}$ of all such loss distributions. Assessment then proceeds by Bayesian inference, using the classifier's observed failure behaviour during testing to produce a so-called *posterior distribution* over $\mathcal{L}$. And this updated assessor ignorance – this posterior distribution – can be used to compute informed estimates of classifier reliability and risk.

Bayesian inference would typically require the specification of a suitable prior distribution over $\mathcal{L}$ – a daunting task. Perhaps a less daunting task would be to specify the prior only partially – e.g. specify a prior probability that the classifier fails its next classification. Then, amongst all prior distributions over $\mathcal{L}$ that share this value for the probability, one determines which of these priors yields the worst-case value for a posterior reliability measure of interest. With this worst-case value, an assessor gains insight into the range of plausible loss distributions (and thus, classifiers) consistent with their prior evidence and the observed failure behaviour of the classifier. By reasoning conservatively – i.e. reasoning in terms of worst-case values for posterior measures of interest – an assessor is actively avoiding dangerously optimistic assessments resulting from using unjustified priors. In this spirit, *conservative Bayesian inference* (CBI) methods have been used in a number of contexts [2,18,30,35–37]. We develop and apply a new variant of these methods to the problem of producing conservative assessments of binary classifiers. In particular, by performing constrained optimisations over the set $\mathcal{D}$ of all *prior* distributions with sample space $\mathcal{L}$, we produce surprisingly simple expressions for conservative estimates of *posterior* classifier reliability, and identify prior distributions that yield these posteriors.

# 3    A Statistical Model of Binary Classification

Consider the set $\Omega$ of all possible classification tasks, i.e. *demands*, that a classifier can be presented with in a given application. When presented with a demand,
a classifier either raises an alarm or not. With ROC methods, one assumes classifiers always give a response – the same response – to a demand. So, imagine the
operational environment of a classifier as a black box, spewing forth demands
from $\Omega$ for the classifier to classify. The interplay – of a classifier's deterministic behaviour and the uncertainty about which demand will be presented by
the environment next – induces uncertainty about whether a classifier will fail
on the next demand. And, uncertainty about whether the next failure will be
a *false-positive* (FP) error – the classifier raises an alarm when it should not
– or a *false-negative* (FN) error – it does not raise an alarm when it should.
These error-types have associated non-zero costs $l_{fp}$ and $l_{fn}$ when they occur.
Assume the cost associated with an error-type is always the same whenever the
error occurs[1]. Costs are determined by the economic impact of classifier failures;
typically, with $l_{fp}$ much smaller than $l_{fn}$. Correct classification incurs no cost.

A classifier has an associated conditional probability $q_{fn}$ of making an FN
error (i.e. $1 - $ TPR), and conditional probability $q_{fp}$ of an FP error (i.e. FPR).
Estimates of $(1 - q_{fn})$ and $(1 - q_{fp})$ are referred to as *sensitivity* and *specificity*.

Define the following indicator function:

$$\mathbf{1}_{fn} := \begin{cases} 1; & \text{if the classifier commits an FN error} \\ 0; & \text{otherwise} \end{cases}$$

and $\mathbf{1}_{fp}$ is similarly defined for FP errors. Then, the loss resulting from a classifier's failure is the random variable $L := l_{fn}\mathbf{1}_{fn} + l_{fp}\mathbf{1}_{fp}$. The *loss distribution*
for $L$ is depicted in Fig. 2a, where $\gamma := P(\text{next demand should cause an alarm})$.

# 4    Loss-Size vs Reliability Trade-Off

Given the loss values $l_{fp}$ and $l_{fn}$, and a best estimate of a classifier's expected loss
$\mathbb{E}[L]$, what can be conservatively claimed about the classifier's failure behaviour?
In theory, there exists a range of possible discrete, 3-point, loss distributions
(all with the same $\mathbb{E}[L]$) that could characterise the occurrence of failures and
losses for this classifier. And, in practice, specifying which of these distributions
conservatively characterises the classifier must ultimately be a judgement call,
dependent on the specifics of the situation. There is a trade-off here – between
how reliable the classifier is, and how large the losses are when failures occur.

We can elucidate this trade-off. As proved in Appendix A, Fig. 2 shows the 3
extremes of the range of possible loss distributions, using a normalised scale for
the losses (i.e. the losses have been divided by the largest possible loss, typically
$l_{fn}$). The smallest probability $\theta$ of correct classification is either $\theta = \frac{l_{fp} - \mathbb{E}[L]}{l_{fp}}$,

---

[1] View this as the conditional expected loss, given the occurrence of the relevant error.

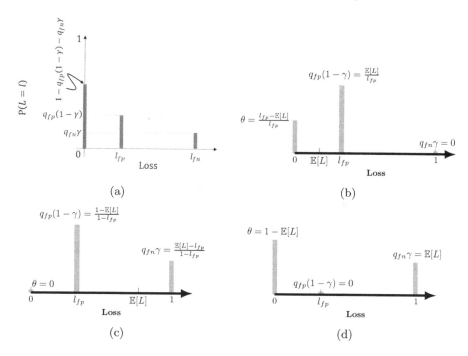

**Fig. 2.** In **(a)** is the distribution of loss $L$ for a classifier. Also shown, in **(b)** **(c)** and **(d)**, are the 3 extremes of the range of all loss distributions that share the same expected loss $\mathbb{E}[L]$. All of the extreme distributions are defined over normalised loss values $\{0, l_{fp}, 1\}$. In particular, the distributions in **(b)** and **(c)** give the smallest values for the probability $\theta$ of correct detection, depending on whether $l_{fp}$ is larger or smaller than $\mathbb{E}[L]$, respectively. These also give the smallest values for the probability $q$ of an FN error. The largest values for both $\theta$ and $q$ are given by the distribution in **(d)**.

when $l_{fp} > \mathbb{E}[L]$ (see Fig. 2b), or it is $\theta = 0$ otherwise (see Fig. 2c). Contrastingly, the largest value of $\theta$ is $\theta = 1 - \mathbb{E}[L]$, given by the loss distribution in Fig. 2d.

What do these extremes represent in practice? In Fig. 2b, the only failures are FPs. While, in Fig. 2c, no correct detection ever occurs – only FPs and FNs. In practice, it is fairly easy to determine that one is not in the extreme situation of Fig. 2c, once the classifiers have been observed to correctly classify *some* demands. In contrast, determining that Fig. 2b is not the situation one faces in practice is more challenging, especially since FNs can be very difficult to identify in certain applications (e.g. cybersecurity, medical diagnosis). Moreover, while Fig. 2b is clearly more preferable than Fig. 2c, the choice between Fig. 2b and Fig. 2d is less clear. Of course, lest one get too excited about the most reliable classifier Fig. 2d, it is sobering that this possibility is also easily excluded in practice, once FPs have been observed. But, the usefulness of thinking in terms of these extremes is not that these are "achievable" in practice, but that classifiers worryingly "close" to these are.

There is another way to view this trade-off. Consider when $l_{fp} = \mathbb{E}[L]$. Both distributions in Figs. 2b and 2c collapse to the same deterministic function – where only failures with associated losses of size $l_{fp} = \mathbb{E}[L]$ occur with probability 1. That is, only demands that should cause no alarms occur, and the classifier raises alarms on all of these. There is no uncertainty – only FP failures *will* occur with accompanying losses of value $\mathbb{E}[L]$. Unlike the distribution in Fig. 2d, which has *the largest amount of variation amongst all of these distributions.* In this sense, the trade-off between the extreme distributions can be viewed as exchanging the certainty of small losses (i.e. losses due to FPs) for an increase in the reliability of the classifiers, but at the added cost of an increased probability of incurring much larger losses when failures occur (i.e. losses due to FNs).

So far, our discussion has focused on the extremes of a range of 3-point loss distributions, while remarking that the distribution Fig. 2d – for the most reliable classifier – possesses the maximum variation amongst all of these distributions. But the following stronger claim is also true (see Appendix A). Amongst *all* loss distributions over *any* (normalised) collection of loss values in the interval $[0, 1]$ (so, not only 3 loss values) – where the distributions all share $\mathbb{E}[L]$ – Fig. 2d is the loss distribution for the most reliable classifiers, and this is also the distribution with the largest possible variation. So, "increased variance" of a loss distribution is the same as "increased reliability and increased losses when failures occur". The proof of this result shares some similarities with the proof of bounds on a system's probability of failure, in [31].[2]

### 4.1 Trade-Off Implications for Randomly Choosing Amongst Diverse Classifiers During Operation

The trade-off becomes more complicated when considering classifiers with different expected losses and variances for their loss distributions. For example, one classifier might have a loss distribution similar to Fig. 2b, while another classifier has a distribution like Fig. 2d, but with a larger expected loss than the first classifier. So, the first classifier has a smaller expected loss, while the second one is noticeably more reliable but perhaps more prone to making FN errors. Consequently, an assessor's preferences for a classifier's failure behaviour may lie somewhere "inbetween" these two classifiers. An "inbetween" classifier can be constructed by a suitable random combination of this pair during operation.

Reducing both expected loss and the probability of failures (i.e. increasing variance) requires multi-objective optimisation techniques. Using expectations and variances together in making multi-objective choices is not a new idea – Markowitz and Sharpe applied this to the finance problem of selecting "efficient" investment portfolios, for which they were awarded the 1990 Nobel prize in Economics [19, 20, 29]. What *is* novel here is the application of these ideas to the problem of choosing optimal configurations of binary classifiers.

---

[2] In [31] their focus was uncertainty about the value of the probability of failure for a system. Contrastingly, our result applies to the uncertainty about whether a given classifier will fail on its next classification task, and the loss incurred if it does.

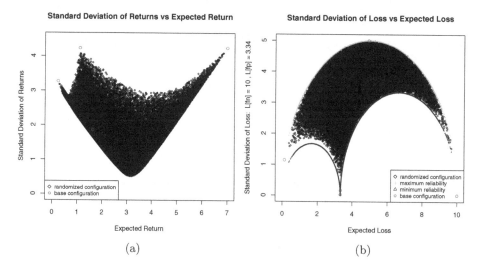

**Fig. 3. (a)** Given 4 distinct risky investments (i.e. "base configurations"), the shaded region contains randomly constructed investment portfolios (i.e. randomised configurations); each point is a convex combination of the investments. Plotted for each portfolio is the standard deviation of returns vs the expected return. All portfolios that lie lower in the region have benefited from diversity reducing uncertainty (i.e. "risk" in this portfolio theory) in the portfolio returns. **(b)** Given 5 distinct classifiers (i.e. "base configurations"), the shaded region contains randomly chosen hybrid classifiers (i.e. randomised configurations). Each hybrid uses a unique distribution for randomly choosing one of the base configurations to exclusively perform the next classification task. Note, one base configuration lies in the bottom right corner of **(b)**. Plotted for each classifier is the standard deviation of losses vs the expected loss, with $l_{fn} = 10$ and $l_{fp} = 3.34$. The classifiers on the boundary have the extreme distributions in Fig. 2, with appropriately scaled losses. All hybrids that lie lower in the region have benefited from diversity reducing the probability of large losses, but also impacting on reliability.

Markowitz's modern portfolio theory shows how combinations of diverse risky investments in a portfolio can lower some of the risk associated with the portfolio, possibly leaving only so-called *undiversifiable risk*. So, with a fixed budget to invest, a desirable portfolio is constructed out of varied investments in carefully chosen proportions. Here, analogous constructions can be made out of classifiers – a convex combination of classifiers defines a hybrid classifier with properties not wholly possessed by any of the constituent classifiers. For other reasons, the works of Scott *et al.* [28], Provost *et al.* [23,25] and Gaffney *et al.* [11,12] have even argued for such convex combinations as a way of creating preferred classifiers out of unsatisfactory ones – these approaches are related to the ROC Convex-Hull theorem (ROCCH) for determining preferred classifiers [8,25].

We construct a convex combination of diverse classifiers as follows. Suppose there are $n$ functionally equivalent classifiers, each with their respective conditional probabilities of FPs, $q_{fp}^1, \ldots, q_{fp}^n$, and FNs, $q_{fn}^1, \ldots, q_{fn}^n$. When a demand from $\Omega$ occurs, with probability $p_i$ it is classified by classifier $i$ exclusively (where $\sum_{i=1}^{n} p_i = 1$). These define a hybrid configuration of $n$ classifiers, with expected loss and variance for the hybrid given as

$$\mathbb{E}[L] = \sum_{i=1}^{n} p_i(l_{fp}q_{fp}^i(1-\gamma) + l_{fn}q_{fn}^i\gamma) \tag{1}$$

$$\mathbb{V}[L] = \sum_{i=1}^{n} p_i(l_{fp}^2 q_{fp}^i(1-\gamma) + l_{fn}^2 q_{fn}^i\gamma) - \left(\sum_{i=1}^{n} p_i(l_{fp}q_{fp}^i(1-\gamma) + l_{fn}q_{fn}^i\gamma)\right)^2 \tag{2}$$

where probability $\gamma$ is defined in Sect. 3. Note, these formulae are fairly general and do not, for instance, assume failures between classifiers are statistically independent. The set of hybrids implied by (1) and (2) is, in a visually striking sense, the "rotation" of the analogous set of portfolios, where both sets are constructed by convex combinations of classifiers/assets respectively. For instance, see the characteristic "aardvark" (or "bullet") silhouettes in Figs. 3a and 3b.

## 4.2   Trade-Off Implications for Optimal Adjudication Amongst Diverse Classifiers

If trustworthy estimates of (conditional) expected loss can be obtained for diverse classifiers, then there are ways of combining the classifier responses into responses that minimise expected loss. A given rule for combining classifier responses into single responses is a so-called *adjudication function*. Minimising expected loss by using an "optimal" adjudication function is possible because: 1) the collective responses of a group of classifiers partition $\Omega$ into disjoint subsets, and 2) on each of these subsets, a response can be chosen that minimizes the conditional expected loss for a subset when demands from this subset occur. In this section, we investigate the relationship between the extreme distributions of Fig. 2 and the loss distribution for a classifier that uses optimal adjudication.

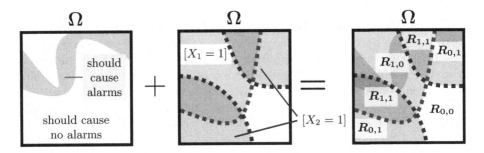

**Fig. 4.** Two classifiers each give a binary response – alarm "1" or no alarm "0" – upon receiving a demand. There are two ways for $\Omega$ to be partitioned into subsets: 1) The demands divide $\Omega$ into two disjoint subsets; 2) The pair of responses $(X_1(\omega), X_2(\omega))$ that classifiers 1 and 2 give for each demand $\omega \in \Omega$ also partitions $\Omega$, into 4 disjoint subsets labelled $R_{i,j} := \{\omega \in \Omega \mid (X_1(\omega), X_2(\omega)) = (i, j)$ where $i, j = 0, 1\}$.

Optimal adjudication has long been advocated in various forms [3,7,11]. To illustrate, consider only two classifiers. Their responses partition $\Omega$ into disjoint subsets (see Fig. 4). Let $X_1$ be classifier 1's response. Then the two events, "no alarm" $[X_1 = 0]$ and "alarm" $[X_1 = 1]$, divide $\Omega$ into two disjoint subsets. Similarly, classifier 2's responses split $\Omega$ into two disjoint subsets. Altogether, $(X_1, X_2)$ divides $\Omega$ into 4 regions – the subsets of $\Omega$ that trigger the responses $(1,1), (1,0), (0,1)$ and $(0,0)$, labelled $R_{1,1}, R_{1,0}, R_{0,1}$ and $R_{0,0}$ respectively.

**Table 1.** The 16 adjudication functions for 2 binary classifiers.

| $R$ | $f_1$ | $f_2$ | $f_3$ | $f_4$ | $f_5$ | $f_6$ | $f_7$ | $f_8$ | $f_9$ | $f_{10}$ | $f_{11}$ | $f_{12}$ | $f_{13}$ | $f_{14}$ | $f_{15}$ | $f_{16}$ |
|---|---|---|---|---|---|---|---|---|---|---|---|---|---|---|---|---|
| $(1,1)$ | 0 | 1 | 0 | 0 | 0 | 1 | 1 | 1 | 0 | 0 | 0 | 1 | 1 | 1 | 0 | 1 |
| $(1,0)$ | 0 | 0 | 1 | 0 | 0 | 1 | 0 | 0 | 1 | 1 | 0 | 1 | 1 | 0 | 1 | 1 |
| $(0,1)$ | 0 | 0 | 0 | 1 | 0 | 0 | 1 | 0 | 1 | 0 | 1 | 1 | 0 | 1 | 1 | 1 |
| $(0,0)$ | 0 | 0 | 0 | 0 | 1 | 0 | 0 | 1 | 0 | 1 | 1 | 0 | 1 | 1 | 1 | 1 |

On each "$R$" the classifiers' responses can be combined into a single response in one of two ways: either issue an alarm or no alarm. An adjudication function is a choice of response on each "$R$". There are 16 possible adjudication functions, $f_1 \ldots f_{16}$ (see Table 1). An "optimal" adjudication function minimises expected loss. Define the expected loss $\mathbb{E}_f[L]$ from adjudication function $f$ as

$$\mathbb{E}_f[L] = l_{fp} \cdot P(f(X_1, X_2) = 1, \text{FP error}) + 1 \cdot P(f(X_1, X_2) = 0, \text{FN error})$$

To determine an optimal $f$, one computes two expectations over each "$R$" – the expected loss due to an FP error, and that due to an FN error. In total, over the 4 regions, 8 expectations are computed[3]. Using these expectations, the optimal $f$ is simply a choice of responses that give the smallest expected losses for each region. The worst adjudication possible is given by choosing responses which give the largest expected losses instead. This can be done using a table such as Table 2. Two numerical examples of Table 2 are illustrated in Table 3. In either example, $l_{fp} = 0.3$, $l_{fn} = 1$ are the normalised losses. These examples are based on the two probability distributions in Table 4 for the various $R$ regions in Fig. 4. They show that $f_{10}$ and $f_7$ are the optimal and worst adjudications for scenario 1, while $f_{13}$ and $f_4$ are the optimal and worst adjudications for scenario 2. The $R$ distributions were generated from a Dirichlet distribution that randomly assigned probabilities to the regions.

In scenario 1 of Table 3, notice how the optimal response for region $R_{11}$ suggests that one should risk FN failures rather than FP ones. Since, from Table 4, the chances of demands causing FN errors is very small (approx. 0.008) compared to the chances of demands causing FP errors (approx. 0.112). However, for scenario 2, the probabilities are roughly the same order of magnitude (approx. 0.07

---

[3] From these, the expected loss for *any* of the adjudication functions may be computed.

**Table 2.** The 8 expected losses for the adjudication functions in Table 1.

| $R$ | |
|---|---|
| $(\mathbf{1,1})$ | $P(f(\mathbf{1,1}) = 1$ & demand should cause no alarm$) \cdot l_{fp}$ |
| | $P(f(\mathbf{1,1}) = 0$ & demand should cause alarm$)$ |
| $(\mathbf{1,0})$ | $P(f(\mathbf{1,0}) = 1$ & demand should cause no alarm$) \cdot l_{fp}$ |
| | $P(f(\mathbf{1,0}) = 0$ & demand should cause alarm$)$ |
| $(\mathbf{0,1})$ | $P(f(\mathbf{0,1}) = 1$ & demand should cause no alarm$) \cdot l_{fp}$ |
| | $P(f(\mathbf{0,1}) = 0$ & demand should cause alarm$)$ |
| $(\mathbf{0,0})$ | $P(f(\mathbf{0,0}) = 1$ & demand should cause no alarm$) \cdot l_{fp}$ |
| | $P(f(\mathbf{0,0}) = 0$ & demand should cause alarm$)$ |

**Table 3.** Two example scenarios of Table 2, with expected losses for each $R$ subset of $\Omega$ (see Fig. 4). In each scenario, for each subset, the adjudication response associated with the smallest expected loss is the optimal response, while the response associated with largest expected loss is the worst response.

| $R$ | Scenario 1 | Scenario 2 |
|---|---|---|
| $(\mathbf{1,1})$ | $3.36e-2$, worst response $= 1$ | $6.781e-2$, optimal response $= 1$ |
| | $8.018e-3$, optimal response $= 0$ | $7.408e-2$, worst response $= 0$ |
| $(\mathbf{1,0})$ | $1.302e-2$, optimal response $= 1$ | $2.081e-2$, optimal response $= 1$ |
| | $2.836e-1$, worst response $= 0$ | $3.248e-2$, worst response $= 0$ |
| $(\mathbf{0,1})$ | $6.436e-2$, worst response $= 1$ | $9.374e-2$, worst response $= 1$ |
| | $4.696e-2$, optimal response $= 0$ | $8.729e-2$, optimal response $= 0$ |
| $(\mathbf{0,0})$ | $5.865e-2$, optimal response $= 1$ | $1.553e-2$, optimal response $= 1$ |
| | $9.599e-2$, worst response $= 0$ | $1.465e-2$, worst response $= 0$ |

and 0.23 respectively). Hence, since $l_{fp}$ is much smaller than $l_{fn}$ in both scenarios, one expects more loss from FN errors than FP errors for $R_{11}$ in scenario 2. This reversal illustrates how optimal adjudication depends on the likelihood of the various regions in Fig. 4, and the relative sizes of $l_{fp}$ and $l_{fn}$.

Are optimal configurations very reliable ones, or do they trade-off reliability in favour of making FNs very unlikely? Figure 5 shows a randomly generated example (i.e. a Dirichlet distributed assignment of the probabilities over the regions in Fig. 4) where the worst adjudication is similar to the most reliable classifier with the same expected loss, while optimal adjudication is similar to the least reliable classifier. But how big can the difference be between an optimal adjudicator and the most reliable classifier with the same expected loss? Fig. 6 depicts empirical distributions of the ratio between the accuracy of a randomly chosen extreme adjudication function (over $\Omega$ in Fig. 4) and the accuracy for an extreme loss distribution with the same expected loss. To generate these empirical ratio distributions, 100,000 (Dirichlet distributed) distributions over the $R$

**Table 4.** Two probability distributions, used respectively to compute the expectations in Table 3, for the regions in Fig. 4.

| $R$ | | Scenario 1 | Scenario 2 |
|---|---|---|---|
| $(\mathbf{1,1})$ | $P(\mathbf{1,1},$ demand should cause no alarm) | 0.112005 | 0.226038 |
| | $P(\mathbf{1,1},$ demand should cause alarm) | 0.008017 | 0.074076 |
| $(\mathbf{1,0})$ | $P(\mathbf{1,0},$ demand should cause no alarm) | 0.043412 | 0.069367 |
| | $P(\mathbf{1,0},$ demand should cause alarm) | 0.283563 | 0.032477 |
| $(\mathbf{0,1})$ | $P(\mathbf{0,1},$ demand should cause no alarm) | 0.214545 | 0.312464 |
| | $P(\mathbf{0,1},$ demand should cause alarm) | 0.046961 | 0.087287 |
| $(\mathbf{0,0})$ | $P(\mathbf{0,0},$ demand should cause no alarm) | 0.195507 | 0.051752 |
| | $P(\mathbf{0,0},$ demand should cause alarm) | 0.095986 | 0.146535 |

regions in Fig. 4 were sampled. In particular, Fig. 6b shows that in applications where FNs are very rare and $l_{fp} \ll l_{fn}$, the most reliable system can be over two orders of magnitude more reliable than the optimal adjudication configuration with the same expected loss.

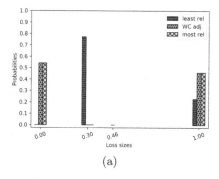
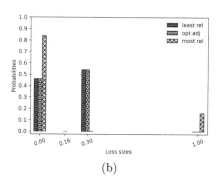

**Fig. 5.** A Dirichlet distributed assignment of probabilities over the $R$ regions (in Fig. 4) produced loss distributions for the worst-case and optimal adjudication functions. These are compared with the extreme loss distributions (in Fig. 2) that have the same expected losses. In **(a)**, with expected loss 0.46, the worst adjudication is identical to the most reliable classifier with the same expected loss. While **(b)**, with expected loss 0.16, shows the optimal adjudication distribution is identical to the distribution for the least reliable classifier. In both plots, $l_{fp} = 0.3$ and $l_{fn} = 1$.

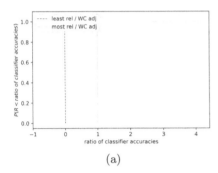
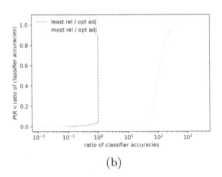

(a)                                    (b)

**Fig. 6.** From 100,000 randomly chosen distributions over the $\boldsymbol{R}$ regions in Fig. 4, with FNs unlikely and $l_{fp} \ll l_{fn}$, **(a)** a randomly chosen worst adjudicator (WC adj) gives approximately the same accuracy as the most reliable classifier with the same expected loss. But, **(b)** the most reliable classifier can have "orders of magnitude" greater accuracy than the corresponding, randomly chosen, optimal adjudicator (opt adj).

# 5    Conservative Bayesian Assessment

In this section, we explicitly account for an assessor's ignorance when assessing a classifier. Our proposed approach is Bayesian, and a novel extension of CBI methods [2, 18, 30, 35–37]. Rather than a completely specified prior distribution, CBI requires only a partial specification of the prior, such as specifying the value of the prior *probability of the classifier passing n tests*. There is a collection of all prior distributions consistent with this probability value. And from this collection, one determines a prior distribution that yields the most undesirable value for a posterior reliability measure of interest (one example measure is the posterior *probability of correctly classifying the next n demands*).

CBI encourages, but does not require, an assessor to be "minimalist" in their application of Bayesian methods. By only providing partial specifications of priors, the assessor can base their assessment only on those properties of a prior they are confident in demonstrating. But, should an assessor completely specify a prior, then CBI becomes identical to traditional Bayesian inference.

Our CBI variant differs from previous ones in two respects. First, it concerns estimating the probability that a known-to-be-imperfect system will have a "perfect run" on the next $n$ demands. Many of the other CBI applications have dealt with systems where *any* failure during testing is unacceptable (i.e. systems with ultra-high reliability requirements). In contrast, classifiers are often known to be imperfect, upon observing them commit FP/FN errors. We determine the worst-case value for the posterior probability of a classifier correctly executing $n$ tasks, after observing the classifier fail "$k$" out of "$k + r$" tasks.

Secondly, we combine CBI analyses and the extremes implied by the loss-size/reliability trade-off, to obtain conservative measures of reliability and risk that account for this trade-off. Previous applications of CBI do not consider the "impact" of failures explicitly – this new variant does.

More formally, consider estimating the unknown probability $U$ that the classifier fails its next classification (i.e. $u = q_{fp}(1 - \gamma) + q_{fn}\gamma$). During operational testing, suppose the classifier fails "$k$" out of "$k + r$" classification tasks. Furthermore, suppose we have some estimate $\mu$ of the probability of seeing these results (e.g. this probability takes the form $\mathbb{E}[U^k(1 - U)^r]$ for a sequence of classifications made by the classifier, where these classification tasks arise according to some Bernoulli process with unknown parameter $U$). Assuming classification tasks follow a Bernoulli process, what is the largest probability of the classifier failing the next classification after operational testing?

**Proposition 1.** *Consider the set $\mathcal{D}$ of all probability distributions over the unit interval, each distribution representing a prior distribution of $U$, where $U$ is the unknown probability of a classification failure. For $\mu \in (0, 1)$ and $k, r > 0$,*

$$\underset{F \in \mathcal{D}}{maximise} \quad \mathbb{E}[\, U \mid k \text{ failures } \& \, r \text{ successes}]$$

$$subject\ to \quad \mathbb{E}[U^k(1 - U)^r] = \mu$$

**Solution**: Appendix B proves the existence of a "single point" prior $F^* \in \mathcal{D}$ – it assigns probability 1 to the unique $u^* \in [\frac{k}{k+r}, 1]$ that satisfies $\mu = (u^*)^k(1-u^*)^r$. Upon using $F^*$ as a prior, $\mathbb{E}[\, U \mid k \text{ failures } \& \, r \text{ successes}]$ attains its maximum value, $u^*$. So, the conservative choice of loss distribution for the classifier would be *any* distribution in $\mathcal{L}$ satisfying $U = u^*$. Note, $F^*$ is unique "*almost everywhere*" [27] (i.e. up to Lebesgue measure zero subsets of $(0, 1)$).    ∎

This result has a number of consequences. The following are limiting cases:

– before any observations (so $k, r = 0$), we must have $\mu = 1$ (i.e. *any* value for the unknown probability of failure $U$ is possible). So, the conservative value $u^*$ must also be 1 – the assessor should conservatively expect the classifier to fail on its next task, in the absence of any evidence to the contrary;
– if no successes are observed (so $r = 0$) then $u^* = 1$: the classifier must be expected to fail its next task, in the absence of any evidence to the contrary;
– if no failures are observed (so $k = 0$) then $u^* = 1 - \mu^{1/r}$. And, as $r \to \infty$, $u^* \to 0$. That is, despite being conservative, with increasing failure free evidence the assessor becomes more convinced that the classifier is "perfect";

Note that the *maximum likelihood estimate* (MLE) for $U$ is $\frac{k}{k+r}$. Since $u^* \in [\frac{k}{k+r}, 1]$, this reassures us that $u^*$ *is* worse than the corresponding MLE. Working backwards, this also means that $\mu \leqslant (\frac{k}{k+r})^k(1-\frac{k}{k+r})^r$ necessarily – a requirement that our assessor can easily use to check the feasibility of their initial $\mu$ estimate.

Using this conservative value $u^*$ in the distributions of Figs. 2b and 2d, we can reason about the possible expected loss values this probability implies. The largest expected loss implied by $u^*$ is $u^*$ (using Fig. 2d). The smallest expected loss is $u^* l_{fp}$ (using Fig. 2b). The case in Fig. 2c applies only if $u^* = 1$; i.e. the classifier *will* fail, the only question is how big the losses will be. Conservatively, one should expect the losses to be as large as possible, so $\mathbb{E}[L] = 1$.

Interestingly, in those cases where we deduced expected losses from Figs. 2b and 2d, the probability of the classifier being correct is $1 - u^*$. This is the smallest plausible value for classifier accuracy, given the assessor's prior evidence $\mu$.

Of course, an assessor might also look to more optimistic assessments. Nevertheless, CBI still offers a useful check against dangerously optimistic assessments. CBI reliability estimates – when compared with estimates from alternative, similarly constrained, approaches – reveal how optimistic these alternatives are. In fact, Appendix C proves the following "most optimistic" posterior expected value for $U$, in a Bernoulli process of classifications.

**Proposition 2.** *Consider the set $\mathcal{D}$ of all probability distributions over $[0, 1]$, each distribution representing a potential prior distribution of $U$, where $U$ is the unknown probability of a classification failure. For $\mu \in (0, 1)$ and $k, r > 0$,*

$$\underset{F \in \mathcal{D}}{minimise} \quad \mathbb{E}[\, U \mid k \, failures \, \& \, r \, successes]$$

$$subject \ to \quad \mathbb{E}[U^k(1 - U)^r] = \mu$$

**Solution**: Appendix C proves the existence of a "single point" prior $F_* \in \mathcal{D}$ – it assigns probability 1 to the unique $u_* \in [0, \frac{k}{k+r}]$ that satisfies $\mu = (u_*)^k(1-u_*)^r$. Upon using $F_*$ as a prior, $\mathbb{E}[\, U \mid k$ failures $\& r$ successes] attains its minimum value, $u_*$. So, the optimistic choice of loss distribution for the classifier would be *any* distribution in $\mathcal{L}$ satisfying $U = u_*$. Estimates of $U$ that are close in value to $u_*$ should be used with caution. Note, $F_*$ is unique "*almost everywhere*". ∎

Finally, using $u^*$ and $u_*$, Appendix D proves the worst-case (i.e. smallest) probability of a classifier correctly classifying the next $n$ tasks, having already observed the classifier correctly classify "$r$" out of "$k + r$" tasks. That is,

**Corollary 1.** *Consider the previous propositions (proved in appendices B and C) which give the largest and smallest values for $\mathbb{E}[\, U \mid k \, failures \, \& \, r \, successes]$, $u^*$ and $u_*$ respectively. Then,*

$$(1 - u^*)^n \leqslant \mathbb{E}[\, (1 - U)^n \mid k \ failures \, \& \, r \ successes] \leqslant 1 - u_*$$

In particular, the lower bound is attained with the prior distribution $F^* \in \mathcal{D}$. So, any loss distribution in $\mathcal{L}$ with $U = u^*$ conservatively characterises the long-run failure behaviour of the classifier. Note that $1 - u^*$ is the smallest plausible (i.e. consistent with the evidence) value for the classifier's unknown accuracy, $\theta$. Consequently, as $n \to \infty$, $(1 - u^*)^n$ tends to zero faster than any other plausible value for accuracy. Given the well-known unsuitability of accuracy as a measure of classifier performance in imbalanced dataset settings [5, 16, 17], conservative long-run success probabilities provide an increasingly stringent (as $n$ increases) alternative measure of classifier performance.

# 6   Conclusions

This work has two main focuses: 1) to explicitly account for various uncertainties inherent in assessing binary classifiers, and 2) to highlight trade-offs between classifier reliability and the size of losses when a classifier fails. These have consequences for deciding which, amongst a collection of classifiers, is most desirable.

Assessment is fraught with uncertainty. And, while ROC techniques address some of these, they do not go far enough. Classifiers fail and incur losses according to some unknown loss distribution, and our work bounds the possible range of such distributions. For example, for classifiers with the same expected loss, the more reliable a classifier is, the larger the losses when it fails – a trade-off.

A consequence of such trade-offs is that hybrid classifiers – i.e. convex combinations of diverse classifiers – can have a reduced risk of a sequence of high-impact failures, but at the expense of reliability. This is akin to how, in modern portfolio theory, diverse risky assets can be combined into one investment portfolio, to reduce investment risk. There are visually arresting parallels (Figs. 3a and 3b) between the sets in these two scenarios. The trade-offs also have implications for "optimal adjudication" schemes that combine the outputs of diverse classifiers to reduce expected loss. Such schemes can be significantly less reliable than the most reliable classifier with the same "optimal" expected loss.

These trade-offs are a strong argument for using multi-objective optimisations during assessment – it is not enough to only consider the well-known trade-off between sensitivity and specificity. In this sense, our approach furthers the benefits of traditional ROC approaches. It also turns out that the loss distributions for the most reliable classifiers (amongst classifiers with a common expected loss) are those with the largest variation. This strongly suggests parallels with mean-variance optimisation methods used in modern portfolio theory. So, our work also complements techniques employed in modern portfolio theory.

Of course, there are significant differences between assets and classifiers that limit the analogy. For example, assets can be leveraged against each other – borrowing on the one hand to buy an investment on the other hand. Such leveraging does not make sense when randomising amongst classifiers. Also, as of this writing, classifiers are typically "indivisible" and cannot be broken up into smaller functionally equivalent artefacts, unlike many investment assets.

Unlike either ROC approaches or modern portfolio theory, with CBI it is fundamental that an assessor explicitly models their uncertainty, about the occurrence of classifier failures. This must be based on justifiable evidence gathered prior to operational testing. Classifiers are then assessed by carrying out "worst-case" inference, using prior distributions in conjunction with observed reliability.

A number of this paper's results (e.g. Propositions 1 and 2) apply more widely (e.g. to multiclass classifiers). But, outstanding challenges remain. For instance, how best to quantify and gain sufficient confidence in prior estimates, such as the prior probability $\mathbb{E}[U^k(1-U)^r]$ of observing classifier failures? Or, investigating whether analogous trade-offs hold for metrics other than $\mathbb{E}[L]$ and reliability, such as F-measures or surrogate estimates of loss [1,6]. Investigating settings where classifications and likelihoods arise according to processes more general than

Bernoulli ones. And, explicitly accounting for how classifier performance can evolve and change over time (e.g. due to "learning" or patching) – currently, the techniques we have outlined apply inbetween changes in classifier performance.

# 7    Appendices and Supplementary Material

For all of the proofs, please see this paper's appendices online, in Springer's Electronic Supplementary Materials (ESM) system.

**Acknowledgment.** This work was supported by the European Commission through the H2020 programme under grant agreement 700692 (DiSIEM). My thanks to the anonymous reviewers for their helpful suggestions for improving the presentation.

# References

1. Bartlett, P., Jordan, M., McAuliffe, J.: Convexity, classification, and risk bounds. J. Am. Stat. Assoc. **101**, 138–156 (2006). https://doi.org/10.1198/016214505000000907
2. Bishop, P., Bloomfield, R., Littlewood, B., Povyakalo, A., Wright, D.: Toward a formalism for conservative claims about the dependability of software-based systems. IEEE Trans. Softw. Eng. **37**(5), 708–717 (2011)
3. Blough, D.M., Sullivan, G.F.: A comparison of voting strategies for fault-tolerant distributed systems. In: Proceedings Ninth Symposium on Reliable Distributed Systems, pp. 136–145 (1990). https://doi.org/10.1109/RELDIS.1990.93959
4. Box, G.E., Tiao, G.C.: Nature of Bayesian Inference, chap. 1, pp. 1–75. Wiley (2011). https://doi.org/10.1002/9781118033197.ch1
5. Buda, M., Maki, A., Mazurowski, M.A.: A systematic study of the class imbalance problem in convolutional neural networks. Neural Netw. **106**, 249–259 (2018). https://doi.org/10.1016/j.neunet.2018.07.011
6. Dembczyński, K., Kotłowski, W., Koyejo, O., Natarajan, N.: Consistency analysis for binary classification revisited. In: Precup, D., Teh, Y.W. (eds.) Proceedings of the 34th International Conference on Machine Learning. Proceedings of Machine Learning Research, PMLR, International Convention Centre, Sydney, Australia, vol. 70, pp. 961–969, 06–11 August 2017. http://proceedings.mlr.press/v70/dembczynski17a.html
7. Di Giandomenico, F., Strigini, L.: Adjudicators for diverse-redundant components. In: Proceedings Ninth Symposium on Reliable Distributed Systems, pp. 114–123, October 1990. https://doi.org/10.1109/RELDIS.1990.93957
8. Fawcett, T.: An introduction to ROC analysis. Pattern Recogn. Lett. **27**(8), 861–874 (2006). http://dx.doi.org/10.1016/j.patrec.2005.10.010
9. Fawcett, T., Flach, P.A.: A response to Webb and Ting's on the application of ROC analysis to predict classification performance under varying class distributions. Mach. Learn. **58**(1), 33–38 (2005). https://doi.org/10.1007/s10994-005-5256-4
10. Flach, P., Shaomin, W.: Repairing concavities in ROC curves. In: Proceedings of the 19th International Joint Conference on Artificial Intelligence (IJCAI 2005), IJCAI, pp. 702–707, August 2005

11. Gaffney, J.E., Ulvila, J.W.: Evaluation of intrusion detectors: a decision theory approach. In: Proceedings of the 2001 IEEE Symposium on Security and Privacy, pp. 50–61. IEEE (2001). http://dl.acm.org/citation.cfm?id=882495.884438
12. Gaffney, J.E., Ulvila, J.W.: Evaluation of intrusion detection systems. J. Res. Natl. Inst. Stand. Technol. **108**(6), 453–473 (2003)
13. Gelman, A., Carlin, J.B., Stern, H.S., Rubin, D.B.: Bayesian Data Analysis, 2nd edn. Chapman and Hall/CRC (2004)
14. Gelman, A., Shalizi, C.R.: Philosophy and the practice of Bayesian statistics. Br. J. Math. Stat. Psychol. **66**(1), 8–38 (2013). https://doi.org/10.1111/j.2044-8317.2011.02037.x
15. Hand, D.J., Till, R.J.: A simple generalisation of the area under the roc curve for multiple class classification problems. Mach. Learn. **45**(2), 171–186 (2001). https://doi.org/10.1023/A:1010920819831
16. Japkowicz, N., Stephen, S.: The class imbalance problem: a systematic study. Intell. Data Anal. **6**(5), 429–449 (2002)
17. Koyejo, O.O., Natarajan, N., Ravikumar, P.K., Dhillon, I.S.: Consistent binary classification with generalized performance metrics. In: Ghahramani, Z., Welling, M., Cortes, C., Lawrence, N.D., Weinberger, K.Q. (eds.) Advances in Neural Information Processing Systems, vol. 27, pp. 2744–2752. Curran Associates, Inc. (2014)
18. Littlewood, B., Salako, K., Strigini, L., Zhao, X.: On reliability assessment when a software-based system is replaced by a thought-to-be-better one. Reliab. Eng. Syst. Saf. **197**, 106752 (2020). https://doi.org/10.1016/j.ress.2019.106752
19. Markowitz, H.M.: Portfolio selection. J. Finan. **7**(1), 77–91 (1952)
20. Markowitz, H.M.: Portfolio Selection, Efficient Diversification of Investments. Wiley, Hoboken (1959)
21. Nayak, J., Naik, B., Behera, D.H.: A comprehensive survey on support vector machine in data mining tasks: applications and challenges. Int. J. Database Theory Appl. **8**, 169–186 (2015). https://doi.org/10.14257/ijdta.2015.8.1.18
22. Pouyanfar, S., et al.: A survey on deep learning: algorithms, techniques, and applications. ACM Comput. Surv. **51**(5) (2018). https://doi.org/10.1145/3234150
23. Provost, F., Fawcett, T.: Robust classification systems for imprecise environments. In: Proceedings of AAAI 1998, pp. 706–713. AAAI press (1998)
24. Provost, F., Fawcett, T.: Analysis and visualization of classifier performance: comparison under imprecise class and cost distributions. In: Proceedings of the Third International Conference on Knowledge Discovery and Data Mining, KDD 1997, pp. 43–48. AAAI Press (1997)
25. Provost, F., Fawcett, T.: Robust classification for imprecise environments. Mach. Learn. **42**(3), 203–231 (2001). https://doi.org/10.1023/A:1007601015854
26. RavinderReddy, R., Kavya, B., Yellasiri, R.: A survey on SVM classifiers for intrusion detection. Int. J. Comput. Appl. **98**, 34–44 (2014). https://doi.org/10.5120/17294-7779
27. Schilling, R.: Measures, Integrals and Martingales, 2nd edn. Cambridge University Press, Cambridge (2017)
28. Scott, M.J.J., Niranjan, M., Prager, R.W.: Realisable classifiers: improving operating performance on variable cost problems. In: Proceedings of the British Machine Vision Conference, pp. 31.1–31.10. BMVA Press (1998). https://doi.org/10.5244/C.12.31
29. Sharpe, W.F.: Capital asset prices: a theory of market equilibrium under conditions of risk. J. Finan. **19**(3), 425–442 (1964)

30. Strigini, L., Povyakalo, A.: Software fault-freeness and reliability predictions. In: Bitsch, F., Guiochet, J., Kaâniche, M. (eds.) SAFECOMP 2013. LNCS, vol. 8153, pp. 106–117. Springer, Heidelberg (2013). https://doi.org/10.1007/978-3-642-40793-2_10
31. Strigini, L., Wright, D.: Bounds on survival probability given mean probability of failure per demand; and the paradoxical advantages of uncertainty. Reliab. Eng. Syst. Saf. **128**, 66–83 (2014). https://doi.org/10.1016/j.ress.2014.02.004
32. Swets, J., Dawes, R., Monahan, J.: Better decisions through science. Sci. Am. **283**, 82–87 (2000). https://doi.org/10.1038/scientificamerican1000-82
33. Swets, J.A.: Measuring the accuracy of diagnostic systems. Science **240**(4857), 1285–93 (1988)
34. Webb, G., Ting, K.: On the application of ROC analysis to predict classification performance under varying class distributions. Mach. Learn. **58**, 25–32 (2005). https://doi.org/10.1007/s10994-005-4257-7
35. Zhao, X., Littlewood, B., Povyakalo, A., Strigini, L., Wright, D.: Modeling the probability of failure on demand (pfd) of a 1-out-of-2 system in which one channel is 'quasi-perfect'. Reliab. Eng. Syst. Saf. **158**, 230–245 (2017)
36. Zhao, X., Littlewood, B., Povyakalo, A., Strigini, L., Wright, D.: Conservative claims for the probability of perfection of a software-based system using operational experience of previous similar systems. Reliab. Eng. Syst. Saf. **175**, 265–282 (2018). https://doi.org/10.1016/j.ress.2018.03.032
37. Zhao, X., Robu, V., Flynn, D., Salako, K., Strigini, L.: Assessing the safety and reliability of autonomous vehicles from road testing. In: The 30th International Symposium on Software Reliability Engineering (ISSRE), Berlin, Germany. IEEE (2019, in press)

# Bayesian Inference by Symbolic Model Checking

Bahare Salmani[✉] and Joost-Pieter Katoen

RWTH Aachen University, Aachen, Germany
{salmani,katoen}@cs.rwth-aachen.de

**Abstract.** This paper applies probabilistic model checking techniques for discrete Markov chains to inference in Bayesian networks. We present a simple translation from Bayesian networks into tree-like Markov chains such that inference can be reduced to computing reachability probabilities. Using a prototypical implementation on top of the Storm model checker, we show that symbolic data structures such as multi-terminal BDDs (MTBDDs) are very effective to perform inference on large Bayesian network benchmarks. We compare our result to inference using probabilistic sentential decision diagrams and vtrees, a scalable symbolic technique in AI inference tools.

## 1 Introduction

*Bayesian Networks.* Bayesian networks (BNs, for short) are one of the most prominent class of probabilistic graphical models [32] in AI. They are used in very different domains, both for knowledge representation and reasoning. BNs represent conditional dependencies between random variables yielding – if these dependencies are sparse – compact representations of their joint probability distribution. Probabilistic inference is the prime evaluation metric on BNs. It amounts to compute conditional probabilities. It is computationally hard: PP-complete [13,14]. A vast amount of inference algorithms exists, both exact ones (possibly tailored to specific graph structures such as bounded tree-width graphs), as well as advanced approximate and simulation algorithms. State-of-the-art symbolic exact inference use different forms of decision diagrams. In particular, sentential decision diagrams (SDDs for short [44]) and their probabilistic extension (PSDDs [30]) belong to the prevailing techniques.

*Probabilistic Model Checking.* Model checking of Markov chains and non-deterministic extensions thereof is a vibrant field of research since several decades. The central problem is computing reachability probabilities, i.e., what is the probability to reach a goal state from a given start state? Algorithms for computing conditional probabilities have been considered in [5]. Efficient model-checking algorithms have been developed and tools such as PRISM [33]

This work is funded by the ERC AdG Projekt FRAPPANT (Grant Nr. 787914).

M. Gribaudo et al. (Eds.): QEST 2020, LNCS 12289, pp. 115–133, 2020.
https://doi.org/10.1007/978-3-030-59854-9_9

and storm [20] have been applied to case studies from several different application areas. Like ordinary model checking, the state-space explosion problem is a major practical obstacle. As for BNs, the use of decision diagrams has attracted a lot of attention since its first usages in probabilistic model checking [2,4] and improvements are still being developed, see e.g., [31]. As demonstrated in the QComp 2019 competition, MTBDD-based model checking prevails on various benchmarks [24].

*Topic of This Paper.* The aim of this work is to investigate to what extent off-the-shelf techniques from probabilistic model checking can be used for exact probabilistic inference in BNs. We are in particular interested to study the usage of MTBDD-based symbolic model checking for inference, and to empirically compare its performance to inference using state-of-the-art decision diagrams in AI such as SDDs and their probabilistic extension. To that end, we define a simple mapping from (discrete) BNs into discrete-time Markov chains (MCs) and relate Bayesian inference to computing reachability probabilities. We report on an experimental evaluation on BNs of the `bnlearn` repository varying in size from small to huge (BN categorization) using a prototypical implementation on top of the storm model checker. Our experiments show that inference using MTBDD-based model checking is quite sensitive to the evidence in the inference query, both in terms of size (i.e., the number of random variables) and depth (i.e., the ordering). For BNs of small to large size, MTBDD-based symbolic model checking is competitive to BN-specific symbolic techniques such as PSDDs whereas for very large and huge BNs, PSDD techniques prevail.

*Contributions.* Our work aimed to reduce the gap between the area of probabilistic model checking and probabilistic inference. Its main contributions are:

- A simple mapping from Bayesian networks to Markov chains.
- A prototypical tool chain to enable model-checking based inference.
- An experimental evaluation to compare off-the-shelf MTBDD-based inference by model checking to tailored PSDD inference.

*Related Work.* There is a large body of related work on exploiting verification and/or symbolic data structures to inference. We here concentrate on the most relevant papers. Deininger *et al.* [21] is perhaps the closest related work. They apply PCTL model checking on factored representations of dynamic BNs and compare an MTBDD-based approach using partitioned representations of the transition probability matrix to monolithic representations. Their experiments show that quantitative model checking does not significantly benefit from partitioned representations. Langmead *et al.* [34,35] employ probabilistic model checking algorithms to perform inference on Dynamic Bayesian Networks. They emphasize on broadening the queries, using temporal logic as the query language. Holtzen *et al.* [25] consider symbolic inference on discrete probabilistic programs. They generalize efficient inference techniques for BNs that exploit the BN structure to such programs. The key is to compile a program into a weighted Boolean formula, and to exploit BDD-based techniques to carry out weighted model counting on such formulas. The works by Darwiche *et al.* [11,15,16,19] compile BNs into arithmetic circuits (via CNF) and perform inference by mainly differentiating these circuits in linear time in terms of the circuit sizes. This method is also applicable to relational BNs [12]. Minato *et al.* [36] propose an approach for exact inference by compiling BNs directly into multiple linear formulas using zero-suppressed BDDs. This differs from Darwiche's approach as it does not require the intermediate compilation into CNF. Shih *et al.* [43] propose a symbolic approach to compile BNs into reduced ordered BDDs in order to verify them against monotonicity and robustness. Sanner and McAllester [40] propose an affine version of algebraic decision diagrams to compactly represent context-specific, additive, and multiplicative structures. They proved that the time and memory footprint of these affine ADDs for inference can be linear in the number of variables in cases where ADDs are exponential. Batz *et al.* [6] use deductive verification to analyse BNs. They show that BNs correspond to a simple form of probabilistic programs amenable to obtaining closed-form solutions for exact inference. They exploited this principle to determine the expected time to get one sample from the BN under rejection sampling. Approximate model checking has been applied to verify dynamic BNs against finite-horizon probabilistic linear temporal properties [37]. Finally, we mention the works [23,42] that use exact symbolic inference methods, so-called PSI tools, on belief networks and probabilistic programs, basically through mapping program values to symbolic expressions.

*Outline.* Section 2 introduces Bayesian networks and probabilistic inference. Section 3 briefly recapitulates Markov chain model checking. Section 4 presents the various symbolic data structures that are relevant to this work. Section 5 details how BNs are mapped onto Markov chains and how inference can be reduced to computing reachability probabilities. Section 6 reports on the experimental results, while Sect. 7 concludes the paper.

## 2    Bayesian Networks

A Bayesian network (BN for short) is a tuple $B = (G, \Theta)$ where $G = (V, E)$ is a directed acyclic graph with finite set of vertices $V$ in which $v \in V$ represents a *random variable* taking values from the finite domain $D$ and edge $(v, w) \in E$ represents the *dependency* of $w$ on $v$. We let $parents(v) = \{w \in V \mid (w, v) \in E\}$. For each vertex $v$ with $k$ parents, the function $\Theta_v : D^k \to Dist(D)$ is the *conditional probability table* of (the random variable corresponding to) vertex $v$. $Dist(D)$ here denotes the set of probability distribution functions on $D$. Figure 1 indicates a small BN, in which all the variables are binary. The DAG indicates the dependencies between the variables. For example, the grade a student gets for an exam depends on whether the exam has been easy or difficult, and additionally on whether she has been prepared for the exam. See Fig. 1.

The conditional probability table $\Theta_v$ (CPT for short) of vertex $v$ defines a probability distribution which determines the evaluation of $v$, given some evaluation of *parents(v)*. For example, according to the CPT of Grade, the probability of a low grade is 0.95 for an easy exam and non-prepared student.

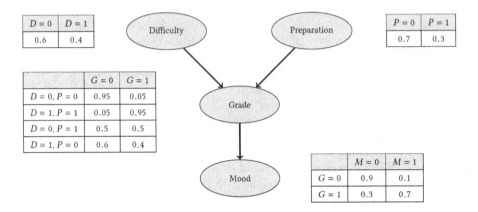

**Fig. 1.** Simple example of Bayesain networks - Student Mood

The semantics of BN $B = (V, E, \Theta)$ is the joint probability function that it defines. Let $W \subseteq V$ be a downward-closed set of vertices where $w \in W$ has value $\underline{w} \in D$. The unique joint probability function of BN $B$ equals:

$$Pr(W = \underline{W}) = \prod_{w \in W} Pr(w = \underline{w} \mid parents(w) = \underline{parents(w)}) \qquad (1)$$

In this paper, we are interested in probabilistic inference. Let $B$ be a BN with set $V$ of vertices, $F \subseteq V$ be the evidence, and $H \subseteq V$ be the hypothesis. The evidence can be simply seen as what we already know and the hypothesis as what we are interested in, given the evidence. The problem of *(exact) probabilistic inference* is to determine the following conditional probability:

$$Pr(H = h \mid F = f) = \frac{Pr(H = h \wedge F = f)}{Pr(F = f)} \qquad (2)$$

In case $Pr(F = f) = 0$, the query is considered ill-conditioned. In the student mood example, shown in Fig. 1, let assume that we are interested to know how likely a student ends up with a bad mood after getting a bad grade for an easy exam, given that she is well prepared. This is defined as:

$$Pr(D = 0, G = 0, M = 0 \mid P = 1) = \frac{Pr(D = 0, G = 0, M = 0, P = 1)}{Pr(P = 1)}$$

$$= \frac{0.6 \cdot 0.5 \cdot 0.9 \cdot 0.3}{0.3} = \frac{0.081}{0.3} = 0.27$$

The decision variant of probabilistic inference can be defined for a given probability $p \in \mathbb{Q} \cap [0, 1)$ as follows:

$$Does \ Pr(H = h \mid F = f) > p?$$

This problem is PP-complete [17]. The *average Markov blanket* of a BN is an indication of the practical complexity of performing inference. The Markov blanket for a vertex $v$ in a BN is the set $\partial_v$ composed of the parents, the children, and the spouses of $v$ [17], where the spouses of $v$ are the nodes that have some common children with $v$. It follows that $Pr(v \mid \partial_v \wedge w) = Pr(v \mid \partial_v)$, for any $w \in V$. The average Markov blanket of a BN, or AMB for short, then is the average size of the Markov blanket of all its vertices, that is, $\frac{1}{|V|} \sum_{v \in V} |\partial_v|$. AMB indicates the average degree of dependence between the random variables in the BN.

# 3    Markov Chain Model Checking

Since in this work we are focused on discrete time BNs, we are interested in discrete-time Markov chains. DTMCs or simply MCs for short, are simple probabilistic models that equip each edge with a probability. An MC $M$ is a tuple $(\Sigma, \sigma_I, P)$ where $\Sigma$ is a countable non-empty set of states, $\sigma_I$ is the initial state, and $P : S \rightarrow Dist(\Sigma)$ is the transition probability function.

In this work we are interested in computing reachability probabilities in an MC. The reachability probability for $G \subseteq \Sigma$ is defined as the probability of finally reaching $G$, starting from the initial state $\sigma_I$. This is denoted by $Pr_M(\Diamond G)$. Computing $Pr_M(\Diamond G)$ can be reduced to computing the unique solution of a linear equation system [29] whose size is linear in $|\Sigma|$. This can be done in a symbolic manner using MTBDDs [4].

# 4    Symbolic Data Structures

The need to represent Boolean functions and probability distributions in a succinct manner has led to various compact representations [3,8,18,22,30]. Symbolic model checking mainly relies on set-based and binary encoding of states and transitions enabling the use of compact representations such as BDDs and MTBDDs. In the following we briefly review the data structures related to this work: BDDs, MTBDDs, vtrees, SDDs and PSDDs. The first two are popular in symbolic model checking while the last three are state-of-the-art in probabilistic inference.

## 4.1    Reduced Ordered Binary Decision Diagrams

ROBDDs or simply BDDs for short, are dominantly-used structures for representing switching functions. BDDs result from compacting binary decision trees mainly by eliminating *don't care* nodes and duplicated subtrees. Essential characteristic of ROBDDs is that they are canonical for a given function and a given variable ordering [8]. Optimal variable orderings can yield very succinct ROB-DDs. Although finding the optimal variable ordering is NP-hard [7], ROBDDs can be very compact in practice [10].

Let $\wp = (z_1, ..., z_m)$ be a (total) variable ordering for $Var = \{z_1, ..., z_m\}$ where $z_1 <_\wp ... <_\wp z_m$. An $\wp$-OBDD is a tuple $\mathfrak{B} = (V, V_I, V_T, succ_0, succ_1, var, val, v_0)$ with the finite set $V$ of nodes, partitioned into inner nodes $V_I$ and terminal nodes $V_T$. $v_0 \in V_I$ is the unique distinguished root node. $succ_0, succ_1 : V_I \to V$ are the successor functions assigning a zero-successor $v_r \in V$ and a one-successor $v_l \in V$ to each $v \in V$. The labelling functions $var : V_I \to Var$ and $val : V_T \to \{0, 1\}$ must satisfy the following equation for $v \in V_I$ and $w \in \{succ_0(v), succ_1(v)\}$:

$$(var(v) = z_i \wedge w \in V_I) \Rightarrow var(w) = z_j \quad with \quad z_i <_\wp z_j. \tag{3}$$

Every inner node $v$ in an OBDD represents a variable from $Var$. The terminal nodes are mapped to 0 or 1. Based on the evaluation of $var(v)$ either to 0 or to 1, the transition from $v$ to the next node is chosen from $\{succ_0(v), succ_1(v)\}$. The semantics of an $\wp$-OBDD is the switching function $f_\mathfrak{B}$ where $f_\mathfrak{B}([z_1 = b_1, ..., z_m = b_m])$ is determined by the value of the resulting leaf obtained by traversing the OBDD starting from the root $v_0$ and branching according to the evaluation $[z_1 = b_1, ..., z_m = b_m]$.

For terminal $v \in V_T$, $f_v$ represents the constant function $f_v$ with value $val(v)$. For $v \in V_I$, $f_v$ is defined based on the Shannon expansion over $v$ as $f_v = (\neg z \wedge f_{succ_0}(v)) \vee (z \wedge f_{succ_1}(v))$, where $z = var(v)$. A $\wp$-OBDD $\mathfrak{B}$ is reduced if for every pair $(v, w)$ of nodes in $\mathfrak{B}$, $v \neq w$ implies $f_v \neq f_w$. An OBDD can be reduced by recursively applying simple reduction rules: the elimination of don't care vertices, and the elimination of isomorphic subtrees.

## 4.2   Multi-terminal BDDs

While BDDs represent Boolean functions, the terminal values in MTBDDs can acquire values from other domains such as real or rational numbers. This allows rational or real functions to be succinctly represented, enabling representing probability distribution functions. The formal definition of MTBDD is not very different from the one for BDD. Let $Var$ and $\wp$ be as before. An MTBDD $\mathfrak{M}$ is the same structure as an OBDD except that (in our setting) the value function $val$ is refined to $val : V_T \to [0, 1]$ assigning each terminal node $v \in V_T$ a probability $val(v)$. The semantics of MTBDD $\mathfrak{M}$ is defined by $f_\mathfrak{M} : Eval(Var) \to [0, 1]$ similarly to $f_\mathfrak{B}$ for BDD $\mathfrak{B}$.

## 4.3   Sentential Decision Diagrams

Sentential Decision Diagrams [18] represent propositional knowledge bases. They are inspired by two concepts: structured decomposability [38] which is based on vtrees, and the generalisation of Shannon decomposition which is strongly deterministic decomposition [39].

*vtree.* A vtree [38] for a set of variables $V$ is a full (but not necessarily complete), rooted binary tree whose leaves represent the variables in $V$. The node $v$ and the subtree rooted at the node $v$ are often called the same. Let $var(v)$ indicate the set of variables stored in the leaves of the subtree rooted at the node $v$. Let $v^l$ and $v^r$ be respectively indicate the left and right children of $v$. A Boolean function $f$ in Decomposable Negation Normal Form (DNNF) is said to *respect* a vtree $T$ if for every conjunction $\alpha \wedge \beta$ in $f$, there is a node $t$ in $T$ such that $var(\alpha) \subseteq var(t^l)$ and $var(\beta) \subseteq var(t^r)$.

*Strongly Deterministic Decomposition.* Let $f$ be a Boolean function with disjoint sets of variables $X$ and $Y$. If $f$ can be written as $(p_1(X) \wedge s_1(Y)) \vee ... \vee (p_n(X) \wedge s_n(Y))$, then $\{(p_1, s_1), ..., (p_n, s_n)\}$ is called an $(X, Y)$-decomposition of $f$ in terms of Boolean functions $p_i$ and $s_i$ on $X$ and $Y$ respectively. Provided that $p_i \wedge p_j = false$ for $i \neq j$, the decomposition is called *strongly deterministic* on $X$. Here, each $p_i$ is called a prime and each $s_i$ a sub. Let $\mathfrak{S}$ be a strongly deterministic $(X, Y)$-decomposition of function $f$. $\mathfrak{S}$ is called an $X$-partition of $f$ iff its primes make a *partition*. This means each prime is consistent (i.e., can be true at some evaluation), every pair of distinct primes are mutually exclusive, and the disjunction of all primes is valid (true).

*SDD.* An SDD can be seen as a recursive $(X, Y)$-strongly deterministic decomposition of a switching function according to a particular vtree, starting from the root node. The semantics of SDD $\mathfrak{S}$ is defined by the switching function $f_{\mathfrak{S}}$ with respect to the vtree $v$. $\mathfrak{S}$ is an SDD respecting vtree $v$ iff (1) $\mathfrak{S} = \perp$ or $\mathfrak{S} = \top$, with the semantics $f_{\perp} = false$ and $f_{\top} = true$, (2) $\mathfrak{S} = X$ or $\mathfrak{S} = \neg X$ and $v$ is a leaf with variable $X$ with the semantics $f_X = X$ and $f_{\neg X} = \neg X$, (3) $\mathfrak{S} = \{(p_1, s_1), ..., (p_n, s_n)\}$, $v$ is internal, $p_1, ..., p_n$ are SDDs that respect subtrees of $v^l$, and $s_1, ..., s_n$ are SDDs that respect subtrees of $v^r$, where $f_{p_1}, ..., f_{p_n}$ makes a partition. In this case, the semantics of SDD $\mathfrak{S}$ is given by $f_{\mathfrak{S}} = \bigvee_{i=1}^{n} (p_i \wedge s_i)$.

An SDD is *compressed* iff all its subs are distinct ($s_i \neq s_j$ for $i \neq j$). Moreover, it is called *trimmed* if it does not contain any decomposition of the form $\{(\top, \mathfrak{S})\}$ or $\{(\mathfrak{S}, \top), (\neg \mathfrak{S}, \perp)\}$. These two properties characterise the canonicity of SDDs as follows. Two SDDs that are compressed and trimmed and respect the same vtree are semantically equivalent if and only if they are equal [18]. Thus, vtrees for SDDs resemble the role of variable ordering for BDDs. In the same manner, compression and trimming resemble reduction in BDDs.

*Example 1.* Figure 2 (right) indicates an SDD for the Student Mood example (see Fig. 1). The underlying vtree is depicted in Fig. 2 (left). The two-parts boxes in SDD visually indicate the prime-sub pairs. The circles indicate decision nodes. Decision node $(p_1, s_1), ..., (p_k, s_k)$ has $k$ outgoing edges and edge $i$ is connected to $(p_i, s_i)$. The SDD respects the vtree in the sense that on each vtree node the leaves in the left subtree determine the primes and the leaves in the right subtree determine the subs. For instance, for node 3 in the vtree, $P$ determines the primes and $G$ and $M$ determine the subs. Since each decision node makes an $(X, Y)$-partition, each variable evaluation holds on exactly one prime.

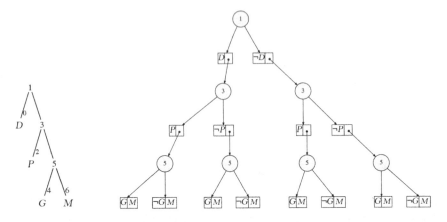

**Fig. 2.** An SDD (right) and the corresponding vtree (left) for Student Mood example

### 4.4  Probabilistic SDDs

PSDDs [30] are recent representations in the domain of reasoning and learning. Similarly to MTBDDs, which are BDDs to represent non-Boolean functions, SDDs are extended to PSDDs in order to particularly represent probability distributions. A single SDD can be parameterized in infinitely many ways, each yielding a probability distribution. This is similar to BNs in a sense that each DAG can be extended to infinitely many $\Theta$s, (i.e., conditional probability tables) where each $\Theta$ specifies a probability distribution. PSDDs are complete in the sense that every distribution can be induced by a PSDD. PSDDs are canonical in the sense that for a given vtree, there is a unique trimmed and compressed PSDD. Interestingly, computing the probability of a term can be done in time linear in the PSDD size [30].

*Syntax.* A PSDD parametrizes an SDD in the following manner: (1) Every decision node $(p_i, s_i), ..., (p_k, s_k)$ and every prime $p_i$ is equipped with a positive parameter $\theta_i$ such that $\theta_1 + ... + \theta_k = 1$ and $\theta_i = 0$ iff $s_i = \bot$. The PSDD decision node is indicated by $(p_1, s_1, \theta_1), ..., (p_k, s_k, \theta_k)$. (2) For each terminal node $\top$, a positive parameter $\theta$ is supplied such that $0 < \theta < 1$. Syntactically, the terminal node $\top$ with parameter $\theta$ is indicated by $x : \theta$, where $x$ is the variable of the vtree leaf node that $\top$ is normalized [18] for. Other terminal nodes ($\bot$, $x$, and $\neg x$) have fixed pre-defined parameters.

*Semantics.* Let $n$ be a PSDD node respecting a vtree node $v$. Node $n$ represents the probability distribution $Pr_n$ over the variables of vtree $v$ defined by:

– If $n$ is a terminal node and $v$ consists of variable $x$, then
  - for $n = x : \theta$, $Pr_n(x) = \theta$ and $Pr_n(\neg x) = 1 - \theta$
  - for $n = \bot$, $Pr_n(x) = 0$ and $Pr_n(\neg x) = 0$
  - for $n = x$, $Pr_n(x) = 1$ and $Pr_n(\neg x) = 0$
  - for $n = \neg x$, $Pr_n(x) = 0$ and $Pr_n(\neg x) = 1$.

– If $n$ is a decision node $(p_1, s_1, \theta_1), ..., (p_k, s_k, \theta_k)$ and $v$, the corresponding vtree node, has $X$ as left variables and $Y$ as right variables, $Pr_n(\underline{X}, \underline{Y}) = Pr_{p_i}(\underline{X}) \cdot Pr_{s_i}(\underline{Y}) \cdot \theta_i$ for $i$ that $\underline{X} \models p_i$. Here, $\underline{X}$ denotes the evaluation of $X$'s variables. The condition $\underline{X} \models p_i$ holds on exactly one of the primes, by definition of $(X, Y)$-decomposition.

*Example 2.* Figure 3 (right) denotes the PSDD for the Student Mood example, with the same underlying vtree as in Example 1. The visual representation of PSDD extends SDD in two manners: (1) edge $i$ directing from the decision node $(p_1, s_1, \theta_1), ..., (p_k, s_k, \theta_k)$ to the pair $(p_i, s_i)$ is labeled with $\theta_i$, (2) terminal node $\top$ is replaced by $x : \theta$, according to the PSDD syntax and semantics. The PSDD in Fig. 3 (right) induces the same probability distribution function induced by the Student Mood BN. For instance, $Pr(\neg D \wedge P \wedge \neg G \wedge \neg M) = 0.6 \cdot 0.3 \cdot (0.5 \cdot (1 - 0.1)) = 0.081$, by the PSDD semantics.

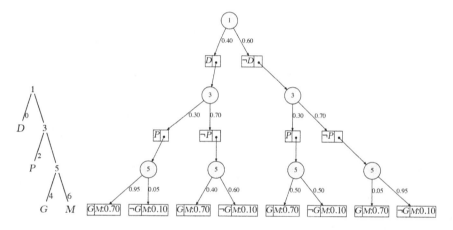

**Fig. 3.** A PSDD (right) and vtree (left) compiled from the Student Mood example

## 5   BN Analysis Using Probabilistic Model Checking

In this section, we are going to explain our approach in detail. First, let us explain some notations. Let $X$ be a set of variables. Let $\underline{X}$ denote the evaluation of $X$'s variables. We use $*$ to denote a don't care value. We use $\mu$ to denote a probability distribution. Let $\prod_{i=1}^{n} X_i = X_1 \times ... \times X_n = \{(x_1, ..., x_n)|x_1 \in X_1, ..., x_n \in X_n\}$ be the Cartesian product over the sets $X_i, ..., X_n$.

The basic idea for the transformation is to map a Bayesian network $B$ onto the Markov chain $M_B$ such that the conditional reachability probabilities in $M_B$ correspond to the conditional probabilistic inference queries in $B$. Colloquially stated:

$$Pr_B(H = h \mid F = f) = Pr_{M_B}(\Diamond(H = h) \mid \Diamond(F = f)). \qquad (4)$$

The definition of MC $M_B$ is as follows.

**Definition 1 *(The Markov chain of a BN)*.**     Let $B = (V, E, \Theta)$ be a BN with $V = \{v_1, ..., v_n\}$ and $dom(v_i) = D_i$ with elements $d_i \in D_i$. For $\varrho = (v_1, ..., v_n)$ a *topological order* on the DAG $(V, E)$, let MC $M_B = (\Sigma, \sigma_I, P)$ be the Markov chain of $B$ where:

- $\Sigma = \prod\limits_{i=1}^{n} \{v_i\} \times (D_i \cup \{*\})$ is the set of *states*,
- $\sigma_I = V \times \{*\}$ is the *initial state*, and
- $P : \Sigma \times \Sigma \to [0, 1]$ is the *transition probability function* defined by the following SOS rules:

$$\frac{\Theta_{v_1} = \mu, \quad \mu(d_1) = p}{\sigma_I \xrightarrow{p} \sigma_{(v_1, d_1)} = ((v_1, d_1), (v_2, *), ..., (v_n, *))} \tag{5}$$

$$\frac{\Theta_{v_i}(\underline{parents(v_i)}) = \mu, \quad \mu(d_i) = p,}{\sigma = ((v_1, d_1), ..., (v_{i-1}, d_{i-1}), (v_i, *), ..., (v_n, *))}$$
$$\frac{parents(v_i) \times \underline{parents(v_i)} \subseteq \{(v_1, d_1), ..., (v_{i-1}, d_{i-1})\}}{\xrightarrow{p} \sigma' = ((v_1, d_1), ..., (v_i, d_i), (v_{i+1}, *), ..., (v_n, *))} \tag{6}$$

$$\sigma = ((v_1, d_1), ..., (v_n, d_n)) \xrightarrow{1} \sigma = ((v_1, d_1), ..., (v_n, d_n)). \tag{7}$$

The states in MC $M_B$ are tuples of pairs in the form of $(v, d)$ where $v$ is a variable of BN $B$ and $d \in dom(v) \cup \{*\}$ is the current value of $v$. The symbol $*$ is used to denote the initial evaluation of a variable. The initial state is $((v_1, *), ..., (v_n, *))$. The transition probability function specifies the probability of evolving between states. These transition probabilities correspond to the values in the conditional probability tables of $B$'s variables. The rule (5) defines the transitions from the initial state to its successors according to $\Theta_{v_1}$. Since $v_1$ is the first variable in the topological order, $parents(v_1) = \emptyset$. If $\Theta_{v_1} = \mu$ and $\mu(d_1) = p$, there is a transition with probability $p$ from the initial state $\sigma_I$ to the state $\sigma$ in which all the variables are $*$ except for $v_1$ which is mapped to $d_1$. Let states in which all the variables have taken values from their domain constitute the final states, $\{(v, d) \mid v \in V, d \neq *\}$. According to the rule (7), the final states are equipped with a 1-probability self-loop. The transitions from the states that are neither initial nor final are formalized in the rule (6). Let $parents(v_i)$ be the evaluation of all the variables in the set $parents(v_i)$. Let $\mu = \Theta(\overline{parents(v_i)})$. If $\mu(d_i) = p$, then from the state $\sigma$ where all variables before $v_i$ based on the ordering $\varrho$ are evaluated to values other than $*$, the transition goes to the state $\sigma'$ where all the variable evaluations remain the same, except for $(v_i, *)$ that changes to $(v_i, d_i)$. The transition can take place only provided that all the variables in $parents(v_i)$ are already evaluated at the state $\sigma$ and their evaluation is consistent with the values in $\overline{parents(v_i)}$. This is ensured by the premise $parents(v_i) \times \underline{parents(v_i)} \subseteq \{(v_1, d_1), ..., (v_{i-1}, d_{i-1})\}$.

*Example 3.* Reconsider the BN Student Mood shown in the Fig. 1. Figure 4 (left) indicates the MC $M_{StudentMood}$ resulting from the above definition. Here, the *don't care* evaluations are omitted from the states and the states related to "Mood" variable are ignored. The probabilities on evolving edges are based on their corresponding values in the conditional probability tables. For instance, for the left most path in the MC this is 0.4, 0.3, and 0.95. The last number is set, for example, according to the Grade's CPT where the probability of $Grade = 1$ equals 0.95, for $Dif = 1$ and $Prep = 1$. The MTBDD corresponding to the probability distribution over the leaves of $M_{StudentMood}$ is shown in Fig. 4 (right), again abstracting from *don't care* values and "Mood" variable. The nodes binary evaluations are coded as the right and left edges in an MTBDD. For example, the left most path indicates all the variables being evaluated to 1. The terminal nodes denote the joint probability, which in this case equals $0.4 \cdot 0.3 \cdot 0.95 = 0.114$.

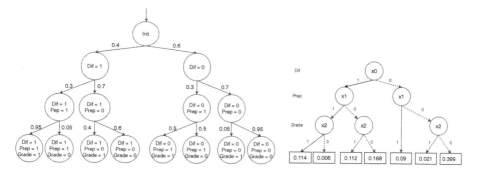

**Fig. 4.** The corresponding MC (left) and MTBDD (right) for Student Mood example

**Proposition 1 (*The size of MC $M_B$*).** *Let $B$ be a BN with $dom(v_i) = D_i$ for each $v_i \in V$ and $|M_B|$ be the number of states in the Markov chain $M_B$. Then,*

$$|M_B| \leq 1 + \sum_{i=1}^{n} \prod_{k=1}^{i} |D_k|. \tag{8}$$

In the special case where all random variables in $B$ have domain $D$

$$|M_B| \leq \sum_{i=0}^{n} |D|^i, \text{ thus } |M_B| \leq \frac{1 - |D|^{n+1}}{1 - |D|}.$$

*Example 4.* The number of states in Fig. 4 (right) is $|M_B| = 2^4 - 1 = 15$.

We now consider the reachability probability as shown in the Eq. (4). By definition of conditional probabilities we have:

$$Pr_{M_B}(\lozenge(H = h) \mid \lozenge(F = f)) = \frac{Pr_{M_B}(\lozenge(H = h) \wedge \lozenge(F = f)))}{Pr_{M_B}(\lozenge(F = f))} \tag{9}$$

To determine the right-hand side we observe that given the tree structure of MC $M_B$ it holds:

**Proposition 2.** *For Markov Chain $M_B$ of BN B:*

$$Pr_{M_B}(\Diamond(H = h) \wedge \Diamond(F = f)) = Pr_{M_B}(\Diamond(H = h \wedge F = f)).$$

From this, it is concluded that inference in BN $B$ can be reduced to computing reachability probabilities in MC $M_B$.

## 6  Experimental Results

*Experimental Setup.* We implemented a prototypical software tool for performing Bayesian inference as an extension of the probabilistic model checker Storm [20]. Our tool takes as input a BN in the Bayesian network Interchange Format [1] (BIF, for short). The BN is translated into a Markov chain as described in Definition 1. The MC is specified using the Jani [9] modelling language, a high-level modelling language in the domain of probabilistic model checking. We evaluated our tool using various BN benchmarks from the Bayesian network repository **bnlearn** [41] that contains several BNs categorized in small, medium, large, very large, and huge networks. Table 1 indicates some statistics of the evaluated BNs from the repository. The first column denotes whether all the variables in the BN are binary. The other statistics are the number of vertices, the number of edges, the maximum in-degree, the maximum domain size of variables, the average Markov blanket, and the number of parameters. The number of parameters is related to the total number of probabilities in all the conditional probability tables. All our experiments were conducted on a 2.3 GHz Intel Core i5 processor with 16 GB of RAM.

We focused our experimental validations on the following three questions:

1. What is the performance of MTBDD-based symbolic probabilistic model checking on Bayesian inference?
2. What is the effect of the number of observations and their depth in the topological ordering on the inference time in our approach?
3. How does inference using MTBDDs compares to PSDD techniques in terms of compilation time and inference time?

*Bayesian Inference Using MTBDD-Based Model Checking.* In order to answer the first question, we have fed the Jani descriptions of the MCs into storm's sparse engine, and storm's bdd engine. The former fully builds a sparse matrix of the Markov chain $M_B$ of BN $B$, while the latter generates an MTBDD representation from the Jani file. The variable ordering of the MTBDD is determined based on the topological order of the BN. Table 2 indicates the size of the resulting data structures and the compilation time. Here $E19$, for instance, denotes $10^{19}$ as the order of magnitude. The inference time on the sparse representation is prohibitive, even for medium-sized BNs, while inference using MTBDDs is mostly a matter of a few seconds or less. The large MC sizes are due to the

**Table 1.** Statistics on the evaluated Bayesian networks in **bnlearn**

| Binary | BN | #Vertices | #Edges | InDegreeMax | Dmax | AMB | #Parameters |
|---|---|---|---|---|---|---|---|
| YES | cancer | 5 | 4 | 2 | 2 | 2.00 | 10 |
| | earthquake | 5 | 4 | 2 | 2 | 2.00 | 10 |
| | asia | 8 | 8 | 2 | 2 | 2.5 | 18 |
| | win95pts | 76 | 112 | 7 | 2 | 5.92 | 574 |
| | andes | 223 | 338 | 6 | 2 | 5.61 | 1157 |
| NO | survey | 6 | 6 | 2 | 3 | 2.67 | 21 |
| | sachs | 11 | 17 | 3 | 3 | 3.09 | 178 |
| | child | 20 | 25 | 2 | 6 | 3.00 | 230 |
| | alarm | 37 | 46 | 4 | 4 | 3.51 | 509 |
| | insurance | 27 | 52 | 3 | 5 | 5.19 | 984 |
| | hepar2 | 70 | 123 | 6 | 4 | 4.51 | 1453 |
| | hailfinder | 56 | 66 | 4 | 11 | 3.54 | 2656 |
| | water | 32 | 66 | 5 | 4 | 7.69 | 10083 |
| | pathfinder | 135 | 200 | 5 | 63 | 3.81 | 72079 |

exponential growth of the state space in the domain of the BN's variables, see Proposition 1. The significant size reduction with MTBDDs is due to the symmetrical and repetitive structure of the $M_B$. Those kind of symmetries and duplicated subtrees are merged in the MTBDD representation. The type of MTBDD shown in Fig. 4 (right) represents a discrete probability distribution. However, the MTBDD generated by storm encodes the Markov chain, i.e. its terminal nodes carry the transition probabilities of the Markov chain. This makes sharing of the subgraphs much more likely.

**Table 2.** Analysis of BN benchmarks using storm symbolic engine compared to sparse

| Binary | BN | Size | | Construction time | |
|---|---|---|---|---|---|
| | | MC (#states) | MTBDD (#nodes) | Sparse engine | Symbolic engine |
| YES | cancer | 63 | 56 | 0.018 s | 0.007 s |
| | earthquake | 63 | 55 | 0.023 s | 0.006 s |
| | asia | 278 | 154 | 0.028 s | 0.011 s |
| | win95pts | E19 | 446752 | >1.5 h | 11 s |
| | andes | E67 | 339485 | >1.5 h | 180 s |
| NO | survey | 238 | 70 | 0.031 s | 0.008 s |
| | sachs | 265720 | 165 | 0.469 s | 0.072 s |
| | child | E9 | 731 | >1.5 h | 0.277 s |
| | alarm | E16 | 2361 | >1.5 h | 1 s |
| | insurance | E11 | 102903 | >1.5 h | 2 s |
| | hepar2 | E42 | 7675 | >1.5 h | 17 s |
| | hailfinder | E17 | 152201 | >1.5 h | 18 s |
| | water | E9 | 64744 | >1.5 h | 20 s |
| | pathfinder | E242 | MO | >1.5 h | – |

*The Influence of Observations.* In order to answer the second question, we have chosen different ways to pick the set of evidences. This is aimed to investigate how the number of evidence nodes and their depth in the topological order affect the verification time. For each BN, three different sets of observations are considered; the evidence nodes at the beginning in the topological order, a random selection, and the last nodes in the topological ordering. We also varied the number of evidence nodes. Figure 5 (in log-log scale) demonstrates the results for two large benchmarks, win95pts (left) and hepar2 (right). The x-axis denotes the number of evidence nodes and the y-axis denotes the model checking time in seconds. For the "first" setting, where $i$ nodes are picked from the beginning of the topological order, the time for performing model checking is relatively small; less than 3.064 s in all the experiments for win95pts and less than 0.682 s in all the experiments for hepar2. The results follow a similar pattern in almost all the other BN benchmarks. The last nodes in the topological order are the highest dependent ones on the other nodes. That explains why model checking is significantly more time-consuming in the "last" setting. The verification time becomes negligible if the number of evidences is large. That is mostly because then the final result of the inference tends more likely to become zero when the number of evidence nodes are high. In this case there are many restrictions to be satisfied, and the zero probability can be computed very fast.

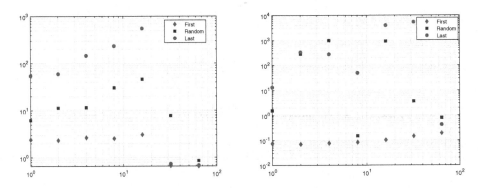

**Fig. 5.** Inference time (in s) for different size and depth of evidence for win95pts (left) and hepar2 (right) - log log scale

*Comparing MTBDD-Based Model Checking to Inference Using PSDDs.* In order to answer our last question, we have conducted a series of experiments to see how our approach performs compared to the recent prominent inference tool based on PSDD. PSDD[1] is a scalable tool for reasoning and learning on belief networks in the AI literature. We have compiled the BNs in the benchmark into MTBDD with storm symbolic engine. To this end, we have converted the benchmark BNs into PSDDs and vtrees using the PSDD-Nips package[2]. Our experiments covers

---

[1] https://github.com/hahaXD/psdd.

[2] https://github.com/hahaXD/psdd_nips.

**Table 3.** Empirical comparison with PSDD based inference regarding different vtree methods - Binary cases

| BN | #Evidence | MTBDD Compilation | MTBDD Inference | PSDD random vtree Compilation | PSDD random vtree Inference | PSDD fixed vtree Compilation | PSDD fixed vtree Inference | PSDD minfill vtree Compilation | PSDD minfill vtree Inference |
|---|---|---|---|---|---|---|---|---|---|
| cancer | 1 | | 0.001 s | | 0.002 s | | 0.003 s | | 0.003 s |
| | 2 | | 0.001 s | | 0.003 s | | 0.003 s | | 0.003 s |
| | 3 | 0.022 s | 0.001 s | 0.016 s | 0.003 s | 0.016 s | 0.003 s | 0.005 s | 0.003 s |
| | 4 | | 0.001 s | | 0.003 s | | 0.002 s | | 0.003 s |
| | 5 | | 0.001 s | | 0.003 s | | 0.003 s | | 0.003 s |
| earthquake | 1 | | 0.001 s | | 0.003 s | | 0.003 s | | 0.003 s |
| | 2 | 0.006 s | 0.001 s | 0.015 s | 0.003 s | 0.016 s | 0.003 s | 0.004 s | 0.003 s |
| | 4 | | 0.001 s | | 0.003 s | | 0.003 s | | 0.003 s |
| asia | 1 | | 0.001 s | | 0.004 s | | 0.003 s | | 0.003 s |
| | 2 | 0.018 s | 0.001 s | 0.026 s | 0.003 s | 0.023 s | 0.003 s | 0.005 s | 0.003 s |
| | 4 | | 0.001 s | | 0.004 s | | 0.004 s | | 0.003 s |
| | 8 | | 0.002 s | | 0.003 s | | 0.003 s | | 0.003 s |
| win95pts | 1 | | 2.409 s | | 0.074 s | | 0.068 s | | **0.042 s** |
| | 2 | | 2.760 s | | 0.066 s | | 0.060 s | | **0.039 s** |
| | 4 | | 2.501 s | | 0.067 s | | 0.067 s | | **0.039 s** |
| | 8 | 11.214 s | 2.452 s | 0.258 s | 0.063 s | 0.233 s | 0.061 s | 0.047 s | **0.039 s** |
| | 16 | | 2.576 s | | 0.056 s | | 0.050 s | | **0.033 s** |
| | 32 | | 0.671 s | | 0.053 s | | 0.046 s | | **0.033 s** |
| | 64 | | 0.658 s | | 0.053 s | | 0.043 | | **0.032 s** |
| andes | 1 | | **1.165 s** | | 5.479 s | | 12.893 s | | 4.863 s |
| | 2 | | **0.989 s** | | 5.824 s | | 12.832 s | | 4.818 s |
| | 4 | | **0.992 s** | | 5.423 s | | 13.312 s | | 4.823 s |
| | 8 | 180 s | **1.144 s** | 12.617 s | 5.620 s | 13.046 s | 12.874 s | 4.724 s | 4.838 s |
| | 16 | | **1.247 s** | | 5.612 s | | 9.921 s | | 4.122 s |
| | 32 | | **1.385 s** | | 5.457 s | | 10.362 s | | 4.120 s |
| | 64 | | **2.538 s** | | 5.552 s | | 8.996 s | | 3.442 s |
| | 128 | | 3.488 s | | 3.656 s | | 8.096 s | | **3.141 s** |

the available vtree methods: random, fixed and minfill. The decisive difference between these methods is the heuristics they employ to triangulate the DAG underlying a BN [26]. Due to the fact that the PSDD packages are inherently limited to perform inference only on BNs with binary variables, we categorize our results into two parts: *binary* BNs and *non-binary* BNs. Table 3 indicates the results for the binary benchmarks. The results include the compilation time and inference time by different methods, taking the same sets of evidence nodes. In each row the minimum inference time is highlighted in bold face. As inference in our tool is applicable to non-binary BNs, we have built a prototypical script to binarize non-binary networks such that they can be fed into the PSDD package. Table 4 indicates the results for these non-binary benchmarks where $\#Nodes$

**Table 4.** Empirical comparison with PSDD based inference regarding different vtree methods - Non-binary cases

| BN | #Nodes | #Evidence | MTBDD Compilation | MTBDD Inference | PSDD random vtree Compilation | PSDD random vtree Inference | PSDD fixed vtree Compilation | PSDD fixed vtree Inference | PSDD minfill vtree Compilation | PSDD minfill vtree Inference |
|---|---|---|---|---|---|---|---|---|---|---|
| survey | 14 | 1 | | 0.002 s | | 0.004 s | | 0.004 s | | 0.004 s |
| | | 2 | 0.018 s | 0.002 s | 0.114 s | 0.004 s | 0.129 s | 0.003 s | 0.019 s | 0.004 s |
| | | 4 | | 0.004 s | | 0.003 s | | 0.003 s | | 0.003 s |
| sachs | 33 | 1 | | 0.002 s | | 0.008 s | | 0.008 s | | 0.010 s |
| | | 2 | 0.076 s | 0.002 s | 0.208 s | 0.008 s | 0.212 s | 0.008 s | 0.096 s | 0.008 s |
| | | 4 | | 0.004 s | | 0.009 s | | 0.009 s | | 0.009 s |
| | | 8 | | 0.004 s | | 0.008 s | | 0.007 s | | 0.008 s |
| child | 60 | 1 | | 0.014 s | | 0.014 s | | 0.012 s | | 0.021 s |
| | | 2 | | 0.004 s | | 0.013 s | | 0.010 s | | 0.018 s |
| | | 4 | 0.273 s | 0.005 s | 0.304 s | 0.013 s | 0.293 s | 0.010 s | 0.191 s | 0.018 s |
| | | 8 | | 0.006 s | | 0.010 s | | 0.010 s | | 0.013 s |
| | | 16 | | 0.008 s | | 0.011 s | | 0.010 s | | 0.016 s |
| alarm | 104 | 1 | | 0.010 s | | 0.014 s | | 0.012 s | | 0.013 s |
| | | 2 | | 0.011 s | | 0.012 s | | 0.013 s | | 0.012 s |
| | | 4 | 1.538 s | 0.013 s | 0.703 s | 0.013 s | 0.685 s | 0.013 s | 0.345 s | 0.012 s |
| | | 8 | | 0.014 s | | 0.013 s | | 0.013 s | | 0.013 s |
| | | 16 | | 0.019 s | | 0.010 s | | 0.011 s | | 0.010 s |
| | | 32 | | 0.031 s | | 0.013 s | | 0.013 s | | 0.012 s |
| insurance | 88 | 1 | | 0.432 s | | 0.011 s | | 0.013 s | | 0.012 s |
| | | 2 | | 0.462 s | | 0.012 s | | 0.012 s | | 0.012 s |
| | | 4 | 2.258 s | 0.461 s | 0.695 s | 0.012 s | 0.672 s | 0.013 s | 0.342 s | 0.012 s |
| | | 8 | | 0.478 s | | 0.013 s | | 0.013 s | | 0.012 s |
| | | 16 | | 0.174 s | | 0.012 s | | 0.011 s | | 0.010 s |
| hepar2 | 162 | 1 | | 0.074 s | | 0.058 s | | 0.054 s | | 0.056 s |
| | | 2 | | 0.069 s | | 0.054 s | | 0.049 s | | 0.054 s |
| | | 4 | | 0.076 s | | 0.052 s | | 0.044 s | | 0.053 s |
| | | 8 | 17 s | 0.083 s | 32.129 s | 0.052 s | 37.466 s | 0.043 s | 12.205 s | 0.043 s |
| | | 16 | | 0.101 s | | 0.045 s | | 0.043 s | | 0.033 s |
| | | 32 | | 0.144 s | | 0.039 s | | 0.036 s | | 0.051 s |
| | | 64 | | 0.191 s | | 0.042 s | | 0.038 s | | 0.051 s |

indicates the number of resulting binary variables. The pre-processing time for conversion and binarization is not included. Due to the large number of parameters in the benchmarks "hailfinder", "water", and "pathfinder" (see Table 1), these cases are computationally hard to binarize. Therefore, these cases are not included in Table 4. The main conclusions of our experimental results are:

1. Inference using MTBDD-based symbolic model checking is competitive to BN-specific symbolic techniques like PSDD for small to large BNs.

2. PSDD techniques outperform our MTBDD-based approach for very large and huge BNs.
3. MTBDD-based inference is quite sensitive to the number and depth (in the topological order) of evidences.

# 7 Conclusions

In this paper, we have investigated MTBDD-based symbolic probabilistic model checking to perform exact inference on Bayesian networks. We have translated Bayesian networks into Markov chains, and have reduced inference to computing reachability probabilities. Our prototypical tool chain built on top of storm [20] is evaluated on BNs from the **bnlearn** repository. We investigated several hypotheses to see which factors are affecting the inference time.

Future work consists of optimizing our implementation and approach, and to consider other metrics on BNs, such as maximum a posteriori (MAP) and the most probable explanation (MPE). We also like to generalize our approach to recursive BNs [27] or dynamic BNs [28], which bring respectively the notion of recursion and time on the table. Moreover, we believe that this work provides a good basis for performing probabilistic model checking on a broader set of graphical models such as Markov networks, which are, unlike Bayesian networks, undirected in nature.

**Acknowledgement.** The authors would like to thank Yujia Shen (UCLA) for his kind support with running the PSDD tools.

# References

1. Bayesian network Interchange Format. http://www.cs.washington.edu/dm/vfml/appendixes/bif.htm. Accessed 2019
2. de Alfaro, L., Kwiatkowska, M., Norman, G., Parker, D., Segala, R.: Symbolic model checking of probabilistic processes using MTBDDs and the Kronecker representation. In: Graf, S., Schwartzbach, M. (eds.) TACAS 2000. LNCS, vol. 1785, pp. 395–410. Springer, Heidelberg (2000). https://doi.org/10.1007/3-540-46419-0_27
3. Bahar, R.I., et al.: Algebraic decision diagrams and their applications. Formal Methods Syst. Des. **10**(2/3), 171–206 (1997)
4. Baier, C., Clarke, E.M., Hartonas-Garmhausen, V., Kwiatkowska, M., Ryan, M.: Symbolic model checking for probabilistic processes. In: Degano, P., Gorrieri, R., Marchetti-Spaccamela, A. (eds.) ICALP 1997. LNCS, vol. 1256, pp. 430–440. Springer, Heidelberg (1997). https://doi.org/10.1007/3-540-63165-8_199
5. Baier, C., Klein, J., Klüppelholz, S., Märcker, S.: Computing conditional probabilities in Markovian models efficiently. In: Ábrahám, E., Havelund, K. (eds.) TACAS 2014. LNCS, vol. 8413, pp. 515–530. Springer, Heidelberg (2014). https://doi.org/10.1007/978-3-642-54862-8_43
6. Batz, K., Kaminski, B.L., Katoen, J.-P., Matheja, C.: How long, O Bayesian network, will I sample thee? In: Ahmed, A. (ed.) ESOP 2018. LNCS, vol. 10801, pp. 186–213. Springer, Cham (2018). https://doi.org/10.1007/978-3-319-89884-1_7

7. Bollig, B., Wegener, I.: Improving the variable ordering of OBDDs is NP-complete. IEEE Trans. Comput. **45**(9), 993–1002 (1996)
8. Bryant, R.E.: Binary decision diagrams. Handbook of Model Checking, pp. 191–217. Springer, Cham (2018). https://doi.org/10.1007/978-3-319-10575-8_7
9. Budde, C.E., Dehnert, C., Hahn, E.M., Hartmanns, A., Junges, S., Turrini, A.: JANI: quantitative model and tool interaction. In: Legay, A., Margaria, T. (eds.) TACAS 2017. LNCS, vol. 10206, pp. 151–168. Springer, Heidelberg (2017). https://doi.org/10.1007/978-3-662-54580-5_9
10. Chaki, S., Gurfinkel, A.: BDD-based symbolic model checking. Handbook of Model Checking, pp. 219–245. Springer, Cham (2018). https://doi.org/10.1007/978-3-319-10575-8_8
11. Chavira, M., Darwiche, A.: Compiling Bayesian networks with local structure. In: IJCAI, pp. 1306–1312. Professional Book Center (2005)
12. Chavira, M., Darwiche, A., Jaeger, M.: Compiling relational Bayesian networks for exact inference. Int. J. Approx. Reason. **42**(1–2), 4–20 (2006)
13. Cooper, G.F.: The computational complexity of probabilistic inference using Bayesian belief networks. Artif. Intell. **42**(2–3), 393–405 (1990)
14. Dagum, P., Luby, M.: Approximating probabilistic inference in Bayesian belief networks is NP-Hard. Artif. Intell. **60**(1), 141–153 (1993)
15. Darwiche, A.: A logical approach to factoring belief networks. In: KR, pp. 409–420. Morgan Kaufmann (2002)
16. Darwiche, A.: New advances in compiling CNF into Decomposable Negation Normal Form. In: ECAI, pp. 328–332. IOS Press (2004)
17. Darwiche, A.: Modeling and Reasoning with Bayesian Networks. Cambridge University Press, Cambridge (2009)
18. Darwiche, A.: SDD: a new canonical representation of propositional knowledge bases. In: IJCAI, pp. 819–826. IJCAI/AAAI (2011)
19. Darwiche, A.: A differential approach to inference in Bayesian networks. CoRR abs/1301.3847 (2013)
20. Dehnert, C., Junges, S., Katoen, J.-P., Volk, M.: A STORM is coming: a modern probabilistic model checker. In: Majumdar, R., Kunčak, V. (eds.) CAV 2017. LNCS, vol. 10427, pp. 592–600. Springer, Cham (2017). https://doi.org/10.1007/978-3-319-63390-9_31
21. Deininger, D., Dimitrova, R., Majumdar, R.: Symbolic model checking for factored probabilistic models. In: Artho, C., Legay, A., Peled, D. (eds.) ATVA 2016. LNCS, vol. 9938, pp. 444–460. Springer, Cham (2016). https://doi.org/10.1007/978-3-319-46520-3_28
22. Fujita, M., McGeer, P.C., Yang, J.C.: Multi-terminal binary decision diagrams: an efficient data structure for matrix representation. Formal Methods Syst. Des. **10**(2/3), 149–169 (1997)
23. Gehr, T., Misailovic, S., Vechev, M.: PSI: exact symbolic inference for probabilistic programs. In: Chaudhuri, S., Farzan, A. (eds.) CAV 2016. LNCS, vol. 9779, pp. 62–83. Springer, Cham (2016). https://doi.org/10.1007/978-3-319-41528-4_4
24. Hahn, E.M., et al.: The 2019 comparison of tools for the analysis of quantitative formal models. In: Beyer, D., Huisman, M., Kordon, F., Steffen, B. (eds.) TACAS 2019. LNCS, vol. 11429, pp. 69–92. Springer, Cham (2019). https://doi.org/10.1007/978-3-030-17502-3_5
25. Holtzen, S., Millstein, T.D., Van den Broeck, G.: Symbolic exact inference for discrete probabilistic programs. CoRR abs/1904.02079 (2019)

26. Hopkins, M., Darwiche, A.: A practical relaxation of constant-factor treewidth approximation algorithms. In: Proceedings of the First European Workshop on Probabilistic Graphical Models. PGM, pp. 71–80. Citeseer (2002)
27. Jaeger, M.: Complex probabilistic modeling with recursive relational Bayesian networks. Ann. Math. Artif. Intell. **32**(1–4), 179–220 (2001)
28. Jensen, F.V.: Bayesian Networks and Decision Graphs. Statistics for Engineering and Information Science. Springer, New York (2001). https://doi.org/10.1007/978-1-4757-3502-4
29. Katoen, J.: The probabilistic model checking landscape. In: LICS, pp. 31–45. ACM (2016)
30. Kisa, D., Van den Broeck, G., Choi, A., Darwiche, A.: Probabilistic sentential decision diagrams. In: KR. AAAI Press (2014)
31. Klein, J., et al.: Advances in symbolic probabilistic model checking with PRISM. In: Chechik, M., Raskin, J.-F. (eds.) TACAS 2016. LNCS, vol. 9636, pp. 349–366. Springer, Heidelberg (2016). https://doi.org/10.1007/978-3-662-49674-9_20
32. Koller, D., Friedman, N.: Probabilistic Graphical Models - Principles and Techniques. MIT Press, Cambridge (2009)
33. Kwiatkowska, M., Norman, G., Parker, D.: PRISM 4.0: verification of probabilistic real-time systems. In: Gopalakrishnan, G., Qadeer, S. (eds.) CAV 2011. LNCS, vol. 6806, pp. 585–591. Springer, Heidelberg (2011). https://doi.org/10.1007/978-3-642-22110-1_47
34. Langmead, C., Jha, S., Clarke, E.: Temporal logics as query languages for dynamic Bayesian networks: application to d. melanogaster embryo development. Technical report, Carnegie Mellon University (2006)
35. Langmead, C.J.: Towards inference and learning in dynamic Bayesian networks using generalized evidence. Technical report, Carnegie Mellon University (2008)
36. Minato, S., Satoh, K., Sato, T.: Compiling Bayesian networks by symbolic probability calculation based on zero-suppressed BDDs. In: IJCAI, pp. 2550–2555 (2007)
37. Palaniappan, S.K., Thiagarajan, P.S.: Dynamic Bayesian networks: a factored model of probabilistic dynamics. In: Chakraborty, S., Mukund, M. (eds.) ATVA 2012. LNCS, pp. 17–25. Springer, Heidelberg (2012). https://doi.org/10.1007/978-3-642-33386-6_2
38. Pipatsrisawat, K., Darwiche, A.: New compilation languages based on structured decomposability. In: AAAI, pp. 517–522. AAAI Press (2008)
39. Pipatsrisawat, T., Darwiche, A.: A lower bound on the size of decomposable negation normal form. In: AAAI. AAAI Press (2010)
40. Sanner, S., McAllester, D.A.: Affine algebraic decision diagrams (AADDs) and their application to structured probabilistic inference. In: IJCAI, pp. 1384–1390. Professional Book Center (2005)
41. Scutari, M.: Bayesian network repository. https://www.bnlearn.com. Accessed 2019
42. Shachter, R.D., D'Ambrosio, B., Favero, B.D.: Symbolic probabilistic inference in belief networks. In: AAAI, pp. 126–131. AAAI Press/The MIT Press (1990)
43. Shih, A., Choi, A., Darwiche, A.: Formal verification of Bayesian network classifiers. In: Proceedings of Machine Learning Research. PGM, vol. 72, pp. 427–438. PMLR (2018)
44. Xue, Y., Choi, A., Darwiche, A.: Basing decisions on sentences in decision diagrams. In: AAAI. AAAI Press (2012)

# Queuing Networks

# CogQN: A Queueing Model that Captures Human Learning of the User Interfaces of Session-Based Systems

Olivia Das[1][(✉)] and Arindam Das[2]

[1] Electrical and Computer Engineering, Ryerson University, Toronto, Canada
odas@ee.ryerson.ca
[2] School of Engineering Technology & Applied Science,
Centennial College, Toronto, Canada

**Abstract.** A session-based system provides various services to its end users through user interfaces. A novice user of a service's user interface takes more think time—the time to comprehend the content, and the layout of graphical elements, on the interface—in comparison to expert users. The think time gradually decreases, as she repeatedly comprehends the same interface, over time. This decrease in think time is the user learning phenomenon. Owing to this learning behavior, the proportion of users—at various learning levels for different services—changes dynamically leading to a difference in the workload. Traditionally though, workload specifications (required for system performance evaluation) never accounted for user learning behavior. They generally assumed a global mean think time, instead. In this work, we propose a novel queueing network (QN) model called CogQN that accounts for *user learning*. It is a multi-class QN model where each service and its learning level constitute a class of users for the service. The model predicts overall mean response times across different learning modes within 10% error in comparison to empirical data.

**Keywords:** Queueing network · Think time · User learning · Response time

## 1 Introduction

Often, a user interface (UI) is associated with a service. A typical user of a UI takes a finite amount of time to comprehend its content, and the layout of its graphical elements. This time is termed as think time. As the user repeatedly uses a service, her think time for the service gradually decreases—this is user learning phenomenon. The existing works on workload generation [1–3, 5] have not considered user learning. In contrast we posit that *user learning*, together with different proportions of users arriving at the system with various learning levels for different services, *is likely* to impact the system's performance.

To this end, we propose a predictive performance model, CogQN. Our model is a multi-class queueing network augmented with two novel parameters related to user learning—the Think Time Matrix and the Arrival Probability Matrix. We define them later in the paper. The CogQN model introduces the concept of a class to represent *a service* invoked by the users who are at a certain *learning level* for that service.

© Springer Nature Switzerland AG 2020
M. Gribaudo et al. (Eds.): QEST 2020, LNCS 12289, pp. 137–143, 2020.
https://doi.org/10.1007/978-3-030-59854-9_10

It further introduces the concept of *learning level-dependent* class-switching probabilities to model the dynamic transition of learning levels of users for different services.
For our convenience, we refer to the traditional queueing network (QN) as *QN-no-
learning* to distinguish it from our CogQN model.

We modify a TPC-W system [4, 7] in terms of user learning included (we call this
*yes-learning* or YL mode), versus not included (we call this *no-learning* or NL mode).
We then run the modified system and collect measured data out of it. Next, we realize
two discrete-event simulation models of the TPC-W system: First, a CogQN model
capturing various YL modes. Second, a *QN-no-learning* model—a traditional QN
model capturing the NL mode. We realize this QN model as a special case of the
CogQN model. Finally, using the models, we predict the overall mean response times
across different YL modes, and the NL mode. We validate the predictions against
measured data within 10% error. The empirical result indicates that one *should not
likely avoid* end-user learning if accurate performance predictions are desired.

## 2    System Description

When a user submits a service request successfully through a user interface-based
client, a response results in form of a user interface (e.g., a web page) corresponding to
that service request. To initiate a user session, the user first submits a request for the
entry service *e* (e.g., Home) through the client. Once the user interface corresponding to
service *e* is obtained as a response, the user has the option of ending the session or
submitting another request through the responded interface. The cycle—request submission; response generation; time for response comprehension (*think time*); next
request submission—continues until the session is ended. Studies in user interfaces
have observed that the think time of a user decreases for a service with its repeated
usage [6, 8–10, 12]. In our context, this happens because the user gets more and more
familiar with the content on the user interface obtained as a response to the service
request. The work in [8] surveys that the learning curves, resulting from change in
think time over repeated usages, can have different shapes—it may be flat showing
small change in think time across consecutive usages; or it may be steep showing sharp
fall in think time in the early stage of usages (number of usages being plotted in x-axis,
think time being plotted in y-axis). The shape of the curve depends on the usability of
the interface in terms of the complexity of its content and the layout of its graphical
elements.

## 3    The CogQN Model

### 3.1    Notations

To develop our CogQN model, we first introduce the notations as follows:

- *N* is the number of nodes (representing devices).
- *R* denotes the number of services provided by an application.
- *L* denotes the number of learning levels a user can be at for a service (Note: At any
  given time-point, the user can be at only one learning level for a given service).

Once a user reaches the last learning level for a service, her learning level does not increase any further even if she continues to submit requests for that service.

- $U$ denotes the number of users in the system. Some of the users may be in the thinking phase and some of them may be in the waiting phase (Users in waiting phase are those who have submitted the request and waiting for system to respond).

- $\lambda$ is the arrival rate of users. A user session begins by requesting for entry service $e$.

- **Think Time Matrix**, $T = [t_{ij}]_{R \times L}$. An element $t_{ij}$ denotes the mean think time of the users who are at learning level $j$ for service $i$. Each row $i$ in this matrix represents the learning curve for service $i$.

- **Arrival Probability Matrix**, $A = [a_{ij}]_{R \times L}$. A user is at different learning levels for different services. An element $a_{ij}$ denotes the probability that an arriving user is at learning level $j$ for service $i$. Given any service $i$, $\sum_{j=1}^{L} a_{ij} = 1$.

- **Service Transition Probability Matrix**, $P = [p_{rs}]_{R \times R}$. An element $p_{rs}$ is the probability that a user will send a request for service $s$ after finishing with a request for service $r$. For our modeling purpose, we assume that whenever the service $s$ is the entry service $e$, the user session ends, and the user leaves the network.

- A **class $rl$** (where $r = 1, 2, ..., R; l = 1, 2, ..., L$) represents a set of users who are in one of the two states:
  - a user has submitted a $rl$ request for service $r$ at learning level $l$.
  - a user has got back the response of her $rl$ request and is comprehending the response for the duration of the relevant think time.

The classes $rl$ and $rl'$ (representing the same service, different learning levels) differ in their think times, but always need the same service demands from a node.

- $C = [c_{rl}]_{R \times L}$. $c_{rl}$ denotes the number of users in class $rl$. $\sum_{r=1}^{R} \sum_{l=1}^{L} c_{rl} = U$.
- $K = [k_{ij}]_{R \times L}$. $k_{ij}$ in this matrix denotes the number of users in the system who are at learning level $j$ for service $i$. The row $i$ shows the distribution of $U$ users who are at various learning levels for the service $i$. $k_{ij}$ includes $c_{ij}$ (the number of users who are in class $ij$) **plus** the number of users who are in other classes but who have the learning level $j$ for service $i$. So, $k_{ij} \geq c_{ij}$. Thus, for each row $i$, $\sum_{j=1}^{L} k_{ij} = U$.
- **Learning Level-dependent Class Switching Probability Matrix**, $S = [s_{rl,r'l'}]_{R.L \times R.L}$. An element $s_{rl,r'l'}$ is the probability that a user will switch from the class $rl$ to the class $r'l'$. The element $s_{rl,r'l'}$ is computed using $C$, $K$ and $P$.
- **Service Demand Matrix**, $D = [d_{rn}]_{R \times N}$. An element $d_{rn}$ is the service demand to process a request for service $r$ on node $n$. For a fixed $r$, all $rl$ requests (where $l = 1, 2, ..., L$) will have the same service demand $d_{rn}$ on node $n$ (where $n = 1, 2, ..., N$).
- **RT** is the overall mean response time of a request regardless of its class.

## 3.2  Model Description

Figure 1 shows our CogQN model. The parameters of this model are Think Time Matrix $T$, Arrival Probability Matrix $A$, Service Transition Probability Matrix $P$, Service Demand Matrix $D$ and the Arrival rate $\lambda$. $T$ and $A$ are related to user learning.

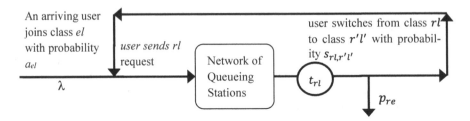

**Fig. 1.** A Cog-QN model for a session-based application system.

Our model has $R \times L$ classes of users. The box in Fig. 1 represents a network of queueing stations i.e. nodes, where a node represents a resource in the real system. Users arrive to the system with an average rate $\lambda$. All the arriving users enter the system by requesting entry service $e$. Once a user arrives at the system, it joins a class $el$ with probability $a_{el}$. A user then submits an $rl$ request ($r$ is $e$ for a newly arrived user) and waits for the response. Once the response is obtained, the user takes $t_{rl}$ amount of time to comprehend the response. The user then either leaves the network with probability $p_{re}$ or switches its class from $rl$ to $r'l'$ with learning level-dependent class switching probability $s_{rl,r'l'}$.

### 3.3   Model Solution

In our model, the system maintains a state on the number of users using a service at a given learning level, and the number of users at each learning level for each service. The former is maintained by the matrix $C$ and the latter is maintained by the matrix $K$. We solve our CogQN model using discrete-event simulation. In our simulation, the probability of switching of a user from one class to another class is computed using these two matrices, and the matrix $P$.

Similar to any queueing model simulation, we first generate the users at an inter-arrival time sampled from an exponential distribution with rate $\lambda$. Once a user joins a class $rl$, it traverses different queueing nodes in the network to obtain service $r$ with service demand $d_{rn}$ on node $n$. The service demands are sampled from an exponential distribution. The user then visits the pure delay server representing the think time. The think time is sampled from an exponential distribution with rate $1/t_{rl}$ depending on the user's class.

Next, we describe how we maintain the matrices $C$ and $K$ as well as how we compute the class switching probability (where relevant) for the three cases—i) when a user arrives; ii) when a user leaves; iii) when a user switches its class.

**Case i):**
When a user arrives, the matrices $C$ and $K$ are updated as follows:

- The user joins a class $el$ with probability $a_{el}$. Consequently, with probability $a_{el}$, the $c_{el}$ element of matrix $C$ as well as the $k_{el}$ element of $K$ are increased by 1.
- For every service $i$ other than $e$, only one $k_{ij}$ element of row $i$ of matrix $K$ is incremented by 1 depending on the probability $a_{ij}$ where $\sum_{j=1}^{L} a_{ij} = 1$.

**Case ii):**

When a user leaves, matrices $C$ and $K$ are updated as follows:

- The user leaves the class $rl$ with probability $p_{re}$. Consequently, with probability $p_{re}$, the $c_{rl}$ element of matrix $C$ as well as the $k_{rl}$ element of matrix $K$ are decremented by 1.
- For every service $i$ other than $r$, only one $k_{ij}$ element of row $i$ of matrix $K$ is decremented by 1 depending on the following probability:

$$\frac{k_{ij} - c_{ij}}{\sum_{j=1}^{L} \left( k_{ij} - c_{ij} \right)}$$

where $k_{ij} - c_{ij}$ is the number of users who are at learning level $j$ for service $i$ but who belong to a class $(m*)$ where $m \neq i$; $*$ implies any learning level.

**Case iii):**

The switching probability of a user $s_{rl,r'l'}$ from class $rl$ to $r'l'$ is determined in two steps given below:

*Step-1.* First, the user goes to class $(r'*)$ depending on the probability $p_{rr'}$. The $*$ implies any learning level.

*Step-2.* Second, the user specifically enters the class $r'l'$ with a certain probability as follows:

- when $r' = r$ and $l < L$, the probability is 1 if $l' = l + 1$; 0 otherwise.
- when $r' = r$ and $l = L$, the probability is 1 if $l' = l$; 0 otherwise.
- when $r' \neq r$, the probability is:

$$\frac{k_{r'l'} - c_{r'l'}}{\sum_{j=1}^{L} \left( k_{r'j} - c_{r'j} \right)}$$

Once *Step-2* is finished, the two matrices $C$ and $K$ are updated as follows:

- $c_{rl}$ element of matrix C is decreased by 1, and its $c_{r'l'}$ element is increased by 1.
- If $l < L$, then $k_{rl}$ of $K$ is decreased by 1, and its $k_{r(l+1)}$ element is increased by 1.

## 4   Model Validation

We build the CogQN model and the *QN-no-learning* model of the modified TPC-W system. We develop the CogQN model (Fig. 1) with appropriate queueing stations for the CPU and the disk being guided by the hardware resource information of the computer where we carried out the measurements. The CogQN model takes as input the service demands incurred by each service request on the CPU and the disk. In a preliminary assessment, we found that the no-load response time of a service is not the same when invoked standalone versus when invoked in a mix with other services. Thus, determining the demands on the resources for each service becomes a chal-lenging task. As an approximation, we estimate the demands based on the assumption that the ratio of the disk demand to the CPU demand for a service stays the same regardless of invoking the service standalone or in a mix with other services. The mix we consider is the ordering-mix of TPC-W [4]. To capture the navigation pattern within a session, we use the Service Transition Probability Matrix of the ordering-mix. This matrix comes with the TPC-W system [7] that we have modified. The arrival rate for the model is set at $\lambda = 10$ users/sec. We feed the model with four YL modes made from two Think Time Matrices and two Arrival Probability Matrices. The four modes are shown in Table 1. One Think Time Matrix is replete with the learning curve LC1 = [70 s, 13 s, 7 s] for every service; the other is replete with LC2 = [50 s, 30 s, 10 s] for every service. Similarly, one Arrival Probability Matrix is replete with the arrival probability set AP1 = [0.3, 0.3, 0.4] for every service; the other is replete with AP2 = [0, 1, 0] for every service. We take a similar approach to parameterize the *QN-no-learning* model, except that it is the case of NL mode where the think time at all learning levels for a given service is the same (30 s—the average of the entries in either LC1 or LC2), and the arrival probabilities are irrelevant.

We implement our models using SimPy 2.3 [11]. We accomplish 5 simulation runs at every mode (NL and four YL). For each run, we set the ramp-up time to 300 s, measurement interval to 300 s, and ramp-down time to 100 s (like empirical settings). We obtain the average of all the runs in a given mode. A model takes a maximum of 6.22 s per run. We validate model predictions against measured data. We have empirically observed an YL mode to be *statistically significantly different* than the NL mode (Mode #3 versus #1; Mode #5 versus #1 of Table 1). We have also observed *statistically significant difference* in a pair of YL modes themselves (Mode #2 versus #3; Mode #4 versus #5 of Table 1). Table 1 shows the percentage error between a model prediction of the overall mean response time and the corresponding measured data. The error values in the table are absolute values. The errors are within 10%.

**Table 1.** % errors in model predictions for overall mean response time RT

| Mode # | NL and YL modes (Arrival rate $\lambda = 10$ users/sec) | Error % |
|---|---|---|
| 1 | NL: Think Time 30 s | 2.8 |
| 2 | YL: LC1 = [70 s, 13 s, 7 s], AP1 = [0.3, 0.3, 0.4] | 2.1 |
| 3 | YL: LC1 = [70 s, 13 s, 7 s], AP2 = [0, 1, 0] | 9.3 |
| 4 | YL: LC2 = [50 s, 30 s, 10 s], AP1 = [0.3, 0.3, 0.4] | 4.2 |
| 5 | YL: LC2 = [50 s, 30 s, 10 s], AP2 = [0, 1, 0] | 4.6 |

## 5   Conclusions

This work introduces a QN model with *user learning*. It models the dynamic transition of learning levels of users for different services using class-switching probabilities. It indicates that user learning behavior *is indeed likely* to impact the performance of session-based systems. The model predicts overall mean response times across different learning modes. The predictions are within 10% error in comparison to measured data.

# References

1. Menasce, D.A., et al.: A methodology for workload characterization of e-commerce sites. In: Proceedings of the 1st Conference on Electronic Commerce (EC), pp. 119–128. ACM (1999)
2. Mi, N., Casale, G., Cherkasova, L., Smirni, E.: Sizing multi-tier systems with temporal dependence: benchmarks and analytic models. Springer J. Internet Services and App. **1**(2), 117–134 (2010). https://doi.org/10.1007/s13174-010-0012-9
3. Vögele, C., van Hoorn, A., Schulz, E., Hasselbring, W., Krcmar, H.: WESSBAS: extraction of probabilistic workload specifications for load testing and performance prediction—a model-driven approach for session-based application systems. Softw. Syst. Model. **17**(2), 443–477 (2018). https://doi.org/10.1007/s10270-016-0566-5
4. Menasce, D.A.: TPC-W: a benchmark for e-commerce. IEEE Internet Comput. **6**(3), 83–87 (2002)
5. Casale, G., et al.: Dealing with burstiness in multi-tier applications: models and their parameterization. IEEE Trans. Software Eng. **38**(5), 1040–1053 (2012)
6. Ritter, F.E., Schooler, L.J.: The learning curve. In: International Encyclopedia of the Social and Behavioral Sciences, vol. 13, pp. 8602–8605 (2001)
7. Zhang, L., Down, Douglas G.: SMVA: a stable mean value analysis algorithm for closed systems with load-dependent queues. In: Puliafito, A., Trivedi, Kishor S. (eds.) Systems Modeling: Methodologies and Tools. EICC, pp. 11–28. Springer, Cham (2019). https://doi.org/10.1007/978-3-319-92378-9_2
8. Ritter, F.E., et al.: Learning and Retention. The Oxford Handbook of Cognitive Engineering. Oxford Press, New York (2013)
9. Cockburn, A., Gutwin, C., Greenberg, S.: A predictive model of menu performance. In: ACM CHI, pp. 627–636 (2007)
10. Ahlstrom, D., et al.: Why it's quick to be square: modelling new and existing hierarchical menu designs. In: ACM CHI, pp. 1371–1380 (2010)
11. SimPy discrete-event simulation framework, version 2.3. https://simpyclassic.readthedocs.io/en/latest/Manuals/Manual.html
12. Das, A., Stuerzlinger, W.: Unified modeling of proactive interference and memorization effort: a new mathematical perspective within act-r theory. In: Proceedings of the Annual Meeting of the Cognitive Science Society (CogSci), pp. 358–363 (2013)

# A Matlab Toolkit for the Analysis of Two-Level Processor Sharing Queues

Andrea Marin$^{(\boxtimes)}$, Sabina Rossi, and Carlo Zen

Università Ca' Foscari Venezia, Venice, Italy
{marin,sabina.rossi}@unive.it, 864429@stud.unive.it

**Abstract.** This paper presents a Matlab toolkit for the numerical analysis of the two-level processor sharing queue (2LPS). The job sizes are expressed in terms of acyclic phase type distributions which can approximate any distribution arbitrary well while arrivals occur according to a homogeneous Poisson process. The toolkit provides a simple yet efficient way to find the optimal parametrization of the 2LPS queueing disciplines given the job size distributions and the intensity of the workload. In practice, the tool can be used to configure the 2LPS scheduler for TCP flows. The time complexity of the solution depends on the cube of the number of phases of the distribution describing the flow sizes.

## 1 Introduction

Two-level processor sharing queues (2LPS) have been widely studied in queueing theory and since the first results due to Kleinrock [4], many other papers appeared and showed when the benefits of this scheduling discipline are prominent with respect to the processor sharing (PS) queue (see, e.g., [1]). From the networking prospective, the introduction of the Least Attained Service (LAS) discipline lead to a new approach in the scheduling of TCP flows thanks to its optimality properties [6]. LAS uses a PS discipline among the jobs that have received the least amount of service. However, the difficulties in practical implementation of LAS shifted the attention to the multi-level PS (MLPS) queues as practical approximations easier to implement. In MLPS, there are several levels of attained service and the jobs belonging to a certain level are served according to PS discipline only if there are no jobs which belong to levels corresponding to a lower amount of attained service. In contrast with LAS, multi-level PS queues require a parametrisation, i.e., the setting of the attained service thresholds. A wrong parametrization may reduce the benefits of this approach and even worsen the performance of the system with respect to the simpler PS scheduling.

In this paper, we provide a set of Matlab tools[1] that, given the distribution of the job sizes expressed by a canonical phase type distribution, allows the analysis of the 2LPS queue and hence the determination of its optimal parametrization for different intensities of the arrival process. Several solution approaches have been

---

[1] http://www.dais.unive.it/~marin/2lpstoolkit.zip.

© Springer Nature Switzerland AG 2020
M. Gribaudo et al. (Eds.): QEST 2020, LNCS 12289, pp. 144–147, 2020.
https://doi.org/10.1007/978-3-030-59854-9_11

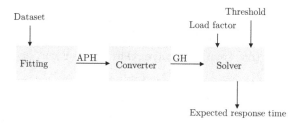

**Fig. 1.** Phases of use of the Matlab Toolkit for the two-level PS queues.

developed by the scientific literature and they follow the lines of the solution proposed in [4] that is however complete only for a specific class of job size distributions. The solution is complicated since it requires the analysis of two queueing systems, one of which is a PS queue with batch arrivals and for this class of queues it is known that the insensitivity property does not hold [2]. Thus, we believe that this Toolkit will be useful in practice for those interested in the configuration of 2LPS schedulers. The algorithms which are implemented are described in [5].

For networking purposes, i.e., scheduling TCP flows, we can safely focus on systems consisting only of two levels (and one threshold) for several reasons. First, it has been shown (see, e.g., [5] and the references therein) that the difference in the expected response time between the 2LPS with optimal configuration and the LAS is small at least for TCP flow distributions taken from real datasets. Second, the increase of the number of levels in a MLPS makes its practical implementation more and more difficult since each level requires a priority queue. Third, the case of two levels introduces only a single parameter to optimize and this makes the computations rather fast and may be done while the scheduler is running on a real system to self-configure its optimal threshold.

## 2   Description of the Analysis Procedure and of the Toolkit

The toolkit has been developed baring in mind the procedure shown in Fig. 1. Algorithms for fitting dataset into phase type distributions have been largely studied. Any acyclic phase-type distribution can be expressed in a certain canonical form and cyclic distributions can be approximated arbitrary well by acyclic ones. Tools like PhFit [3] returns the description of the canonical phase type distribution fitting the dataset in terms of their generating matrices. In general, to fit heavy-tailed distribution without using too many phases, the fitting of the body and the tail of the distributions are kept separate. Thus, we have an acyclic phase type (APH) with $n$ phases for the body and $m$ for the tail. The Matlab Toolkit transforms this APH into a generalized hyperexponential distribution (GH) using the algorithm described in [5]. GH distributions do not have the Markovian interpretation of APH but they are merely an algebraic instrument whose expression of their Laplace-Stieltjes transform allows for efficient computations of the steady-state performance measures of the queueing system. The transformation is performed by function:

```
function [p,u] = convertPhfit(a, B, n, m)
```

where `a` is the vector of the initial probabilities, `B` is the sub-infinitesimal generator (i.e., the infinitesimal generator without the column associated with the absorbing state) and `n`, `m` are the number of phases of the body and tail of the distribution, respectively. The function returns the (non-Markovian) description of the GH distribution. The other key-function of the tool is

```
function resp = averageR(u, p, l, a)
```

which takes as input the description of the GH distribution thanks to parameters `u` and `p`, the intensity of the arrival process `l` and the threshold `a`. `averageR` returns the expected response time of the 2LPS with those parameters.

Finally, the optimization of the parametrization can be performed with

```
function [th,resp] = optimalThreshold(a, B, body, tail, thmin,
                     thmax, rho, tol)
```

that takes as input the description `a`, `B` of the APH describing the job size distribution, where `body` and `tail` indicate the number of phases used for the body and the tail of the distribution. `thmin` and `thmax` are parameters used to specify the range within the search of the optimum is done. `rho` is the load factor of the queue and finally `tol` is the tolerance of the numerical search of the minimum. The function implements a *golden section search*. It returns the optimal threshold `th` and the corresponding expected response time `resp`.

## 3   Example

In this example, we have fitted the distribution of the TCP flow sizes collected at the data center of the University Ca' Foscari of Venice in November 2019. The vector of initial probabilities `A` and the sub-infinitesimal generator `B` are shown in the file `main.m` of the downloadable toolkit. The TCP flows have an average size of 88.63 packets. The main file performs several experiments with a load factor of 0.9, 0.7 and 0.5. Figure 2 shows the Matlab plot of the expected response time as function of the chosen threshold. The optimal thresholds found by the optimizer are 1029, 853 and 769 for the workload intensities 0.9, 0.8, 0.7, respectively and correspond with those shown by the plot.

## 4   Final Remarks

2LPS scheduling plays an important role especially in networking applications because it can heavily reduce the expected response time of the system with respect to a PS scheduler while being blind with respect to the prior knowledge of the flow sizes. While 2LPS has slightly lower performance than LAS, its implementation is less resource demanding and complex than LAS.

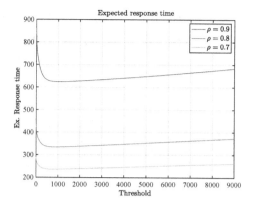

**Fig. 2.** Example of output.

However, 2LPS requires a parametrization that, if wrongly done, could even worsen the performance with respect to a standard PS scheduling. The Matlab toolkit that we have presented solves this optimization problem without the need for the practitioner to choose one of the available (and usually quite complicated) solution methods and reimplement it. Moreover, the toolkit is compatible with PhFit, a popular tool for fitting datasets into phase type distributions and this may turn to be an important feature for its practical use. Future works include a comparison with the performance of the scheduling proposed in [7].

# References

1. Aalto, S., Ayesta, U.: Mean delay analysis of multi level processor sharing disciplines. In: Proceedings of the IEEE of the 25th Annual Joint Conference of the IEEE Computer and Communications Society (INFOCOM) (2006)
2. Bansal, N.: Analysis of the M/G/1 processor-sharing queue with bulk arrivals. Oper. Res. Lett. **31**(3), 401–405 (2003)
3. Horváth, A., Telek, M.: PhFit: a general phase-type fitting tool. In: Field, T., Harrison, P.G., Bradley, J., Harder, U. (eds.) TOOLS 2002. LNCS, vol. 2324, pp. 82–91. Springer, Heidelberg (2002). https://doi.org/10.1007/3-540-46029-2_5
4. Kleinrock, L.: Queueing Systems, volume II: Computer Applications. Wiley, Hoboken (1976)
5. Marin, A., Rossi, S., Sottana, M., Zen, C.: Theoretical and experimental evaluation of the two-level processor sharing discipline for TCP flows. In: 27th IEEE International Symposium on Modeling, Analysis, and Simulation of Computer and Telecommunication Systems, MASCOTS 2019, pp. 94–106 (2019)
6. Righter, R., Shanthikumar, J.: Scheduling multiclass single server queueing systems to stochastically maximize the number of successful departures. Probab. Eng. Inf. Sci. **3**, 323–333 (1989)
7. Van Houdt, B., Van Velthoven, J., Blondia, C.: QBD Markov chains on binomial-like trees and its application to multilevel feedback queues. Ann. Oper. Res. **160**(1), 3–18 (2008). https://doi.org/10.1007/s10479-007-0288-8

# M/M/1 Vacation Queue with Multiple Thresholds: A Fluid Analysis

Mehmet Akif Yazici[1]([✉]) [iD] and Tuan Phung-Duc[2] [iD]

[1] Informatics Institute, Istanbul Technical University, Istanbul, Turkey
yazicima@itu.edu.tr
[2] Faculty of Engineering, Information and Systems, University of Tsukuba,
1-1-1 Tennodai, Tsukuba, Ibaraki 305-8573, Japan
tuan@sk.tsukuba.ac.jp

**Abstract.** We propose an analytical method for an M/M/1 vacation queue with workload dependent service rates. We obtain the distribution of the workload in the system, and consider a power-saving and performance trade-off problem. Numerical experiments reveal that square root service rate function has lower cost than that of linear and quadratic service functions under certain scenarios.

**Keywords:** Data center · Variable service rate · Power-saving · Fluid model · Vacation queue

## 1 Introduction

Cloud computing is supported by data centers with a large number of servers and a huge amount of energy consumption. This calls for energy saving mechanisms in data centers while keeping service level high. A natural approach to this problem is to adjust the processing rate of data centers according to the workload level in the system so as to balance the energy consumption and performance. This can be realized by turning the servers on and off (ON-OFF policy) in data centers [3], frequency scaling or dynamic voltage and frequency scaling (DVFS) [5].

In this paper, we model power-saving in data centers by a single server queueing system with vacation and workload-dependent service rate. We are able to obtain the probability distribution and relevant statistics of the workload in the system. This allows us to consider the energy-performance trade-off problem and to investigate optimal service rate function as well as vacation policy.

As related work, Yajima and Phung-Duc [5] consider an M/M/1 system where the service rate is proportional to the number of jobs in the system and analyze the response time distribution. Marin et al. [2] consider an M/M/1 system with SRPT scheduling policy and $K$ speeds. In these papers, the service rate depends on the number of jobs in the system. In contrast, in our present paper, we consider the workload in the system instead of the number of jobs. As a closely related work, Sakuma et al. [4] consider the same model and analyzed it using

© Springer Nature Switzerland AG 2020
M. Gribaudo et al. (Eds.): QEST 2020, LNCS 12289, pp. 148–152, 2020.
https://doi.org/10.1007/978-3-030-59854-9_12

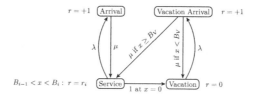

**Fig. 1.** The modulating CTMC of the fluid model, and the corresponding drift rates.

renewal theory and level-crossing approach. Yazici and Akar [6] analyzed the MAP/PH/1 queue with workload-dependent behavior.

In this paper, we approach the problem using fluid queues. We numerically solve the exact distribution of the workload, and analytically compute relevant statistics. One of the advantages of our approach is that it could easily be extended to analyze models with MAP arrivals and PH-type service times, which is the main difference between our work and [4]. The remainder of the paper is organized as follows. In Sect. 2, we describe our model in detail. Section 3 shows some numerical results while concluding remarks are presented in Sect. 4.

## 2   System Model

We consider an infinite-capacity vacation queue with Poisson arrivals, whose intensity is $\lambda$, and exponentially distributed job sizes with mean $1/\mu$. The service rate depends on the instantaneous workload, $x$, through a piecewise-constant function, i.e. the rate is $r_i$ when $B_{i-1} < x <= B_i$ where $B_0 = 0$, $B_K = \infty$ and $i \in \{1, \ldots, K\}$. The server enters vacation when the workload hits 0, and returns from vacation when the workload reaches $B_V$. Without loss of generality, we assume $B_V \in \{B_1, \ldots, B_{K-1}\}$. We model the workload as a fluid and thus, the system can be described as a multi-regime feedback fluid queue [1] due to the piecewise dependence of the service rate to the workload. The modulating continuous-time Markov chain (CTMC) is given in Fig. 1, along with the associated drift rates for each state. Notice that transition rates also depend on the workload, hence producing a multi-regime fluid model.

To obtain the numerical results, we employed the methodology described in [1]. One important detail worth mentioning is that the drift rate in the vacation regime is 0, and this needs special treatment. In general, the pdf of the regimes with 0 drifts can be expressed as a linear combination of the pdf's of the remaining states; see equations (20)–(22) in [1] and the explanations therein. Furthermore, *Vacation* state does not exist beyond $x > B_V$.

After the distribution of the fluid in each state is obtained, the *Arrival* and *Vacation Arrival* states are censored out, as the linear increases in these two states represent the abrupt increases in the workload due to job arrivals and in reality, the system does not spend any time in either of these states.

To study the effect of the rate function and the vacation threshold, $B_V$, we consider a cost function as follows [2]:

$$C = \left( \sum_{i=1}^{K} p_i r_i^2 \right) + p_0 c_0 + c_h E[V] + c_s \lambda V(0), \tag{1}$$

where $p_i$ is the probability that the server works at speed $r_i$, $p_0$ is the probability that the server is on vacation, $c_0$ is the power consumption when the server is on vacation, $c_h$ is the weight placed on the mean workload, i.e., performance, $E[V]$ is the mean workload, $c_s$ is the switching cost, and $V(0)$ is the probability that the workload is 0. Here, the product $\lambda V(0)$ is equal to the reciprocal of the mean cycle time from the beginning of one vacation to the next [4], and hence represents switching frequency of the server from OFF to ON.

Following the definitions in [1], the pdf of the workload is of the form

$$f^{(i)}(x) = a_0^{(i)} L_0^{(i)} + a_-^{(i)} \exp\left(A_-^{(i)}(x - B_{i-1})\right) L_-^{(i)} + a_+^{(i)} \exp\left(-A_+^{(i)}(B_i - x)\right) L_+^{(i)},$$
$$B_{i-1} < x < B_i, \ i \in \{1, \ldots, K\}, \tag{2}$$

where $A_-^{(i)}$ and $A_+^{(i)}$ are blocks of a matrix obtained through a similarity transform on $A^{(i)} = Q^{(i)}(R^{(i)})^{-1}$, $Q^{(i)}$ and $R^{(i)}$ being the infinitesimal generator of the CTMC and the diagonal drift matrix, respectively, for each regime, $L_0^{(i)}$, $L_-^{(i)}$, and $L_+^{(i)}$ are blocks of the inverse of the aforementioned similarity transform matrix, and $\left[ a_0^{(i)}, a_-^{(i)}, a_+^{(i)} \right]$ are coefficients obtained through a set of boundary conditions [1]. Hence, the required statistics can be obtained as

$$p_i = \int_{B_{i-1}}^{B_i} f_S^{(i)}(x)\, dx, \quad \text{and} \quad E[V] = \sum_{i=1}^{K} \int_{B_{i-1}}^{B_i} x\, f_S^{(i)}(x)\, dx, \tag{3}$$

where $f_S(x)$ is the pdf of the workload in *Service* state after *Arrival* and *Vacation Arrival* states are censored out. Considering that the pdf expression in (2) contain matrix exponentials only, it is clear that the integrals in (3) can be evaluated analytically (We omit the exact expressions due to space limitation.).

## 3   Numerical Results

We obtained numerical results with $\lambda = 1$, $\mu = 1$, $B_i = (i/4)$, $i \in \{1, \ldots, 40\}$ via implementation in Matlab. The parameters $c_0$, $c_h$, $c_s$, and $B_V$ are varied. The service rate functions we experimented with are $r_i^{sr} = \sqrt{B_{i-1}} + 1$, $r_i^{lin} = B_{i-1} + 1$, $r_i^{sq} = (B_{i-1})^2 + 1$, representing square root, linear, and square dependence, respectively, on the threshold values. We first give pdf plots of the workload in Fig. 2 for $B_V \in \{2, 4, 6\}$ and $r_i = r_i^{lin}$. This illustrates the dynamics of the workload and the effect of the selection of vacation threshold, $B_V$.

Next, we compare the mentioned rate functions under various operating scenarios with respect to $c_0$, $c_h$, $c_s$, and $B_V$. We plot in Fig. 3 normalized service

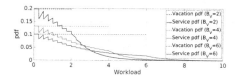
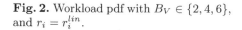

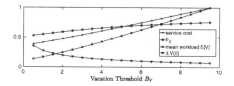

**Fig. 2.** Workload pdf with $B_V \in \{2, 4, 6\}$, and $r_i = r_i^{lin}$.

**Fig. 3.** Cost components against $B_V$. Service cost and $E[V]$ are normalized, with maximum values of 5.8137 and 4.0301, respectively.

cost ($C$ values for $c_0 = c_h = c_s = 0$), $p_0$, normalized $E[V]$, and $\lambda V(0)$, with $r_i = r_i^{lin}$. As $B_V$ is increased, all but $\lambda V(0)$ increase monotonically. Hence, we conclude that $c_s$ is the critical component of the cost against other coefficients. In Fig. 4, we plot the cost for $c_s \in \{10, 30, 50\}$ with $r_i = r_i^{lin}$. We observe that the cost function turns out to be convex in this scenario, and there exist optimum $B_V$ values for each $c_s$ value, which are marked with asterisks on the plots. Finally, we compare the rate functions with $c_0 = c_h = 1$, $c_s = 30$ in Fig. 5. Again, we observe a similar dynamic with respect to $B_V$.

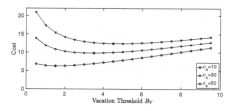

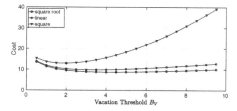

**Fig. 4.** Costs for different values of $c_s$ with $r_i = r_i^{lin}$, $c_0 = c_h = 1$.

**Fig. 5.** Costs for different rate functions, $c_0 = c_h = 1$, $c_s = 30$.

## 4   Conclusion

In this study, we model the M/M/1 vacation queue for the purpose of performance analysis and optimization of cloud data centers. The main mathematical tool we use for our model is multi-regime fluid queues. We observe that the cost is sensitive to the selection of several parameters, as well as the rate function. We quantitatively demonstrate the behavior of the cost function with respect to vacation threshold. It is clear that further analysis is necessary to determine realistic values for the cost coefficients. Hence, future studies will comprise extensive experimentation with regards to the cost parameters, and improvement of the model by considering finite buffer systems and more complicated inter-arrival time and job size distributions.

# References

1. Kankaya, H.E., Akar, N.: Solving multi-regime feedback fluid queues. Stochast. Models **24**(3), 425–450 (2008)
2. Marin, A., Mitrani, I., Elahi, M., Williamson, C.: Control and optimization of the SRPT service policy by frequency scaling. In: McIver, A., Horvath, A. (eds.) QEST 2018. LNCS, vol. 11024, pp. 257–272. Springer, Cham (2018). https://doi.org/10.1007/978-3-319-99154-2_16
3. Phung-Duc, T.: Exact solutions for M/M/c/Setup queues. Telecommun. Syst. **64**(2), 309–324 (2016). https://doi.org/10.1007/s11235-016-0177-z
4. Sakuma, Y., Boxma, O., Phung-Duc, T.: A single server queue with workload-dependent service speed and vacations. In: Phung-Duc, T., Kasahara, S., Wittevrongel, S. (eds.) QTNA 2019. LNCS, vol. 11688, pp. 112–127. Springer, Cham (2019). https://doi.org/10.1007/978-3-030-27181-7_8
5. Yajima, M., Phung-Duc, T.: Batch arrival single-server queue with variable service speed and setup time. Queueing Syst. **86**(3–4), 241–260 (2017)
6. Yazici, M.A., Akar, N.: The workload-dependent MAP/PH/1 queue with infinite/finite workload capacity. Perform. Eval. **70**(12), 1047–1058 (2013)

# Markov Processes

# Bounding Mean First Passage Times in Population Continuous-Time Markov Chains

Michael Backenköhler[1,2]([✉]), Luca Bortolussi[1,3], and Verena Wolf[1]

[1] Saarland University, Saarbrücken, Germany
michael.backenkoehler@uni-saarland.de
[2] Saarbrücken Graduate School of Computer Science, Saarbrücken, Germany
[3] University of Trieste, Trieste, Italy

**Abstract.** We consider the problem of bounding mean first passage times and reachability probabilities for the class of population continuous-time Markov chains, which capture stochastic interactions between groups of identical agents. The quantitative analysis of such models is notoriously difficult since typically neither state-based numerical approaches nor methods based on stochastic sampling give efficient and accurate results. Here, we propose a novel approach that leverages techniques from martingale theory and stochastic processes to generate constraints on the statistical moments of first passage time distributions. These constraints induce a semi-definite program that can be used to compute exact bounds on reachability probabilities and mean first passage times without numerically solving the transient probability distribution of the process or sampling from it. We showcase the method on some test examples and tailor it to models exhibiting multimodality, a class of particularly challenging scenarios from biology.

**Keywords:** Population continuous-time Markov chains · Semi-definite programming · Exit time distribution · Reachability probability · Markov population models

## 1 Introduction

Population Continuous-Time Markov Chains (PCTMCs) provide a widely used framework to capture stochastic interactions between groups of identical agents. This subclass of Continuous-Time Markov Chains (CTMCs) is used to describe the stochastic dynamics of systems in various domains. Prominent applications are chemical reaction networks in quantitative biology [55], epidemic spreading [46], performance analysis of technical and information systems [11,22] as well as the behavior of collective adaptive systems [9].

**Electronic supplementary material** The online version of this chapter (https://doi.org/10.1007/978-3-030-59854-9_13) contains supplementary material, which is available to authorized users.

© Springer Nature Switzerland AG 2020
M. Gribaudo et al. (Eds.): QEST 2020, LNCS 12289, pp. 155–174, 2020.
https://doi.org/10.1007/978-3-030-59854-9_13

For the quantitative analysis of CTMCs, many approaches have been developed, where properties of interest are often expressed in terms of temporal logics such as CSL [2,5,6], MTL [14], and timed-automata specifications [15,41]. In addition, there exist efficient software tools [17,31,38] that can be used to analyze and verify system properties. The computation of reachability probabilities is a central problem in this context.

Popular exact methods for CMTCs rely on numerical approaches that explicitly consider each system state individually. A major problem is that these methods cannot scale in the context of population models with large copy numbers of agents. A popular alternative to tackle this problem is statistical model checking, which is based on stochastic simulation [16]. For PCTMCs arising in the context of chemical reaction networks, trajectories of the process are usually generated using the Stochastic Simulation Algorithm (SSA) [25]. However, since the number of possible interactions grows with the number of agents, stochastic simulations of PCTMCs are time-consuming. Moreover, they are subject to inherent statistical uncertainty and give only statistically estimated bounds.

As an alternative, recent work concentrates on numerical methods that approximate the statistical moments of the system without the need to compute the probability of each state. For groups of identically behaving agents, it is possible to derive systems of differential equations for the evolution of the statistical population moments [10,12,21,22,50,51]. However, as the system of exact moment equations is infinite-dimensional, approximation schemes typically rely on certain assumptions about the underlying probability distribution to truncate it. For example, one might employ a "low dispersion closure" which assumes that higher-order moments are the same as those of a normal distribution [30]. Such approximations are, by nature, ad-hoc and do not come with any guarantees.

Moment-based methods often scale well in terms of population sizes. However, it is not possible to control the effects of the introduced approximations, which in some cases can lead to large errors [50]. This issue reverberates on the application of these methods to compute reachability probabilities and mean first passage times [12,13,28]. Moreover, they can suffer from numerical instabilities, in particular, when the maximum order of the considered moments has to be increased to more appropriately describe the underlying distribution.

Here, we put forward a method based solely on moments that gives *exact bounds* for Mean First Passage Times (MFPTs) and reachability probabilities in PCTMCs. For a set of states, the MFPT within a fixed time-horizon $T$ directly characterizes the probability of reaching that set within $T$ time units. Thus, safe upper and lower bounds on MFPTs can constitute a core component for the verification of properties in PCTMCs. Our approach extends recent work on moment bounds [20,47] and it is based on a martingale formulation of the stopped process that we derive from the exact moment equations. From this formalization, we deduce a set of linear moment constraints from which we derive upper and lower moment bounds using semi-definite programming (SDP). Monotone sequences of both upper and lower bounds can be obtained by increasing the order of the relaxation. Crucially, no closure approximations are introduced. Therefore the bounds are exact up to the numerical accuracy of the SDP solver.

To experimentally validate our method in terms of accuracy and feasibility, we run some tests on examples from biology, leveraging an existing SDP solver and obtaining encouraging results. Comparing with other moment-based methods, our approach is not based on approximations due to closure schemes, thus providing guarantees on the bounds up to the numerical accuracy of the computations. However, similarly to other moment-based methods, we also found the insurgence of numerical instabilities because moments of higher order tend to span over many orders of magnitude. We ameliorate this problem by considering scaling strategies that reduce such variability. We also extend our approach to deal with PCTMCs exhibiting strong multimodal behavior, due to the presence of populations having low copy numbers. This extension exploits some ideas from hybrid moment closures [34].

In summary, this paper presents the following novel contributions:

- the derivation of moment constraints, based on a martingale formulation, for bounding first passage times and reachability probabilities using a convex programming scheme;
- the extension of this scheme using hybrid moment conditions to systems exhibiting multimodal behavior;

The paper is structured as follows: Sect. 2 covers work related to the analysis of first passage times in PCTMCs and recent work on moment bounds. Section 3 introduces the PCTMC framework and its semantics. In Sect. 4 we derive a martingale from the moment dynamics of a PCTMC. Based on this process, in Sect. 5 we formulate linear and semi-definite constraints to state a semi-definite program to compute bounds on the MFPT and reachability probabilities. In Sect. 6, we discuss the practical considerations of the SDP implementation and provide results on a set of case studies. Finally, in Sect. 7 we provide concluding remarks and directions of future work.

## 2  Related Work

Considerable effort has been directed at the analysis of first passage time distributions in PCTMCs. Most works can either focus on an explicit state-space analysis [7,36,37,43] or employ approximation techniques for which, in general, no error bounds can be given [13,28,49]. For some model classes such as kinetic proofreading, analytic solutions are possible [8,32,43].

Barzel and Biham [7] propose a recursive scheme that consists of one equation for each state, expressing the average time the system needs to transition from that state to the target state. Kuntz et al. [36] propose to employ moment bounds in a linear programming approach to compute exit time distribution using state-space truncation schemes. In Ref. [37] the authors propose a finite state-space projection scheme to bound first passage time distributions.

Hayden et al. [28] use moment closure approximations and Chebychev's inequality to gain an understanding of first passage time dynamics. Schnoerr et al. [49] also employ a moment closure approximation and further approximate threshold functions to derive an approximate first passage time distribution. Bortolussi and Lanciani [13] use a mean-field approximation which is required to reach the target region.

Recently, several groups independently suggested the use of semi-definite optimization for the computation of moment bounds for the limiting distribution [19,23,35,47]. In this approach, the differential equations describing the moment dynamics are set to zero and form linear constraints [3]. Alongside, semi-definite constraints can be placed on the *moment matrices*. These give a semi-definite program that can be solved efficiently.

This approach has been extended to the transient case [20,48]. The approach is similar in both works and is a cornerstone of the MFPT analysis presented here. They differ mainly by the fact that Sakurai and Hori apply a polynomial time-weighting [48], while Dowdy and Barton use an exponential one [20]. We adopt the former approach because it can be naturally adapted to the description of densities over time. The resulting forms can also be adapted to statistical estimation problems [4].

Semi-definite programming has been applied to a wide range of problems, including stochastic processes in the context of financial mathematics [33,40]. For good introductions and overviews of application areas, we refer the reader to Parrilo [45] and, more recently, Lasserre [39].

Particularly relevant for this work is the application of convex optimization to first passage times. Helmes et al. [29] formulated a linear program using the Hausdorff moment conditions to bound moments of the first passage time distribution in Markovian processes. Semi-definite optimization has been successfully applied in financial mathematics by Kashima and Kawai [33], as well as Lasserre et al. [40] to bound prices of exotic options.

## 3    Preliminaries

A Population Continuous-Time Markov Chain (PCTMC) describes the interactions among a set of agents of $n_S$ types $S_1, \ldots, S_{n_S}$ in a well-stirred reactor. In the sequel, we will also use other letters than $S_i$ as agent types. Since we assume that all agents are equally distributed in space, we only keep track of the overall copy number of agents for each type. Therefore the state-space is $\mathcal{S} \subseteq \mathbb{N}^{n_S}$. The interactions are expressed as *reactions* with a certain gain and loss of agents, given by the non-negative integer vectors $\boldsymbol{v}_j^-$ and $\boldsymbol{v}_j^+$ for some reaction $j$, respectively. Such a reaction is denoted as

$$\sum_{i=1}^{n_S} v_{ji}^- S_i \xrightarrow{a_j} \sum_{i=1}^{n_S} v_{ji}^+ S_i. \tag{1}$$

The reaction rate constant $a_j > 0$ determines the propensity function $\alpha_j$ of the reaction. If just a constant is given, *mass-action* propensities are assumed, where for $\boldsymbol{x} \in \mathcal{S}$ we define

$$\alpha_j(\boldsymbol{x}) := a_j \prod_{i=1}^{n_S} \binom{x_i}{v_{ji}^-}. \tag{2}$$

This choice of propensity function is natural, since it is proportional to the number of reactant combinations. The system's behavior is described by a stochastic process $\{X_t\}_{t\geq 0}$. We denote the abundance of a given agent type $S_i$ in $X_t$ by $X_t^{(S_i)}$. The propensity $\alpha_j(x)$ gives the infinitesimal probability of a reaction occurring, given a state $x$. That is, for $v_j = v_j^+ - v_j^-$ and a small time step $\Delta t > 0$,

$$\Pr(X_{t+\Delta t} = x + v_j \mid X_t = x) = \alpha_j(x)\Delta t + o(\Delta t). \tag{3}$$

Therefore, given a system of $n_R$ reactions, the semantics of $X_t$ is given by a continuous-time Markov chain (CTMC) on $S$ Accordingly, given an initial distribution on $S$, the time-evolution of the process' distribution is given by the Kolmogorov forward equation. For a single state, in the context of quantitative biology, it is commonly referred to as the *chemical master equation* (CME)

$$\frac{d\pi}{dt}(x,t) = \sum_{j=1}^{n_R} (\alpha_j(x - v_j)\pi(x - v_j, t) - \alpha_j(x)\pi(x, t)), \tag{4}$$

where $\pi(x,t) = \Pr(X_t = x)$ and $\Pr(X_0 = x) = \pi(x, 0)$.

Consider the following simple PCTMC with non-linear propensities as an example.

**Model 1 (Dimerization).** *We first examine a simple dimerization model on an unbounded state-space with reactions*

$$\varnothing \xrightarrow{\lambda} M, \quad 2M \xrightarrow{\delta} D$$

*and initial condition $X_0^{(M)} = X_0^{(D)} = 0$. The semantics is given by a CTMC $X_t = (X_t^{(M)}, X_t^{(D)})^\top$, where $(S_1, S_2) = (M, D)$. The reaction propensities according to (2) are $\alpha_1(x) = \lambda$ and $\alpha_2(x) = \delta\, x^{(M)}(x^{(M)} - 1)/2$. The change vectors $v_1^- = (0,0)^\top$, $v_1^+ = (1,0)^\top$, $v_2^- = (2,0)^\top$, and $v_2^+ = (0,1)^\top$. Consequently, $v_1 = (1,0)^\top$ and $v_2 = (-2,1)^\top$.*

This explicit representation of state probabilities is often not possible, because there are infinitely many states. Usually the state-space is truncated to contain all relevant states [1] or one switches to an approximation such as the mean-field [11].

In this work, we are interested in *first passage times* of such processes. That is the time, the process first enters a set of target states $B \subseteq S$. Naturally, the analysis of first passage times is equivalent to the analysis of times at which the process exits the complement $S \setminus B$. More formally, the first passage time $\tau$ for some target set $B$ is defined as the random variable

$$\tau = \inf\{t \geq 0 \mid X_t \in B\}. \tag{5}$$

In this example, we are interested in the time at which the number of type $M$ agents exceed some threshold $H$. With the framework presented in the sequel, one can bound the expected value of this time using semi-definite programming.

Further, it is possible to impose a time-horizon $T$, and find bounds on the probability of $X_t^{(M)} \geq H$ for some $0 \leq t \leq T$. The employed framework is centered around semi-definite relaxations of the generalized moment problem [39]. These require linear constraints on the moments of measures. In the following section, we derive such constraints.

## 4    Martingale Formulation

Next, we will discuss the ordinary differential equations for the evolution of the statistical moments of the process. The moments over the state-space are then used to derive temporal moments, i.e. moments of measures over both the state-space and the time. This extended description results in a process with the martingale property. This property can be used to formulate linear constraints on the temporal moments and, as a special case, the mean first-passage time. In combination with semi-definite properties of moment matrices, we can formulate mathematical programs that yield upper and lower bounds on mean first passage times.

We start with the description of the *raw moments* dynamics. In particular, a raw moment is $\mathbb{E}(\boldsymbol{X}^{\boldsymbol{m}}) = \mathbb{E}(\prod_{i=1}^{n_S} X_i^{m_i})$, $\boldsymbol{m} \in \mathbb{N}^{n_S}$ with respect to some probability measure. The order of a moment $\mathbb{E}(\boldsymbol{X}^{\boldsymbol{m}})$ is given by the sum of its exponents, i.e. $\sum_i m_i$. Note that the notion of expected value can be generalized to any measure $\mu$ on a Borel-measurable space $(E, \mathcal{B}(E))$, where the $\boldsymbol{m}$-th raw moment is $\int_E \boldsymbol{x}^{\boldsymbol{m}} d\mu(\boldsymbol{x})$. Throughout we assume that moments of arbitrary order remain finite over time, i.e. $\mathbb{E}(|\boldsymbol{X}_t^{\boldsymbol{m}}|) < \infty$, $t \geq 0$. In Ref. [26] the authors propose a framework to verify this property for a given model.

Let $f$ be a polynomial function, $t \geq 0$. Using the CME (4), we can derive ordinary differential equations (ODEs) describing the dynamics of $\mathbb{E}(f(\boldsymbol{X}_t))$ [21]. Specifically,

$$\frac{d}{dt}\mathbb{E}(f(\boldsymbol{X}_t)) = \sum_{j=1}^{n_R} \mathbb{E}((f(\boldsymbol{X}_t + \boldsymbol{v}_j) - f(\boldsymbol{X}_t))\alpha_j(\boldsymbol{X}_t)) . \tag{6}$$

Let us consider Model 1 as an example and agent type $M$. Further, let $X_t = X_t^{(M)}$ for ease of exposition. When choosing $f(X_t) = X_t^m$, $m = 1$ and $m = 2$ we obtain two differential equations describing the change of the first two moments of species $M$, $\mathbb{E}(X_t)$ and $\mathbb{E}(X_t^2)$, respectively.

$$\frac{d}{dt}\mathbb{E}(X_t) = \lambda\mathbb{E}(X_t^0) - 2\delta(\mathbb{E}(X_t^2) - \mathbb{E}(X_t)) \tag{7}$$

$$\frac{d}{dt}\mathbb{E}(X_t^2) = \lambda(2\mathbb{E}(X_t) + 1) - 4\delta(\mathbb{E}(X_t^3) - 2\mathbb{E}(X_t^2) + \mathbb{E}(X_t)) . \tag{8}$$

Fixing initial moments, the ODE system describes the moments over time exactly. However, these ODEs cannot be integrated because the system is not closed. The right-hand side for moment $\mathbb{E}(X_t^m)$ always contains $\mathbb{E}(X_t^{m+1})$. To solve the initial value problem, one typically resorts to ad-hoc approximations of the highest order moments to close the system. Here we do *not* need such approximations because we do not numerically integrate the moment equations. Instead we adopt an approach [20, 48] that extends the description of state-space moments to a temporal one.

This is achieved by the introduction of a time-dependent polynomial $w(t)$ that is multiplied to (6). An integration by parts on $[0, T]$ yields [20, 48]

$$w(T) \, \mathbb{E}\left(f(\boldsymbol{X}_T)\right) - w(0) \, \mathbb{E}\left(f(\boldsymbol{X}_0)\right) - \int_0^T \frac{dw(t)}{dt} \mathbb{E}\left(f(\boldsymbol{X}_t)\right) \, dt$$
$$= \sum_{j=1}^{n_R} \int_0^T w(t) \, \mathbb{E}\left((f(\boldsymbol{X}_t + \boldsymbol{v}_j) - f(\boldsymbol{X}_t)) \, \alpha_j(\boldsymbol{X}_t)\right) \, dt \,. \tag{9}$$

Starting from this equation, it is possible to derive a martingale process, i.e. a process that has an expected value of 0, regardless of time. In Appendix A we describe give its derivation in detail. When choosing $w(t) = t^k$ with $k \in \mathbb{N}$ and $f(\boldsymbol{X}) = \boldsymbol{X}^m$ this process takes the form

$$Z_T^{(m,k)} = T^k \boldsymbol{X}_T^m - 0^k \boldsymbol{X}_0^m + \sum_i c_i \int_0^T t^{k_i} \boldsymbol{X}_t^{m_i} \, dt \tag{10}$$

where $(\boldsymbol{m}_i)_i$, $(k_i)_i$, and $(c_i)_i$ are finite sequences resulting from the substitution of $f$ and $w$. This choice allows to naturally characterize the behavior in time and state-space as moments, because the expected value of (10) then becomes a linear form of moments. We will use these as constraints in the semi-definite program used to bound MFPTs.

If we apply this to our previous example (7), letting $m = 1$ and $k = 1$ we obtain the following process for Model 1.

$$Z_T^{(1,1)} = T X_T - \int_0^T X_t \, dt - \lambda \int_0^T t \, dt - 2\delta \int_0^T t X_t \, dt + 2\delta \int_0^T t X_t^2 \, dt,$$

where the sequences above are $(m_i)_i = (1, 0, 1, 2)$, $(k_i)_i = (0, 1, 1, 1)$, and $(c_i)_i = (-1, -\lambda, -2\delta, 2\delta)$.

## 5  Bounds for Mean First Passage Times

We now turn to the analysis of first passage times within some time-bound $T > 0$. Given some subset of the state-space $B \subseteq \mathcal{S}$ the first passage time is given by the continuous random variable

$$\tau = \inf\{t \geq 0 \mid \boldsymbol{X}_t \in B\} \wedge T$$

where $a \wedge b := \min\{a, b\}$. For this work, we only look at threshold hitting times, i.e. we set a threshold $H$ for species $S$ and thus $B = \{\boldsymbol{x} \mid x^{(S)} \geq H\}$. Note, that this framework allows for a more general class of target sets, which are discussed in Appendix C. In the sequel, we will use $\tau$ as a stopping time in our martingale formulation and consider $Z_\tau^{(m,k)}$ instead of $Z_T^{(m,k)}$. Since (10) defines a martingale, $Z_\tau^{(m,k)}$ remains a martingale by Doob's optional sampling theorem [24]. In particular, this implies that $\mathbb{E}(Z_\tau^{(m,k)}) = 0$ for all moment orders $m$ and degrees $k$ in the weighting function $w(t)$.

## 5.1   Linear Moment Constraints

To simplify our presentation, we fix an initial state $x_0$, i.e. $P(X_0 = x_0) = 1$.
Using $\mathbb{E}(Z_\tau^{(m,k)}) = 0$ and the form (10) for $Z_\tau^{(m,k)}$ yields the following linear
constraint on expected values.

$$0 = \mathbb{E}\left(\tau^k X_\tau^m\right) - 0^k x_0^m + \sum_i c_i \mathbb{E}\left(\int_0^\tau t^{k_i} X_t^{m_i}\, dt\right), \tag{11}$$

where $0^0 = 1$. Hence, we have established a relationship between the process
dynamics up to the hitting time via expected values of the time-integrals and
the final process state at the hitting time via $\mathbb{E}\left(\tau^k X_\tau^m\right)$.

For the ease of exposition, we now turn to the analysis of first passage times
in one-dimensional processes w.r.t. an upper threshold $H$. In particular, we will
consider moments $X^m$ of a one-dimensional process for $m = 0, 1, 2 \ldots$. The
approach proposed in the sequel, however, can be extended to multi-dimensional
processes and more complex target sets $B$.

Consider again Model 1 and assume that we are interested in the time at
which species $M$ exceeds threshold $H$ while fixing the considered time-horizon
to $T = 4$. That is, we are interested in the stopping time $\tau = \inf\{t \geq 0 \mid$
$X_t \geq 10\} \wedge 4$. Since the abundance of $D$ does not influence $M$, we can ignore
species $D$ and treat the process as one-dimensional. Figure 1 shows three example
trajectories: Two reach an upper threshold $H = 10$, while one reaches the final
time-horizon $T = 4$ The figure also illustrates another aspect present in (11).
It gives a connection between the terminal distribution, i.e. the distribution of
$X_\tau$, and the dynamic behavior up to $\tau$. The statistics at $\tau$ are described by a
distribution whose moments are represented by the $\mathbb{E}\left(\tau^k X_\tau^m\right)$ term in (11).
This distribution corresponding two moments encompasses both cases of how $\tau$
can be reached. In the first case threshold $H$ is reached and the second case the
process reaches the time-horizon $T$. In the following we will define the interplay
between these measures more formally.

Therefore we can view (11) as the description of a relationship between two
measures [39, Chapter 9.2]:

- *Expected Occupation Measure* $\xi$ supported on $[0, H] \times [0, T]$:

$$\xi(A \times C) := \mathbb{E}\left(\int_{[0,\tau] \cap C} 1_{\in A}(X_t)\, dt\right), \tag{12}$$

- *Exit Location Probability* supported on $(\{H\} \times [0, T]) \cup ([0, H] \times \{T\})$:

$$\nu(A \times C) := \Pr((X_\tau, \tau) \in A \times C), \tag{13}$$

where $A \times C$ is a measurable set, i.e. $A$ and $C$ are elements of the Borel $\sigma$-algebras
on $[0, H]$ and $[0, T]$, respectively.

Using Fig. 1, one can gain an intuition for these two measures. The expected
occupation measure is shaded in blue. As the name implies $\xi(A \times C)$ tells us how

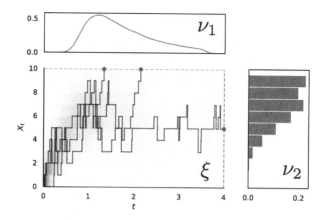

**Fig. 1.** The relationship between the occupation measure $\xi$ and the exit location probability measures $\nu_1$ and $\nu_2$. The shaded area indicates the structure of the occupation measure. Three example trajectories are additionally plotted with their exit location highlighted. The plots are based on 10,000 sample trajectories.

much time the process spends in $A$ up to $\tau$ restricting to the time instants belonging to $C$. In particular, $\xi([0, H] \times [0, T]) = \mathbb{E}(\tau)$. The exit location probability $\nu$, while being a two-dimensional distribution, can be viewed as a composition of a density describing the time at which the process reaches $H$ (if it does) and a probability mass function on the states of the process if the time-horizon is reached without exceeding $H$. We partition the measure $\nu$ into $\nu_1$ and $\nu_2$ by conditioning on $\tau = T$. Thus,

$$\nu_1(C) := \Pr(\tau \in C, \tau < T) \quad \text{and} \quad \nu_2(A) := \Pr(X_T \in A, \tau = T)$$

and hence $\nu(A \times C) = \nu_1(C) + \nu_2(A)$. To refer to the moments of these measures, we define *partial moments*

$$\mathbb{E}(g(X); f(Y) = y) := \mathbb{E}(g(X) \mid f(Y) = y) \Pr(f(Y) = y),$$

for some polynomial $g$ and some indicator function $f$. Then

$$\mathbb{E}\left(\tau^k X_\tau^m\right) = T^k \mathbb{E}\left(X_\tau^m; \tau = T\right) + H^m \mathbb{E}\left(\tau^k; \tau < T, X_\tau = H\right).$$

Therefore the linear moment constraints have the form

$$0 = T^k \mathbb{E}\left(X_\tau^m; \tau = T\right) + H^m \mathbb{E}\left(\tau^k; \tau < T, X_\tau = H\right)$$
$$- 0^k x_0^m + \sum_i c_i \mathbb{E}\left(\int_0^\tau t^{k_i} X_t^{m_i} \, dt\right). \tag{14}$$

Next, we consider infinite sequences of partial moments $\boldsymbol{y}_1 = (y_{1k})_{k \in \mathbb{N}}$, $\boldsymbol{y}_2 = (y_{2m})_{m \in \mathbb{N}}$, and $\boldsymbol{z} = (z_{mk})_{(m,k)^\top \in \mathbb{N}^2}$ of $\nu_1$, $\nu_2$, and $\xi$, respectively.

$$y_{1k} := \mathbb{E}\left(\tau^k; \tau < T\right), \quad y_{2m} := \mathbb{E}\left(X_\tau^m; \tau = T\right), \quad z_{km} := \mathbb{E}\left(\int_0^\tau t^k X_t^m \, dt\right)$$

## 5.2   Objective

Given the above measures and their corresponding moments, we can now identify the moments we are particularly interested in. We formulate an optimization problem with variables corresponding to the moments defined above. The MFPT is exactly the zeroth moment of $\xi$,

$$z_{00} = \mathbb{E}\left(\int_0^\tau 1_{\leq H}(X_t)\, dt\right) = \mathbb{E}\left(\tau\right).$$

Therefore $z_{00}$ corresponds to the objective of the optimization problem that gives bounds for the MFPT. Furthermore, we can easily change the objective to the zeroth moment of $\nu_1$, $y_{10} = \mathbb{E}\left(\tau^0; \tau < T\right) = \Pr(\tau < T)$. This moment is the probability of reaching threshold $H$ before reaching time-horizon $T$. Since the target set can be more complex, this formulation can be used to perform model checking on a wide variety of properties.

Moreover, it is possible to formulate objectives not directly corresponding to a raw moment such as the variance [19,48].

## 5.3   Semi-definite Constraints

The linear constraints alone are not sufficient to identify moment bounds. We further leverage the fact that a necessary condition for a positive measure that the *moment matrices* are positive semi-definite. A matrix $M \in \mathbb{R}^{n \times n}$ is positive semi-definite, denoted by $M \succeq 0$ if and only if

$$\boldsymbol{v}^T M \boldsymbol{v} \geq 0 \quad \forall \boldsymbol{v} \in \mathbb{R}^n.$$

As an example, let us consider a one-dimensional random variable $Z$ with moment sequence $\boldsymbol{z}$. For moment order $r$, the entries of the $(r+1) \times (r+1)$ moment matrix $M_r(\boldsymbol{x})$ are given by the raw moments. In particular,

$$(M_r(\boldsymbol{z}))_{ij} = z_{i+j-2} = \mathbb{E}\left(Z^{i+j-2}\right)$$

for $i, j \in \mathbb{N}_r$ where $\mathbb{N}_r = \{0, 1, \ldots, r\}$ and the maximum order in the matrix is $2r$. For instance,

$$M_1(\boldsymbol{x}) = \begin{bmatrix} x_0 & x_1 \\ x_1 & x_2 \end{bmatrix} \tag{15}$$

needs to be positive semi-definite. By Sylvester's criterion this means $\det M_1 \geq 0$ and $x_0 \geq 0$. We can easily see that in this case this entails

$$\det M_1 = x_0 x_2 - x_1^2 = \mathbb{E}\left(X^2\right) - \mathbb{E}\left(X\right)^2 = \mathrm{Var}(X) \geq 0.$$

This restriction is natural since the variance is always non-negative. This gives us the following restrictions on the moment matrices.

$$M_r(\boldsymbol{z}) \succeq 0, \quad M_r(\boldsymbol{y_1}) \succeq 0, \quad \text{and} \quad M_r(\boldsymbol{y_2}) \succeq 0 \tag{16}$$

for arbitrary orders $r$, providing a first tranche of moment constraints.

Furthermore, we need to enforce the restriction of the measures $\xi$, $\nu_1$, and $\nu_2$ to their supports. This can be done, by defining non-negative polynomials on the intended support of the measure. For example, $\nu_2$ has support $[0, H]$. We can now define

$$u_H(t, x) = Hx - x^2, \quad x \in \mathbb{R}$$

as a polynomial that is non-negative on $[0, H]$. Using such polynomials, we can construct *localizing matrices*, which have to be positive semi-definite [39]. Applying $u_H$ to the moment matrix of measure $\nu_2$, i.e. $M_1(\boldsymbol{y_2})$

$$M_1(u_H, \boldsymbol{y_2}) = \begin{bmatrix} Hy_{21} - y_{22} & Hy_{22} - y_{23} \\ Hy_{22} - y_{23} & Hy_{23} - y_{24} \end{bmatrix}$$

with the constraint $M_1(u_H, \boldsymbol{y_2}) \succeq 0$, where the application of a polynomial such as $u_H$ to a moment matrix is formally defined for the multidimensional case in Appendix C. Similarly, let $u_T(t, x) = Tt - t^2$ to restrict $\nu_1$ to $[0, T)$. The expected occupation measure $\xi$ is constrained similarly to its domain $[0, H] \times [0, T]$. This gives us the following restrictions on the moment matrices.

$$M_r(u_T, \boldsymbol{z}) \succeq 0, \quad M_r(u_H, \boldsymbol{z}) \succeq 0, \quad M_r(u_T, \boldsymbol{y_1}) \succeq 0, \quad M_r(u_H, \boldsymbol{y_2}) \succeq 0. \quad (17)$$

## 5.4   A Semi-definite Program to Bound MFPTs

With the linear constraints given in (11) and the semi-definite constraints (16) and (17) discussed in the previous sections, we can now formulate a semi-definite program (SDP) for any relaxation order $0 < r < \infty$. With each moment sequence $\boldsymbol{x}$ we associate a sequence proxy variables $\boldsymbol{x'}$ used in the optimization problem.

$$
\begin{aligned}
\min / \max \quad & z'_{00} \\
\text{such that} \quad & M_r(\boldsymbol{z'}) \succeq 0, M_r(u_T, \boldsymbol{z'}) \succeq 0, M_r(u_H, \boldsymbol{z'}) \succeq 0 \\
& M_r(\boldsymbol{y_1'}) \succeq 0, M_r(u_T, \boldsymbol{y_1'}) \succeq 0 \\
& M_r(\boldsymbol{y_2'}) \succeq 0, M_r(u_H, \boldsymbol{y_2'}) \succeq 0 \\
& 0 = y'_{1k}H^m - y'_{2m}T^k - 0^k x_0^m + \sum_i c_i z'_{k_i m_i}, \quad \forall m, k
\end{aligned}
\quad (18)
$$

This SDP can be compiled to the canonical form (see Appendix (B.1)). To this end, the moment matrices can be arranged in a block-diagonal form and the localizing constraints (17) can be encoded by the introduction of new variables and appropriate equality constraints. This transformation can be done automatically using modeling frameworks such as CVXPY [18]. We therefore only give the SDP in the more intuitive format. This problem can be solved using off-the-shelf SDP solvers such as MOSEK [42], CVXOPT [56], or SCS [44].

In principle, we can choose an arbitrarily large order $r$ for the moment matrices and their corresponding constraints, because there are infinitely many moments. In practice, however, the order is bounded by practical issues such as the program size (number of constraints and variables) and numerical issues. These issues are discussed in Sect. 6 in more detail. Choosing a finite $r$ is a relaxation of the problem since it removes constraints regarding higher-order moments.

# 6    Implementation and Evaluation

The implementation of the SDP (18) is straightforward using modeling frameworks and off-the-shelf solvers. However, as noted in previous work [19,20,47,48] on moment-based SDPs the direct implementation of the problem may lead to difficulties for the solver. A source of these is that moments of various orders by nature may differ by many orders of magnitude. A re-scaling of the moments [19,48] such that moments only vary by few orders of magnitude may alleviate this problem. In other scenarios such as the bounding of general transient or steady-state moments, the scaling can be particularly difficult, because the magnitude of moments is generally not known a priori. In the context of MFPTs with a finite time-horizon moments are trivially bounded. The resulting scaling scheme is outlined in Appendix D.

## 6.1    Case Studies

We implemented and solved the SDP programs described above using optimization suite MOSEK [42] (version 9.1.2) via the CVXPY interface [18] (version 1.0.24).

**Dimerization.** As a first case study, we use Model 1 with parameters $\lambda = 100$ and $\delta = 0.2$. In this model, we are interested in the time at which the number of agents of type $M$ surpasses a threshold of 25 before some time-horizon $T$, i.e. $\tau = \inf\{t \geq 0 \mid X_t \geq 25\} \wedge T$. First, we set no finite time-horizon $T$, i.e. $T = \infty$. This is achieved by dropping the moments $\boldsymbol{y}_2$ of measure $\nu_2$ in the linear constraints (18). This can be done because the threshold on $M$ makes the state-space finite and therefore the first passage time distribution is a phase-type distribution which possesses finite moments [54, Chapter 7.6]. The empirical FPT distribution based on 100,000 SSA simulations is given in Fig. 2a and the bounds, given different moment orders, are given in Fig. 2b. As we can see in Fig. 2b, the bounds capture the MFPT precisely for orders 5, 6. The difference between upper and lower bound decreases roughly exponentially with increasing relaxation order $r$. We found that this trend was consistent among the case studies presented here (cf. Fig. 4).

Next, we look at first passage times within a finite time-horizon $T$. In Fig. 3a we summarize the bounds obtained for the MFPT over $T$. While low-order relaxations (light) give rather loose bounds, the bounds are already fairly tight when using $r = 4$. In many cases, hitting probabilities, that is, the probability of reaching the threshold before time $T$, are of particular interest. This is done by switching the optimization objective in (18) from the mass of the expected occupation measure $\xi$ to the mass of $\nu_1$. In terms of moments, the objective changes from $z_{00}$ to $y_{10}$. The need for such a scenario often arises in the context of model checking, where one might be interested in the probability of a population exceeding a critical threshold. By varying the time-horizon, we are able to recover bounds on the cumulative density $F(t) = \Pr(X_s = H \mid s < t)$ of the first passage time (Fig. 3b).

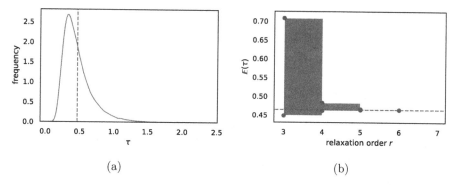

**Fig. 2.** First passage times for Model 1 with $\tau = \inf\{t \geq 0 \mid X_t \geq 10\} \wedge \infty$. The dashed red line denotes the sampled MFPT. (a) The distribution of $\tau$ estimated based on 100,000 SSA samples. (b) The bounds based on the SDP in (18) with different moment orders.

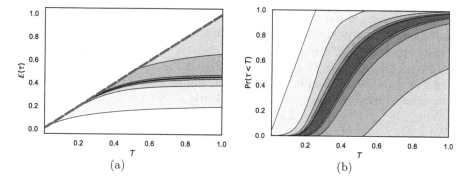

**Fig. 3.** First passage times for the dimerization model with $\tau = \inf\{t \geq 0 \mid X_t \geq 25\} \wedge T$. The results for SDP relaxations of orders 1 (light) to 6 (dark) are shown. (a) The bounds on the MFPT for differing time-horizons $T$. (b) Bounds on the probability to reach the threshold before time $T$.

Finally, we look at turn to the dimer species $D$ that is synthesized by the combination of two monomers $M$. Here, we look at the time until the agents of type $D$ exceed a threshold of five with a time-horizon $T = 1$. Note that we do not limit the number of $M$ agents. Therefore the analyzed state-space is countably infinite. As in the previous two examples, we observe a roughly exponential decrease in interval size with increasing relaxation order $r$ (cf. Fig. 4 and Table 1).

**Parallel Dimerizations.** As a second study, we consider a 2-dimensional model by combining two independent dimerizations.

**Model 2 (Parallel independent dimerizations).**

$$\varnothing \xrightarrow{10^4} M_1, \quad 2M_1 \xrightarrow{0.1} D_1, \quad \varnothing \xrightarrow{10^4} M_2, \quad 2M_2 \xrightarrow{0.1} D_2$$

As a FPT we consider the time at which either $M_1$ or $M_2$ surpasses a threshold of 200 or a time-horizon of $T = 10$ is reached, i.e.

$$\tau = \inf\{t \geq 0 \mid X_t^{(M_1)} \geq 200\} \wedge \inf\{t \geq 0 \mid X_t^{(M_2)} \geq 200\} \wedge 10.$$

As before, we ignore the product species $D_1$ and $D_2$ since they do not influence $\tau$. The SSA (using $n = 10{,}000$ runs) gives the estimate $\mathbb{E}(\tau) \approx 0.028378$ which is captured tightly by the SDP bounds (cf. Table 1). For higher relaxation orders $r \geq 5$ numerical issues prevented the solution of the corresponding SDPs.

## 6.2  Hybrid Models and Multi-modal Behavior

The analysis of switching times is a particularly interesting case of FPTs that arises in many contexts. Often mode switching in such systems can be described a modulating Markov process whose switching rates may depend on the system state (e.g. the population sizes). In biological applications, mode switching often describes a change of the DNA state [27,53] and the analysis of switching time distribution is of particular interest [7,52]. In the context of PCTMCs, typically the state-space $\mathcal{S} = \mathbb{N}^{\tilde{n}_S} \times \{0,1\}^{\hat{n}_S}$. This state is modeled by $\hat{n}_S$ population variables with binary domains. Therefore, at each time point, the state of these modulator variables is given by a set of Bernoulli random variables. When considering the moments of such a variable $X$, clearly $\mathbb{E}(X^m) = \mathbb{E}(X) = \Pr(X = 1)$ for all $m \geq 1$.

We apply a split of variables $\boldsymbol{X}_t$ into the high count part $\tilde{\boldsymbol{X}}_t$ and the binary part $\hat{\boldsymbol{X}}_t$ to the expectations in (6). Similarly, we split $\boldsymbol{v}_j$ and with a case distinction over the mode variable, we arrive at a similar result as in [27]:

$$\frac{d}{dt}\mathbb{E}\left(\tilde{\boldsymbol{X}}_t^m 1_{=\boldsymbol{y}}(\hat{\boldsymbol{X}}_t)\right) = \sum_{j=1}^{n_R} \mathbb{E}\left(\left(\tilde{\boldsymbol{X}}_t + \tilde{\boldsymbol{v}}_j\right)^m \alpha_j(\tilde{\boldsymbol{X}}_t, \boldsymbol{y} - \hat{\boldsymbol{v}}_j) 1_{=\boldsymbol{y} - \hat{\boldsymbol{v}}_j}(\hat{\boldsymbol{X}}_t)\right)$$
$$- \sum_{j=1}^{n_R} \mathbb{E}\left(\tilde{\boldsymbol{X}}_t^m \alpha_j(\tilde{\boldsymbol{X}}_t, \boldsymbol{y}) 1_{=\boldsymbol{y}}(\hat{\boldsymbol{X}}_t)\right). \tag{19}$$

Similarly to the general moment case, we can derive a constraint, by multiplying with a time-weighting factor and integrating.

For simplicity, here we assume $\tilde{n}_S = \hat{n}_S = 1$. Fixing appropriate sequences $(c_i)_i$, $(m_i)_i$, $(k_i)_i$, and $(y_i)_i$ the constraint has the following form.

$$\sum_{y \in \{0,1\}} H^m \mathbb{E}\left(\tau^k; \hat{X}_\tau = y, \tau < T\right) + T^k \mathbb{E}\left(\tilde{X}_T^m; \hat{X}_T = y, \tau = T\right)$$
$$= 0^k \tilde{x}_0^m 1_{=y}(\hat{x}_0) + \sum_i c_i \mathbb{E}\left(\int_0^\tau t^{k_i} \tilde{X}_t^{m_i} \, dt; \hat{X}_t = y_i\right) \tag{20}$$

This way we can decompose the moment matrices such that for each mode $y \in \{0,1\}$, we have moment matrices composed of the respective partial moments.

**Table 1.** MFPT bounds on Models 1, 2, and 3

| Model | | Relaxation order $r$ | | | | |
|---|---|---|---|---|---|---|
| | | 1 | 2 | 3 | 4 | 5 |
| Dimerization (Model 1) | Lower | 0.0909 | 0.2661 | 0.2845 | 0.2867 | 0.2871 |
| $X_t^{(D)} \geq 5$, $T = 1$ | Upper | 1.0000 | 0.3068 | 0.2932 | 0.2886 | 0.2875 |
| Double dim. (Model 2) | Lower | 0.0010 | 0.0250 | 0.0275 | 0.0280 | 0.0280 |
| | Upper | 10.0000 | 0.0575 | 0.0323 | 0.0299 | 0.0290 |
| Gene expression (Model 3) | Lower | 4.0000 | 6.0028 | 6.2207 | 6.3377 | 6.3772 |
| | Upper | 10.7179 | 6.4619 | 6.4079 | 6.4004 | 6.3835 |

To this end, let $z_m^{(y)}$ be the partial moment w.r.t. $\hat{X} = y$. The moment constraint over the partial moments has a linear structure:

$$0 = y_{1k} H^m - y_{2m} T^k - 0^k x_0^m + \sum_i c_i z_{k_i m_i}^{(y_i)}. \qquad (21)$$

**Gene Expression with Negative Feedback.** As an instance of a multimodal system, we consider a simple gene expression with self-regulating negative feedback which is a common pattern in many genetic circuits [53].

**Model 3 (Negative self-regulated gene expression).** *This model consists of a gene state that is either on or off, i.e. $X_t^{D_{on}} + X_t^{D_{off}} = 1$, $\forall t \geq 0$. Therefore the system has two modes.*

$$D_{on} \xrightarrow{\tau_0} D_{off}, \quad D_{off} \xrightarrow{\tau_1} D_{on}, \quad D_{on} \xrightarrow{\rho} D_{on} + P,$$

$$P \xrightarrow{\delta} \varnothing, \quad P + D_{on} \xrightarrow{\gamma} D_{off}$$

*The model parameters are $(\tau_0, \tau_1, \rho, \delta, \gamma) = (10, 10, 2, 0.1, 0.1)$ and $X_0^{(D_{off})} = 1$, $X_0^{(P)} = 0$ a.s.*

As a first passage time we consider

$$\tau = \inf\{t \geq 0 \mid X_t^{(P)} \geq 5\} \wedge 20.$$

The results are summarized in Table 1. The estimated MFPT based on 100,000 SSA samples is $\mathbb{E}(\tau) \approx 6.37795 \pm 0.02847$ at 99% confidence level. Note that our SDP solution for $r = 5$ yields tighter moment bounds than the statistical estimation.

In Fig. 4 we summarize our results about the decrease of the interval widths for increasing relaxation order $r$ by plotting them on a log-scale. We see an approximately exponential decrease with increasing $r$. The semi-definite programs above were all solved within at most a few seconds.

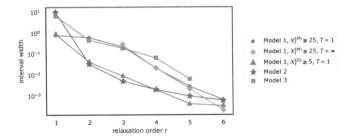

**Fig. 4.** The interval width, i.e. the difference between upper and lower bound, for different case studies and targeted first passage times against the order $r$ of the SDP relaxation.

# 7   Conclusion

Numerical methods to compute reachability probabilities and first passage times for continuous-time Markov chains that are based on an exhaustive exploration of the state-space are exact up to numerical precision. Such methods, however, do not scale and cannot be efficiently applied to models with large or infinite state-spaces, an issue exacerbated in population models. Moment-based methods offer an alternative analysis approach for PCTMCs, which scales with the number of different populations in the system but are approximations with little or no control of the error. In this paper, we bridge this gap by proposing a rigorous approach to derive bounds on first passage times and reachability probabilities, leveraging a semi-definite programming formulation based on appropriate moment constraints.

The method we propose is shown to be accurate in several examples. It does, however, suffer, like all moment-based methods, from numerical instabilities in the SDP solver, caused by the fact that moments typically span several orders of magnitude. We proposed a scaling of moments to mitigate this effect. However, the scaling only addresses the moment matrices but not the linear constraints which still contain values with varying orders of magnitudes. Therefore, we plan as future work to investigate an appropriate scaling for the linear constraints or to redefine the moment constraints (e.g. using an exponential time weighting [20]). Based on this investigation, we expect to make this approach applicable to more problems including, for example, the computation of bounds of rare event probabilities. We also expect that the development of more sophisticated scaling techniques will improve approximate moment-based methods.

Furthermore, moment-based analysis approaches have shown to be successful in a wide range of applications such as optimal control problems or the estimation of densities [39]. We expect that our proposed ideas can be adapted to a wider range of stochastic models such as stochastic hybrid systems, exhibiting partly deterministic dynamics.

**Acknowledgements.** We would like to thank Andreas Karrenbauer for helpful comments on the usage of SDP solvers and Gerrit Großmann for the valuable comments on this manuscript. This work is supported by the DFG project "MULTIMODE", and partially supported by the italian PRIN project "SEDUCE" n. 2017TWRCNB.

# References

1. Andreychenko, A., Mikeev, L., Spieler, D., Wolf, V.: Parameter identification for Markov models of biochemical reactions. In: Gopalakrishnan, G., Qadeer, S. (eds.) CAV 2011. LNCS, vol. 6806, pp. 83–98. Springer, Heidelberg (2011). https://doi.org/10.1007/978-3-642-22110-1_8
2. Aziz, A., Sanwal, K., Singhal, V., Brayton, R.: Verifying continuous time Markov chains. In: Alur, R., Henzinger, T.A. (eds.) CAV 1996. LNCS, vol. 1102, pp. 269–276. Springer, Heidelberg (1996). https://doi.org/10.1007/3-540-61474-5_75
3. Backenköhler, M., Bortolussi, L., Wolf, V.: Moment-based parameter estimation for stochastic reaction networks in equilibrium. IEEE/ACM Trans. Comput. Biol. Bioinform. **15**(4), 1180–1192 (2017)
4. Backenköhler, M., Bortolussi, L., Wolf, V.: Control variates for stochastic simulation of chemical reaction networks. In: Bortolussi, L., Sanguinetti, G. (eds.) CMSB 2019. LNCS, vol. 11773, pp. 42–59. Springer, Cham (2019). https://doi.org/10.1007/978-3-030-31304-3_3
5. Baier, C., Haverkort, B., Hermanns, H., Katoen, J.P.: Model-checking algorithms for continuous-time Markov chains. IEEE Trans. Softw. Eng. **29**(6), 524–541 (2003)
6. Baier, C., Haverkort, B., Hermanns, H., Katoen, J.-P.: Model checking continuous-time Markov chains by transient analysis. In: Emerson, E.A., Sistla, A.P. (eds.) CAV 2000. LNCS, vol. 1855, pp. 358–372. Springer, Heidelberg (2000). https://doi.org/10.1007/10722167_28
7. Barzel, B., Biham, O.: Calculation of switching times in the genetic toggle switch and other bistable systems. Phys. Rev. E **78**(4), 041919 (2008)
8. Bel, G., Munsky, B., Nemenman, I.: The simplicity of completion time distributions for common complex biochemical processes. Phys. Biol. **7**(1), 016003 (2009)
9. Bernardo, M., De Nicola, R., Hillston, J. (eds.): Formal Methods for the Quantitative Evaluation of Collective Adaptive Systems. LNCS, vol. 9700. Springer, Cham (2016). https://doi.org/10.1007/978-3-319-34096-8
10. Bogomolov, S., Henzinger, T.A., Podelski, A., Ruess, J., Schilling, C.: Adaptive moment closure for parameter inference of biochemical reaction networks. In: Roux, O., Bourdon, J. (eds.) CMSB 2015. LNCS, vol. 9308, pp. 77–89. Springer, Cham (2015). https://doi.org/10.1007/978-3-319-23401-4_8
11. Bortolussi, L., Hillston, J., Latella, D., Massink, M.: Continuous approximation of collective system behaviour: a tutorial. Perform. Eval. **70**(5), 317–349 (2013)
12. Bortolussi, L., Lanciani, R.: Model checking Markov population models by central limit approximation. In: Joshi, K., Siegle, M., Stoelinga, M., D'Argenio, P.R. (eds.) QEST 2013. LNCS, vol. 8054, pp. 123–138. Springer, Heidelberg (2013). https://doi.org/10.1007/978-3-642-40196-1_9
13. Bortolussi, L., Lanciani, R.: Stochastic approximation of global reachability probabilities of Markov population models. In: Horváth, A., Wolter, K. (eds.) EPEW 2014. LNCS, vol. 8721, pp. 224–239. Springer, Cham (2014). https://doi.org/10.1007/978-3-319-10885-8_16

172     M. Backenköhler et al.

14. Chen, T., Diciolla, M., Kwiatkowska, M., Mereacre, A.: Time-bounded verification of CTMCs against real-time specifications. In: Fahrenberg, U., Tripakis, S. (eds.) FORMATS 2011. LNCS, vol. 6919, pp. 26–42. Springer, Heidelberg (2011). https://doi.org/10.1007/978-3-642-24310-3_4

15. Chen, T., Han, T., Katoen, J.P., Mereacre, A.: Quantitative model checking of continuous-time Markov chains against timed automata specifications. In: 2009 24th Annual IEEE Symposium on Logic In Computer Science, pp. 309–318. IEEE (2009)

16. David, A., Larsen, K.G., Legay, A., Mikučionis, M., Poulsen, D.B., Sedwards, S.: Statistical model checking for biological systems. Int. J. Softw. Tools Technol. Transf. **17**(3), 351–367 (2015)

17. Dehnert, C., Junges, S., Katoen, J.-P., Volk, M.: A **STORM** is coming: a modern probabilistic model checker. In: Majumdar, R., Kunčak, V. (eds.) CAV 2017. LNCS, vol. 10427, pp. 592–600. Springer, Cham (2017). https://doi.org/10.1007/978-3-319-63390-9_31

18. Diamond, S., Boyd, S.: CVXPY: a Python-embedded modeling language for convex optimization. J. Mach. Learn. Res. **17**(83), 1–5 (2016)

19. Dowdy, G.R., Barton, P.I.: Bounds on stochastic chemical kinetic systems at steady state. J. Chem. Phys. **148**(8), 084106 (2018)

20. Dowdy, G.R., Barton, P.I.: Dynamic bounds on stochastic chemical kinetic systems using semidefinite programming. J. Chem. Phys. **149**(7), 074103 (2018)

21. Engblom, S.: Computing the moments of high dimensional solutions of the master equation. Appl. Math. Comput. **180**(2), 498–515 (2006)

22. Gast, N., Bortolussi, L., Tribastone, M.: Size expansions of mean field approximation: transient and steady-state analysis. Perform. Eval. **129**, 60–80 (2019). https://doi.org/10.1016/j.peva.2018.09.005

23. Ghusinga, K.R., Vargas-Garcia, C.A., Lamperski, A., Singh, A.: Exact lower and upper bounds on stationary moments in stochastic biochemical systems. Phys. Biol. **14**(4), 04LT01 (2017)

24. Gihman, I., Skorohod, A.: The Theory of Stochastic Processes II. Springer, Heidelberg (1975)

25. Gillespie, D.: Exact stochastic simulation of coupled chemical reactions. J. Phys. Chem. **81**(25), 2340–2361 (1977)

26. Gupta, A., Briat, C., Khammash, M.: A scalable computational framework for establishing long-term behavior of stochastic reaction networks. PLoS Comput. Biol. **10**(6), e1003669 (2014)

27. Hasenauer, J., Wolf, V., Kazeroonian, A., Theis, F.J.: Method of conditional moments (MCM) for the chemical master equation. J. Math. Biol. **69**(3), 687–735 (2014)

28. Hayden, R.A., Stefanek, A., Bradley, J.T.: Fluid computation of passage-time distributions in large Markov models. Theor. Comput. Sci. **413**(1), 106–141 (2012)

29. Helmes, K., Röhl, S., Stockbridge, R.H.: Computing moments of the exit time distribution for Markov processes by linear programming. Oper. Res. **49**(4), 516–530 (2001)

30. Hespanha, J.: Moment closure for biochemical networks. In: 2008 3rd International Symposium on Communications, Control and Signal Processing, pp. 142–147. IEEE (2008)

31. Hinton, A., Kwiatkowska, M., Norman, G., Parker, D.: PRISM: a tool for automatic verification of probabilistic systems. In: Hermanns, H., Palsberg, J. (eds.) TACAS 2006. LNCS, vol. 3920, pp. 441–444. Springer, Heidelberg (2006). https://doi.org/10.1007/11691372_29

32. Iyer-Biswas, S., Zilman, A.: First-passage processes in cellular biology. Adv. Chem. Phys. **160**, 261–306 (2016)
33. Kashima, K., Kawai, R.: Polynomial programming approach to weak approximation of lévy-driven stochastic differential equations with application to option pricing. In: 2009 ICCAS-SICE, pp. 3902–3907. IEEE (2009)
34. Kazeroonian, A., Theis, F.J., Hasenauer, J.: Modeling of stochastic biological processes with non-polynomial propensities using non-central conditional moment equation. IFAC Proc. Vol. **47**(3), 1729–1735 (2014)
35. Kuntz, J., Thomas, P., Stan, G.B., Barahona, M.: Rigorous bounds on the stationary distributions of the chemical master equation via mathematical programming. arXiv preprint arXiv:1702.05468 (2017)
36. Kuntz, J., Thomas, P., Stan, G.B., Barahona, M.: Approximation schemes for countably-infinite linear programs with moment bounds. arXiv preprint arXiv:1810.03658 (2018)
37. Kuntz, J., Thomas, P., Stan, G.B., Barahona, M.: The exit time finite state projection scheme: bounding exit distributions and occupation measures of continuous-time Markov chains. SIAM J. Sci. Comput. **41**(2), A748–A769 (2019)
38. Kwiatkowska, M., Norman, G., Parker, D.: PRISM 4.0: verification of probabilistic real-time systems. In: Gopalakrishnan, G., Qadeer, S. (eds.) CAV 2011. LNCS, vol. 6806, pp. 585–591. Springer, Heidelberg (2011). https://doi.org/10.1007/978-3-642-22110-1_47
39. Lasserre, J.B.: Moments, Positive Polynomials and Their Applications, vol. 1. World Scientific, Singapore (2010)
40. Lasserre, J.B., Prieto-Rumeau, T., Zervos, M.: Pricing a class of exotic options via moments and sdp relaxations. Math. Finance **16**(3), 469–494 (2006)
41. Mikeev, L., Neuhäußer, M.R., Spieler, D., Wolf, V.: On-the-fly verification and optimization of DTA-properties for large Markov chains. Form. Methods Syst. Des. **43**(2), 313–337 (2013)
42. MOSEK ApS: MOSEK Optimizer API for C 8.1.0.67 (2018). https://docs.mosek.com/8.1/capi/index.html
43. Munsky, B., Nemenman, I., Bel, G.: Specificity and completion time distributions of biochemical processes. J. Chem. Phys. **131**(23), 12B616 (2009)
44. O'Donoghue, B., Chu, E., Parikh, N., Boyd, S.: SCS: splitting conic solver, version 2.1.0, November 2017. https://github.com/cvxgrp/scs
45. Parrilo, P.A.: Semidefinite programming relaxations for semialgebraic problems. Math. Program. **96**(2), 293–320 (2003)
46. Porter, M.A., Gleeson, J.P.: Dynamical Systems on Networks. FADSRT, vol. 4. Springer, Cham (2016). https://doi.org/10.1007/978-3-319-26641-1
47. Sakurai, Y., Hori, Y.: A convex approach to steady state moment analysis for stochastic chemical reactions. In: 2017 IEEE 56th Annual Conference on Decision and Control (CDC), pp. 1206–1211. IEEE (2017)
48. Sakurai, Y., Hori, Y.: Bounding transient moments of stochastic chemical reactions. IEEE Control. Syst. Lett. **3**(2), 290–295 (2019)
49. Schnoerr, D., Cseke, B., Grima, R., Sanguinetti, G.: Efficient low-order approximation of first-passage time distributions. Phys. Rev. Lett. **119**, 210601 (2017). https://doi.org/10.1103/PhysRevLett.119.210601
50. Schnoerr, D., Sanguinetti, G., Grima, R.: Comparison of different moment-closure approximations for stochastic chemical kinetics. J. Chem. Phys. **143**(18), 185101 (2015). https://doi.org/10.1063/1.4934990

51. Schnoerr, D., Sanguinetti, G., Grima, R.: Approximation and inference methods for stochastic biochemical Kinetics'a tutorial review. J. Phys. Math. Theor. **50**(9), 093001 (2017). https://doi.org/10.1088/1751-8121/aa54d9
52. Spieler, D., Hahn, E.M., Zhang, L.: Model checking CSL for Markov population models. arXiv preprint arXiv:1111.4385 (2011)
53. Stekel, D.J., Jenkins, D.J.: Strong negative self regulation of prokaryotic transcription factors increases the intrinsic noise of protein expression. BMC Syst. Biol. **2**(1), 6 (2008)
54. Stewart, W.J.: Probability, Markov Chains, Queues, and Simulation: the Mathematical Basis of Performance Modeling. Princeton University Press, Princeton (2009)
55. Ullah, M., Wolkenhauer, O.: Stochastic approaches for systems biology. Wiley Interdiscip. Rev. Syst. Biol. Med. **2**, 385–97 (2009). https://doi.org/10.1002/wsbm. 78
56. Vandenberghe, L.: The CVXOPT linear and quadratic cone program solvers (2010). http://cvxopt.org/documentation/coneprog.pdf

# Markovian Arrival Processes
# in Multi-dimensions

Andreas Blume, Peter Buchholz$^{(\boxtimes)}$, and Clara Scherbaum

Informatik IV, Technical University of Dortmund,
44221 Dortmund, Germany
{andreas.blume,peter.buchholz,clara.scherbaum}@cs.tu-dortmund.de

**Abstract.** Phase Type Distributions (PHDs) and Markovian Arrival Processes (MAPs) are established models in computational probability to describe random processes in stochastic models. In this paper we extend MAPs to Multi-Dimensional MAPs (MDMAPs) which are a model for random vectors that may be correlated in different dimensions. The computation of different quantities like joint moments or conditional densities is introduced and a first approach to compute parameters with respect to measured data is presented.

**Keywords:** Input modeling · Markovian arrival process ·
High-dimensional data · Stochastic processes

## 1 Introduction

In many application areas like computer networks [15], supply chains [26] or dependable systems [18], high dimensional data plays an important role in understanding, analyzing and improving the behavior of contemporary systems. Currently available data is mainly analyzed offline to monitor or predict the behavior of complex systems. However, it is known that model-based approaches are often necessary to understand and analyze large systems. In simulation models [23] and also in models based on Markov chains [12], multi-dimensional data is usually described by independent data streams, where at most the elements in one stream are correlated. In practice, multi-dimensional data is correlated in several dimensions and this correlation cannot be neglected in realistic models. The necessary models to describe such a behavior are denoted as multivariate input models.

In multivariate input models one usually distinguishes between random vectors which describe $K$-dimensional vectors of random variables that are correlated. Subsequent vectors are assumed to be independent. Random variables

**Electronic supplementary material** The online version of this chapter (https://doi.org/10.1007/978-3-030-59854-9_14) contains supplementary material, which is available to authorized users.

© Springer Nature Switzerland AG 2020
M. Gribaudo et al. (Eds.): QEST 2020, LNCS 12289, pp. 175–192, 2020.
https://doi.org/10.1007/978-3-030-59854-9_14

that are correlated over time are described by stochastic processes. The combination of both results in multivariate time series. Multivariate input models are mainly considered in simulation, for an overview of available approaches see [5]. Although different approaches for multivariate stochastic processes exist their practical applicability is limited mainly due to very specific structures that capture only parts of the observed behavior, complex methods for parameter fitting, complex methods to generate random variates and the limiting possibilities to perform numerical or analytical analysis of the models. Most promising approaches seem to be VARTA processes [6] and copula-based models [3]. Both approaches are restricted to specific marginal distributions, like normal distributions or distributions of the Johnson-type. The VARTA approach has been extended to phase type distributions in [20].

In computational probability [28,30,33] input models based on Markov processes like phase type distributions (PHDs) or Markovian arrival processes (MAPs) are very popular because they allow one to model a wide variety of behaviors and they can also be analyzed by numerical techniques and not only by simulation. MAPs are a model to describe correlated univariate processes and are therefore an alternative to time series but they cannot be applied to describe multivariate processes. In this paper we extend Markov models like PHDs and MAPs to the multivariate case. This results in a new stochastic model which is an alternative to VARTA processes and similar models. Some older approaches to extend phase type distributions to multivariate phase type distributions exist [1,22]. However, the models defined in these papers differ from our model in that they describe an absorbing Markov process with a multi-dimensional reward structure to generate random vectors. Here, we consider parallel running absorbing Markov processes which are coupled by the initial distributions. This model allows us to generate random vectors with correlated components and correlation between subsequent realizations.

The structure of the paper is as follows. In the following section we introduce the notation and Markov input models. Afterwards, in Sect. 3, we define the multi-dimensional stochastic model and define afterwards multi-dimensional Markovian arrival processes (MDMAPs). In Sect. 4, the analysis of MDMAPs is presented, followed by a first approach to fit the parameters according to some quantities like joint moments or values of the conditional probability distribution function. In Sect. 6 first examples are presented and then the paper is concluded. Proofs of the theorems and major equations can be found in an online appendix.

## 2   Background

We first introduce some notation and define afterwards the basic models used in this paper.

### 2.1   Basic Notation

Matrices and vectors are denoted by bold face small and capital letters. $\mathbb{1}$ is a column vector of 1s, all other vectors are row vectors. $\mathbf{0}$ is a matrix or vector containing only 0 elements, $I$ is the identity matrix, $e_i$ is the $i$th unit row vector. $a^T$ describes the transposed of vector $a$ and diag($a$) denotes a diagonal matrix with elements $a(i)$ on the main diagonal. $\mathbb{R}^{n \times n}$ is the set of $n \times n$ matrices. $Q(i\bullet)$ and $Q(\bullet i)$ describe the $i$th row and column of matrix $Q$. A generator is a matrix $Q \in \mathbb{R}^{N \times N}$ with row sum zero (i.e., $Q\mathbb{1} = 0$) and $Q(i,j) \geq 0$ for $i \neq j$. $Q$ is a sub-generator if $Q\mathbb{1} \leq 0$ and some $i \in \{0,\ldots,n-1\}$[1] exists such that $Q(i\bullet)\mathbb{1} < 0$. In the sequel we assume that all sub-generators we consider in this paper are non-singular which means that the inverse exists and is non-positive. $Q$ is irreducible if between every pair of states $i,j$ a path $i = i_0, i_1, \ldots, i_k = j$ exists such that $Q(i_{h-1}, i_h) > 0$ for $h = 1, \ldots, k$.

## 2.2   Markov Input Models

In stochastic modeling *input modeling* describes the generation of appropriate, usually stochastic, models to represent the input parameters based on measured data from some real process [23]. In simulation, traditionally standard distributions or stochastic processes have been used for this purpose. More recently input models based on Markov processes like phase type distributions and Markovian arrival processes gained much attention. These models are flexible and can be used in simulation as well as in combination with numerical analysis techniques. We use the following traditional definitions [12,28].

**Definition 1.** *A Phase Type Distribution (PHD) is defined by $(\pi, D)$ where $\pi$ is the initial distribution and $D$ is a sub-generator of an absorbing Markov chain.*

A PHD is characterized by the time to absorption of the absorbing Markov chain described by $(\pi, D)$. A Markovian Arrival Process [25,27] is an extension of a PHD.

**Definition 2.** *A Markovian Arrival Process (MAP) is described by two matrices $(D, C)$[2] where $D$ is a sub-generator, $C \geq 0$ and $Q = D + C$ is an irreducible generator.*

The interpretation of the behavior of a MAP is as follows. The process performs transitions as described by the matrices $D$ and $C$ and whenever a transition from $C$ occurs, an event is triggered. Let $d = C\mathbb{1} = -D\mathbb{1}$ and $P = (-D)^{-1} C$ be the transition matrix of the embedded process at event times. Since $Q$ is irreducible, it has a unique stationary vector observing $\phi Q = 0$ and $\phi\mathbb{1} = 1$. Then $\pi = \phi C / (\phi C \mathbb{1})$, $\pi P = \pi$ and $\pi$ describes the stationary vector

---

[1] In an $n$-dimensional space elements are always numbered from 0 through $n-1$ because this numbering is more appropriate for mapping multi-dimensional spaces into a single space.

[2] We use the names $D$ and $C$ rather than $D_0$ and $D_1$ for the matrices of a MAP because the numbers in the postfix are later used to denote matrices of different MAPs or PHDs.

of the MAP immediately after an event. $(\boldsymbol{\pi}, \boldsymbol{D})$ is the embedded PHD of the MAP. We define the following two sets

$$inp\,(\boldsymbol{\pi}, \boldsymbol{D}) = inp\,(\boldsymbol{\pi}) = \{i|\boldsymbol{\pi}(i) > 0\}\,, outp\,(\boldsymbol{\pi}, \boldsymbol{D}) = outp\,(\boldsymbol{D}) = \{i|\boldsymbol{d}(i) > 0\} \tag{1}$$

of input and output states. $n^c = |inp\,(\boldsymbol{\pi}, \boldsymbol{D})|$ and $n^r = |outp\,(\boldsymbol{\pi}, \boldsymbol{D})|$ are the cardinalities of the sets. A PHD is *input flexible* if $n^c > 1$ and it is output flexible if $n^r > 1$. If we assume that each input and output state describes an individual stochastic behavior, then an input flexible PHD allows one to choose a specific behavior by selecting the input state, an output flexible PHD allows one to interpret the previous behavior by considering the output state. To expand a PHD to a MAP (see [12,19] for details), the PHD has to be input and output flexible to specify correlation. A MAP $(\boldsymbol{D}, \boldsymbol{C})$ can also be represented as $(\boldsymbol{D}, \boldsymbol{G})$ where $\boldsymbol{C} = \mathrm{diag}(\boldsymbol{d})\boldsymbol{G}$ and $\boldsymbol{G}$ is a matrix with $\boldsymbol{G}(i\bullet)\mathbf{1} = 1$ for $i \in outp(\boldsymbol{D})$ and 0 otherwise.

PHDs and MAPs can be easily analyzed according to several quantities including moments, probability density and, in case of MAPs, joint moments or joint densities. For details we refer to the literature [12]. Parameter fitting for a stochastic model describes the process of finding good or optimal parameters such that the stochastic model mimics the behavior of a real process for which data is available. Parameter fitting for PHDs or MAPs is more complex than for many other stochastic models because both models have a highly redundant representation [31]. In principle two approaches can be applied. First, some derived measures can be computed, like moments or joint moments and a least squares approach is used to fit the parameters in such a way that the quantities of the measured data are approximated by the PHD or MAP. Alternatively, maximum likelihood estimators for the parameters can be used which are usually based on the EM algorithm. For details about the corresponding algorithms we refer to the literature [12].

## 3   Multi-dimensional Data and Stochastic Models

We first introduce the basic setting for multivariate distributions and random vectors. Afterwards we present a Markov model to describe those quantities.

### 3.1   Multi-dimensional Data

Let $\boldsymbol{X} = (X_1, \ldots, X_K)$ be a random vector where each $X_k$ is a random variable. We assume that all random variables are non-negative and the underlying distribution functions have an infinite support. We denote by $K$ the number of dimensions or components of the random vector. If the random variables are mutually independent, each $X_k$ can be modeled by a MAP, if subsequent realizations of $X_k$ are also independent, a PHD is sufficient. Here we consider the case that various dependencies exist between the random variables and subsequent realizations. Thus, $\boldsymbol{X}^{(t)}$ is the vector observed at time $t\,(= 1, 2, \ldots)$ and $\boldsymbol{X}^{(t+h)}$

is the vector $h$ steps later. In general $X_k^{(t)}$ and $X_l^{(t)}$ as well as $X_k^{(t)}$ and $X_l^{(t+h)}$ may be correlated.

We assume that the stochastic structure of $\boldsymbol{X}$ is unknown but we can observe realizations of $\boldsymbol{X}$. Let $\boldsymbol{x}^{(i)} = \left( x_1^{(i)}, \ldots, x_K^{(i)} \right)$ be the $i$th realization of $\boldsymbol{X}$ and $x_k^{(i)}$ is the $i$th realization of $X_k$. From a sequence of observations $\boldsymbol{x}^{(1)}, \ldots, \boldsymbol{x}^{(L)}$ various quantities can be estimated.

$$\hat{X}_k^j = \frac{1}{L} \sum_{i=1}^{L} \left( x_k^{(i)} \right)^j \text{ and } \hat{\sigma}_k^2 = \frac{1}{L-1} \sum_{i=1}^{L} \left( x_k^{(i)} - \hat{X}_k^1 \right)^2 \tag{2}$$

are estimates for the $j$th moments and the variance of the random variables $X_k$. We denote by $\hat{\boldsymbol{R}}_h$ the correlation matrix of elements $h$ steps apart which contains the correlation coefficients. Elements of the correlation matrix are estimated by

$$\hat{\boldsymbol{R}}_h(k,l) = \frac{1}{(L-h-1)\hat{\sigma}_k\hat{\sigma}_l} \sum_{i=1}^{L-h} \left( x_k^{(i)} - \hat{X}_k^1 \right) \left( x_l^{(i+h)} - \hat{X}_l^1 \right). \tag{3}$$

The definition can be extended to higher order joint moments as follows.

$$\hat{\boldsymbol{J}}_h^{m,n}(k,l) = \frac{1}{(L-h)} \sum_{i=1}^{L-h} \left( x_k^{(i)} \right)^m \left( x_l^{(i+h)} \right)^n \tag{4}$$

where $k,l \in \{1, \ldots, K\}$ and $n, m \geq 1$. Similarly, the distribution function for one or some dimensions of the random vector can be estimated. All presented estimators are consistent. Like for joint moments in (4) we consider especially dependencies between two components $k$ and $l$ which are described in the following joint dependencies.

$$\hat{\boldsymbol{F}}_h^{y,z}(k,l) = \frac{1}{L-h} \sum_{i=1}^{L-h} \delta\left( x_k^{(i)} \leq y \right) \delta\left( x_l^{(i+h)} \leq z \right) \tag{5}$$

## 3.2    Multi-dimensional Markov Models

The available Markov models are not able to describe multi-dimensional data. Therefore we propose an extended model which consists of $K$ absorbing Markov chains that run in parallel, the absorption time of the $k$th chain determines the value of the $k$th random variable. After absorption of all chains, they are restarted according to a joint probability distribution which depends on the states immediately before absorption. The later concept is a direct extension of the idea that is used in MAPs to describe correlation. The following definition formalizes the model.

**Definition 3.** *A Multi-Dimensional Markovian Arrival Process (MDMAP) is defined by $K$ sub-generators $\boldsymbol{D}_k$ of order $n_k$ ($k = 1, \ldots, K$) and a coupling matrix $\boldsymbol{G}$.*

$K$ is the dimension of the MDMAP. Matrix $G$ is an $n_{1:K} \times n_{1:K}$ matrix ($n_{1:K} = \prod_{k=1}^{K} n_k$) matrix where state vector $(i_1, \ldots, i_K)$ ($i_k \in \{0, \ldots, n_k - 1\}$) is mapped onto index $i_{1:K} = \sum_{k=1}^{K} i_k \cdot n_{k+1:K}$ (where $n_{l:k} = \prod_{i=l}^{k} n_k$ for $k \geq l$ and 1 for $l > k$). $G \geq 0$, $G(i_{1:K}\bullet) = 0$ if $i_k \notin outp(D_k)$ for some $k$ and $G(i_{1:K}\bullet)\mathbf{1} = 1$ otherwise. $G(i_{1:K}, j_{1:K}) > 0$ implies $i_k \in outp(D_k)$ and $j_k \in inp(\pi_k)$ for all $k = 1, \ldots, K$ which is denoted as $i_{1:K} \in outp_{1:K}$ and $j_{1:K} \in inp_{1:K}$, respectively. The notations may be restricted to subsets of indices $k : l$ for $k \leq l$ or subset $\mathcal{K} \subseteq \{1, \ldots, K\}$. For the cardinalities of the sets we use the following notations $n_{\mathcal{K}}^r = |outp_{\mathcal{K}}|$ and $n_{\mathcal{K}}^c = |inp_{\mathcal{K}}|$. Let $\mathbf{1}_{n_{1:K}}^{inp}$ be a vector of length $n_{1:K}$ where $\mathbf{1}_{n_{1:K}}^{inp} = 1$ if $i \in inp_{1:K}$ and 0 otherwise. Similarly $\mathbf{1}_{n_{1:K}}^{outp}$ is defined. Then $G\mathbf{1}_{n_{1:K}} = G\mathbf{1}_{n_{1:K}}^{inp} = \mathbf{1}_{n_{1:K}}^{outp}$.

The behavior of an MDMAP is as follows. Each of the $K$ absorbing Markov chains generates a non-negative value, the exit states $i_k$ are kept and finally row $G(i_{1:K}\bullet)$ defines a probability distribution over the input states of each chain. Dependencies between successive events of one chain and between chains are realized by the relation between input and output states.

Let $\pi$ be a vector of length $n_{1:K}$ which contains the distribution immediately before the next event starts. Let $d_k = -D_k\mathbf{1}$ and the stochastic matrix $H_k = (-D_k)^{-1} \operatorname{diag}(d_k)$. Observe that $H_k(i_k, j_k) > 0$ implies $j_k \in outp_k$. Vector $\pi$ can be computed from the following set of linear equations.

$$\pi \left( \left( \bigotimes_{k=1}^{K} H_k \right) G \right) = \pi \quad \text{and} \quad \pi\mathbf{1} = 1, \tag{6}$$

if the matrix in brackets contains a single irreducible subset of states which will be assumed for the moment. For some vector $\pi \in \mathbb{R}^{n_{1:K}}$ we define the mapping onto the $k$th dimension as vector $\pi_k \in \mathbb{R}^{n_k}$ with

$$\pi_k(i_k) = \sum_{i_1=0}^{n_1-1} \cdots \sum_{i_{k-1}=0}^{n_{k-1}-1} \sum_{i_{k+1}=0}^{n_{k+1}-1} \cdots \sum_{i_K=0}^{n_K-1} \pi(i_{1:K}) \tag{7}$$

This mapping can be computed by right multiplication of $\pi$ with the following matrix.

$$\pi_k = \pi V_k \quad \text{where} \quad V_k = \bigotimes_{l=1}^{K} id_l \quad \text{and} \quad id_l = \begin{cases} \mathbf{1}_{n_l} & \text{if } l \neq k, \\ I_{n_k} & \text{if } k = l \end{cases} \tag{8}$$

where $\otimes$ is the Kronecker product. Obviously $V_k = \mathbf{1}_{n_{1:k-1}} \otimes I_{n_k} \otimes \mathbf{1}_{n_{k+1:K}}$. The mapping can be extended to subsets of components. Let $\mathcal{K} \subseteq \{1, \ldots, K\}$ and

$$V_{\mathcal{K}} = \bigotimes_{l=1}^{K} id_l \quad \text{where} \quad id_l = \begin{cases} \mathbf{1}_{n_l} & \text{if } l \notin \mathcal{K}, \\ I_{n_l} & \text{if } l \in \mathcal{K}. \end{cases} \tag{9}$$

$\pi_{\mathcal{K}} = \pi V_{\mathcal{K}}$ is the embedded initial vector mapped onto the subset $\mathcal{K}$. For notational convenience we write $V_{kl}$ for $V_{\{k,l\}}$.

For an initial vector $\boldsymbol{\pi}$, the exit vector $\boldsymbol{\psi}$ is given by

$$\boldsymbol{\psi} = \boldsymbol{\pi} \left( \bigotimes_{k=1}^{K} \boldsymbol{H}_k \right) \text{ and } \boldsymbol{\psi}_k = \boldsymbol{\pi}_k \boldsymbol{H}_k. \tag{10}$$

$\boldsymbol{\psi}$ and $\boldsymbol{\psi}_k$ contain the probabilities of absorption from state $i_{1:k}$ and $i_k$, respectively. Obviously, $\boldsymbol{\psi}_k(i_k) = 0$ for $i_k \notin outp_k$. We assume that $\boldsymbol{\psi}_k(i_k) > 0$ for $i_k \in outp_k$ otherwise the corresponding state would not be reachable from an initial state and can therefore removed from the PHD.

The mapping of matrix $\boldsymbol{G}$ on the state space of some components is defined according to some exit vector $\boldsymbol{\psi}$ using the following matrix

$$\boldsymbol{W}_{\mathcal{K}}[\boldsymbol{\psi}] = \text{diag}\,(\boldsymbol{\psi} \boldsymbol{V}_{\mathcal{K}})^+ \, \boldsymbol{V}_{\mathcal{K}}^T \, \text{diag}(\boldsymbol{\psi}) \text{ and } \boldsymbol{G}_{\mathcal{K}}[\boldsymbol{\psi}] = \boldsymbol{W}_{\mathcal{K}}[\boldsymbol{\psi}] \boldsymbol{G} \boldsymbol{V}_{\mathcal{K}} \tag{11}$$

where $\boldsymbol{A}^+$ is the pseudo-inverse of matrix $\boldsymbol{A}$ which can be computed for the diagonal matrix $\text{diag}\,(\boldsymbol{\psi} \boldsymbol{V}_{\mathcal{K}})$ by substituting non-zero diagonal elements by the inverse and leaving zero diagonal elements unchanged. This is the usual way of aggregation in multi-dimensional Markov models (see e.g. [13] for details).

**Theorem 1.** *For some MDMAP with $K$ components, coupling matrix $\boldsymbol{G}$, initial vector $\boldsymbol{\pi}$ and any subset $\mathcal{K} \subseteq \{1, \ldots, K\}$, the initial vector $\boldsymbol{\pi}_{\mathcal{K}}$ of the MDMAP restricted to the components from $\mathcal{K}$ is the solution of*

$$\boldsymbol{\pi}_{\mathcal{K}} \left( \left( \bigotimes_{k \in \mathcal{K}} \boldsymbol{H}_k \right) \boldsymbol{G}_{\mathcal{K}}[\boldsymbol{\psi}] \right) = \boldsymbol{\pi}_{\mathcal{K}} \text{ and } \boldsymbol{\pi}_{\mathcal{K}} \mathbb{I} = 1$$

*where $\boldsymbol{\psi} = \boldsymbol{\pi} \otimes_{k=1}^{K} \boldsymbol{H}_k$.*

The number of parameters to represent all matrices $\boldsymbol{D}_k$ only linearly with $K$ and quadratic with $n_k$. This does not hold for the number of entries in $\boldsymbol{G}$ which may grow with $n_{1:K}^c n_{1:K}^r$. Therefore we consider MDMAPs of *rank $R$* with product form, that can be represented as follows

$$\boldsymbol{G} = \sum_{r=1}^{R} \lambda^{(r)} \bigotimes_{k=1}^{K} \boldsymbol{G}_k^{(r)} \tag{12}$$

where $\lambda^{(r)} > 0$, $\sum_{r=1}^{R} \lambda^{(r)} = 1$, $\boldsymbol{G}_k^{(r)} \geq 0$ and $\boldsymbol{G}_k^{(r)} \mathbb{I}_{n_k} = \mathbb{I}_{n_k}^{outp}$.

**Theorem 2.** *For an MDMAP of rank $R$ with product form and some set $\mathcal{K} \subseteq \{1, \ldots, K\}$ vector $\boldsymbol{\pi}_{\mathcal{K}} = \boldsymbol{\pi} \boldsymbol{V}_{\mathcal{K}}$ is the solution of*

$$\boldsymbol{\pi}_{\mathcal{K}} \left( \bigotimes_{k \in \mathcal{K}} \boldsymbol{H}_k \right) \left( \sum_{r=1}^{R} \lambda^{(r)} \bigotimes_{k \in \mathcal{K}} \boldsymbol{G}_k^{(r)} \right) = \boldsymbol{\pi}_{\mathcal{K}} \text{ and } \boldsymbol{\pi}_{\mathcal{K}} \mathbb{I}_{n_{\mathcal{K}}} = 1.$$

Theorem 2 holds in particular for sets $\mathcal{K} = \{k\}$. This implies that in a product form MDMAP each component behaves locally like a MAP$(\boldsymbol{D}_k, \sum_{r=1}^{R} \lambda^{(r)} \boldsymbol{G}_k^{(r)})$.

# 4  Analysis of MDMAPs

Analysis of MDMAPs can be performed according to one dimension of the random vector or according to the joint distribution.

## 4.1  Analysis of a Single Vector Component

An MDMAP can be easily mapped on a MAP $(\boldsymbol{D}_k, \boldsymbol{G}_k)$ for one vector component $k$ that neglects all other dimensions. If the MDMAP is of rank $R$ and product form, then $\boldsymbol{G}_k = \sum_{r=1}^{R} \lambda^{(h)} \boldsymbol{G}_k^{(r)}$, otherwise $\boldsymbol{G}_k = \boldsymbol{W}_k[\boldsymbol{\psi}]\boldsymbol{G}\boldsymbol{V}_k$. The resulting MAP can then be analyzed with the available methods (see e.g. [12, Sect. 4]).

## 4.2  Analysis of Joint Measures

In the following we consider mainly dependencies between two dimensions $k$ and $l$. In the equations we assume $k < l$, $k > l$ requires a different ordering of the matrices in the equations, but does, of course, not change the general structure. The case $k = l$ describes a single component and is mentioned above. Equations are formulated for MDMAPs of rank $R$ with product form, matrices $\boldsymbol{G}$ for the general case are written underneath the rank $R$ representation. Let $\boldsymbol{J}_1^{m,n}(k,l) = E\left[\left(X_k^{(t)}\right)^m, \left(X_l^{(t+1)}\right)^n\right]$, the joint moment of order $m, n$ for dimension $k$ and dimension $l$, $h$ steps apart. For $\boldsymbol{J}_0^{m,n}(k,l)$ we obtain

$$\boldsymbol{J}_0^{m,n}(k,l) = m!n!\boldsymbol{\pi}_{kl}\left((\boldsymbol{M}_k)^m \otimes (\boldsymbol{M}_l)^n\right)\mathbb{1}_{n_k n_l} \tag{13}$$

where $\boldsymbol{M}_k = (-\boldsymbol{D}_k)^{-1}$. The joint moment for $h = 1$ and MDMAPs for rank $R$ with product form is given by

$$\boldsymbol{J}_1^{m,n}(k,l) = m!n!\boldsymbol{\pi}_{k,l}\left((\boldsymbol{M}_k)^m \otimes \boldsymbol{H}_l\right)\underbrace{\sum_{r=1}^{R} \lambda^{(r)}\left(\mathbb{1}_{n_k} \otimes \boldsymbol{G}_l^{(r)}\right)}_{\boldsymbol{G}_{kl}\left(\mathbb{1}_{n_l} \otimes I_{n_l}\right)}(\boldsymbol{M}_l)^n \mathbb{1}_{n_l} \tag{14}$$

For two components $k$ and $l$ the joint distribution is given by

$$\boldsymbol{F}_0^{x,y}(k,l) = \boldsymbol{\pi}_{kl}\left(\int_0^x e^{\tau \boldsymbol{D}_k}\boldsymbol{d}_k d\tau \otimes \int_0^y e^{\tau \boldsymbol{D}_l}\boldsymbol{d}_l d\tau\right) \tag{15}$$

For successive observations. $\boldsymbol{F}_1^{x,y}(k,l)$ denotes the conditional probability that we observe values $\leq x$ for $k$ and for the next observation $z_l$ of dimension $l$ $z_l \leq y$ holds. The function can be computed using the following equation.

$$\boldsymbol{F}_1^{x,y}(k,l) = \boldsymbol{\pi}_{kl}\left(\int_0^x e^{\tau \boldsymbol{D}_k}\operatorname{diag}(\boldsymbol{d}_k)\,d\tau \otimes \boldsymbol{H}_l\right)\underbrace{\sum_{r=1}^{R}\lambda^{(r)}\mathbb{1}_{n_k}^{outp} \otimes \boldsymbol{G}_l^{(r)}}_{\boldsymbol{G}_{kl}\left(\mathbb{1}_{n_l} \otimes I_{n_l}\right)}\int_0^y e^{\tau \boldsymbol{D}_l}\boldsymbol{d}_l d\tau \tag{16}$$

Observe that (13) and (15) as well as (14) and (16) are of an identical structure. Therefore we define a common notation which allows us to handle joint moments and values of the distributions functions in a common framework. We denote these measures as zero or first order quantities, respectively. For zero order joint moments and distribution functions we have

$$\boldsymbol{\Theta}_0^{\alpha,\beta}(k,l) = \boldsymbol{\pi}_{kl}\left(\boldsymbol{\phi}_k^{\alpha} \otimes \boldsymbol{\phi}_l^{\beta}\right) \tag{17}$$

where $\boldsymbol{\phi}_l^{\alpha}$ equals $\alpha! \boldsymbol{M}_l^{\alpha} \mathbb{1}_{n_l}$ or $\int_0^{\alpha} e^{\tau \boldsymbol{D}_0} \boldsymbol{d}_l d\tau$ and $\boldsymbol{\Theta}_0^{\alpha,\beta}(k,l)$ equals $\boldsymbol{J}_0^{\alpha,\beta}(k,l)$ or $\boldsymbol{F}_0^{x,y}(k,l)$, $\alpha,\beta \in \mathbb{N}$ for joint moments and $\alpha,\beta \in \mathbb{R}_{>0}$ for joint densities. $\hat{\boldsymbol{\Theta}}_0^{\alpha,\beta}(k,l)$ is then the estimated value for $\boldsymbol{\Theta}_0^{\alpha,\beta}(k,l)$. Similarly we can define a common description of first order joint moments or values of the distribution function.

$$\boldsymbol{\Theta}_1^{\alpha,\beta}(k,l) = \boldsymbol{\pi}_{kl}\left(\boldsymbol{\Xi}_k^{\alpha} \otimes \boldsymbol{H}_l\right) \boldsymbol{G}_{k,l}\left(\mathbb{1}_{n_l} \otimes \boldsymbol{\phi}_l^{\beta}\right) \tag{18}$$

where $\boldsymbol{\Theta}_0^{\alpha,\beta}(k,l)$ equals $\boldsymbol{J}_1^{\alpha,\beta}(k,l)$ or $\boldsymbol{F}_1^{x,y}(k,l)$, $\hat{\boldsymbol{\Theta}}_1^{\alpha,\beta}(k,l)$ is the corresponding estimate and $\boldsymbol{\Xi}_k$ equals $\alpha! \boldsymbol{M}_k^{\alpha} \boldsymbol{H}_k$ or $\int_0^{\alpha} e^{\tau \boldsymbol{D}_k} \operatorname{diag}(\boldsymbol{d}_k) d\tau$.

## 5    Moment-Based Parameter Fitting

We consider different approaches to determine the parameters of an MDMAP based on moments, joint moments and joint values of the probability densities. For the methods we distinguish between general MDMAPs and MDMAPs of rank $R$ with product form. The approaches are based on algorithms that have been proposed for MAPs and MMAPs [9,12,19]. In all cases we start with the computation of a PHD $(\boldsymbol{\pi}_k, \boldsymbol{D}_k)$ from the observations $x_k^{(l)}$ $(l = 1, \dots, L)$. For this purpose any algorithm for parameter fitting of PHDs can be applied, the resulting PHD can be further transformed to increase the number of input and output states using equivalence transformations [8]. The corresponding approach is denoted as two-phase fitting approach [19] and sometimes becomes a three-phase approach in this paper. The computation in different phases allows one to formulate the resulting optimization problems as non-negative least squares problem with linear constraints that can be solved efficiently. Furthermore, it is a common approach used for the parameter fitting of multivariate distributions in general [5]. Most fitting methods for multivariate distributions use $\hat{\boldsymbol{J}}_0^{1,1}(k,l)$ and $\hat{\boldsymbol{J}}_1^{1,1}(k,l)$ as measures to be matched by the multivariate distribution which is often a multivariate normal distribution. This is sometimes criticized [3,4] and other measures like the joint tail behavior of two components are considered. In the following approaches measures such as $\hat{\boldsymbol{J}}_0^{m,n}(k,l)$, $\hat{\boldsymbol{F}}_0^{x,y}(k,l)$ (i.e., $\hat{\boldsymbol{\Theta}}_0^{\alpha,\beta}(k,l)$) and $\hat{\boldsymbol{J}}_1^{m,n}(k,l)$, $\hat{\boldsymbol{F}}_1^{x,y}(k,l)$ (i.e., $\hat{\boldsymbol{\Theta}}_1^{\alpha,\beta}(k,l)$) are incorporated in the fitting process. We do not consider dependencies of lags larger than 1 like $\hat{\boldsymbol{J}}_p^{1,1}(k,l)$ $(p > 1)$ which are used in VARTA processes [6,20].

## 5.1   Dependencies in a Single Component

For a single component the zero and first order quantities are given by

$$\Theta_0^\alpha(k,k) = \pi_k \phi_k^\alpha \text{ and } \Theta_1^{\alpha,\beta}(k,k) = \xi_k^\alpha G_k \phi_k^\beta \tag{19}$$

where $\xi_k^\alpha = \pi_k \Xi_k^\alpha$. Observe that $\Theta_0^\alpha(k,k)$ has only a single parameter $\alpha$ and is completely determined by the PHD $(\pi_k, D_k)$. To describe $\Theta_1^{\alpha,\beta}(k,k)$ we expand the PHD into a MAP $(D_k, G_k)$ (see [12,19]). Now assume that we have $H_k$ estimates $\hat{\Theta}_1^{\alpha_i,\beta_i}(k,k)$ which should be approximated by $(D_k, G_k)$. Computation of matrix $G_k$ results then in the following *Non-Negative Least Squares Problem with Linear Constraints* (NNLSPLC) [24].

$$\min_{G_k \geq 0} \left( \sum_{i=1}^{H_k} \mu_i \left( \hat{\Theta}_1^{\alpha_i,\beta_i}(k,k) - \xi_k^{\alpha_i} G_k \phi_k^{\beta_i} \right)^2 \right) \tag{20}$$
$$\text{subject to } G_k \mathbf{1}_{n_k}^{inp} = \mathbf{1}_{n_k}^{outp}, \psi_k G_k = \pi_k$$

where $\mu_i$ are non-negative weights for the different joint moments/densities. The problem has $n_k^c n_k^r$ variables and $n_k^r + n_k^c$ constraints. The number of non-zero elements in $G_k$ is at most $n_k^r n_k^c$ but often the optimal solution describes a corner case with less non-zero elements. In some situations, it is better to have some more non-zero elements to allow more flexibility for following optimization steps. This can be achieved by adding a penalty term $\lambda \|G_k\|_2$ to the objective function. In this case a matrix with more and smaller non-zero elements results in a smaller two norm. This step can be applied in all NNSPLCs we present in the following paragraphs.

## 5.2   Joint Moment Fitting for General MDMAPs

For general MDMAPs we put no restriction on matrix $G$ which means that $n_{1:K}^c n_{1:K}^r$ variables are available. However, some constraints exist. First, for the row sums $G(i_{1:K}\bullet)\mathbf{1} = 1$ for $i_{1:K} \in out_{1:K}$ and 0 otherwise has to hold. Furthermore, $G$ determines $\pi$ because (6) has to hold for given matrices $H_k$. Additionally, the availability of the distributions $(\pi_k, D_k)$ implies that $\pi V_k = \pi_k$ has to hold.

The parameter fitting is done in two steps. First, vector $\pi$ is determined to approximate quantities $\hat{\Theta}_0^{\alpha_i,\beta_i}(k_i, l_i)$ $(i = 1, \ldots, I_0)$. Then an appropriate matrix $G$ is determined to approximate additional values $\hat{\Theta}_1^{\alpha_i,\beta_i}(k_i, l_i)$ $(i = 1, \ldots, I_1)$.

We begin with the computation of $\pi$ from the zero order quantities. Let $u^{(i)} = V_{kl} \left( \phi_{k_i}^{\alpha_i} \otimes \phi_{l_i}^{\beta_i} \right)$. E.g., if all first joint moments $J_0^{1,1}(k,l)$ are considered, then $(K-1)K/2$ vectors $V_{kl} \left( m_k^1 \otimes m_l^1 \right)$ $(k < l)$ are used. Then $\Theta_0^{\alpha_i,\beta_i} = \pi u^{(i)}$. With these notations we can set up the following NNLSPLC.

$$\min_{\pi \geq 0} \left( \sum_{i=1}^{I_0} \mu_i \left( \hat{\Theta}_0^{\alpha_i,\beta_i}(k_i, l_i) - \pi u^{(i)} \right)^2 \right) \tag{21}$$
$$\text{subject to } \pi \mathbf{1} = 1, \pi \geq 0 \text{ and } \pi V_k = \pi_k \text{ for all } k = 1, \ldots, K$$

Again $\mu_i$ are appropriate non-negative weights. If the minimum of the objective function becomes 0, then all joint moments and conditional values of the distribution function are matched exactly. The result is a set of PHDs coupled via initial vector $\boldsymbol{\pi}$ that generate random vectors. Vector $\boldsymbol{\pi}$ contains $n_{1:K}$ elements of which are at most $n_{1:K}^r$ are non-zero.

To match estimated values $\hat{\boldsymbol{\Theta}}_1^{\alpha_i,\beta_i}(k_i, l_i)$ $(i = 1, \ldots, I_1)$, we assume that the vector $\boldsymbol{\pi}$ is available (e.g. from (21)). This implies that $\boldsymbol{\psi} = \boldsymbol{\pi} \otimes_{k=1}^K \boldsymbol{H}_k$ is also available. The optimization problem for general matrices $\boldsymbol{G}$ results in the following NNLSPLC.

$$\min_{\boldsymbol{G} \geq \boldsymbol{0}} \left( \sum_{i=1}^{I_1} \mu_i \left( \hat{\boldsymbol{\Theta}}_1^{\alpha_i,\beta_i}(k_i, l_i) - \boldsymbol{w}^{(i)} \boldsymbol{G} \boldsymbol{v}^{(i)} \right)^2 \right) \tag{22}$$
$$\text{subject to } \boldsymbol{G} \mathbb{1}_{n_{1:k}}^{inp} = \mathbb{1}_{n_{1:K}}^{outp} \text{ and } \boldsymbol{\psi} \boldsymbol{G} = \boldsymbol{\pi}$$

where

$$\boldsymbol{w}^{(i)} = \boldsymbol{\pi} \left( \boldsymbol{I}_{n_{1:k_i-1}} \otimes \boldsymbol{\Xi}_{k_i}^{\alpha_i} \otimes \boldsymbol{I}_{n_{k_i+1:K}} \right) \prod_{j=1, j \neq k_i}^{K} \left( \boldsymbol{I}_{n_{1:j-1}} \otimes \boldsymbol{H}_j \otimes \boldsymbol{I}_{n_{j+1:K}} \right)$$
$$\boldsymbol{v}^{(i)} = \mathbb{1}_{n_{1:l_i-1}} \otimes \boldsymbol{\phi}_{l_i}^{\beta_i} \otimes \mathbb{1}_{n_{l_i+1:K}}$$

The problem contains $n_{1:K}^c n_{1:K}^r$ non-zero variables, after removing zero elements from $\boldsymbol{G}$, but has relatively simple equality constraints.

## 5.3  Joint Moment Fitting for MDMAPs of Rank $R$ with Product Form

We begin with the generation of product form MDMAPs of rank $R$. As long as we consider the approximation of quantities $\hat{\boldsymbol{\Theta}}_0^{\alpha,\beta}(k, l)$ Theorem 2 applies and allows us to compute the distribution $\boldsymbol{\pi}_{kl}$ from the matrices for components $k$ and $l$, independently of the remaining components. Unfortunately, the joint computation of the matrices $\boldsymbol{G}_k^{(r)}$ and $\boldsymbol{G}_l^{(r)}$ results in a non-linear optimization problem which is hard to solve. To keep the optimization manageable, we use *Alternating Least Squares* (ALS) [21] which is a common approach applied in many areas including the solution of partial differential equations [16,17] or performance models [10]. The basic idea of the approach is fairly simple. It is assumed that matrices $\boldsymbol{G}_l^{(r)}$ $(l \in \{1, \ldots, K\} \setminus \{k\}, r = 1, \ldots, R)$ are known when matrices $\boldsymbol{G}_k^{(r)}$ are computed. Then new matrices are computed for $k = 1, \ldots, K$ and the iteration is repeated until convergence is observed. Some results about local convergence of the approach exist [29] and also hold in our setting, but will not be further analyzed.

To start with the computation we assume that initial matrices $\boldsymbol{G}_k^{(r)}$ are available. Matrices $\boldsymbol{G}_k$ result from the solution of (20) or are initialized as $\mathbb{1}_{n_k}^{outp} \boldsymbol{\pi}_k$. Then a random distribution $(\lambda^{(1)}, \ldots, \lambda^{(R)})$ with $0 < \lambda^{(r)} < 1$ and $\sum_{r=1}^R \lambda^{(r)} = 1$ is generated. The result is an MDMAP of rank $R$ with product form but different components are uncorrelated.

To introduce correlation between two components $k$ and $l$, we consider quantities $\hat{\Theta}_0^{\alpha_i,\beta_i}(k,l)$ $(i = 1, \ldots, I_0^{kl})$ which results in the following NNLSPLC.

$$\min_{\pi_{kl} \geq 0} \left( \sum_{i=1}^{I_0^{kl}} \mu_i \left( \hat{\Theta}_0^{\alpha_i,\beta_i}(k,l) - \pi_{kl} \left( \phi_k^{\alpha_i} \otimes \phi_l^{\beta_i} \right) \right)^2 \right) \tag{23}$$
$$\text{subject to } \pi_{kl}\mathbf{1} = 1, \pi_{kl}\left(\mathbb{1}_{n_k} \otimes I_{n_l}\right) = \pi_l \text{ and } \pi_{kl}\left(I_{n_k} \otimes \mathbb{1}_{n_k}\right) = \pi_l$$

Up to $K(K-1)/2$ NNLSPLCs of the above type have to be solved. From the resulting vectors $\pi_{kl}$ the vectors $\psi_{kl} = \pi_{kl}\left(H_k \otimes H_l\right)$ can be computed.

In the next step matrices $G_k^{(r)}$ have to be found such that the vectors computed in (23) are the embedded stationary vectors of the two components. Due to the product form it is sufficient to consider only the relation between two components if we restrict dependencies to joint moments or densities between two components. Let $\bar{G}_k^{(r)} = \lambda^{(r)}G_k^{(r)}$. If we consider the local optimization problem, where matrices $\bar{G}_k^{(r)}$ are unknown and matrices $G_l^{(r)}$ $(l \neq k)$ are known, we have to find matrices such that

$$\psi_{kl}\left(\sum_{r=1}^R \bar{G}_k^{(r)} \otimes G_l^{(r)}\right) = \pi_{kl} \ (k < l) \text{ and } \psi_{lk}\left(\sum_{r=1}^R G_l^{(r)} \otimes \bar{G}_k^{(r)}\right) = \pi_{kl} \ (k > l) \tag{24}$$

This can be describes in the following NNLSPLC.

$$\min_{\bar{G}_k^{(1)},\ldots,\bar{G}_k^{(R)},\lambda^{(1)},\ldots,\lambda^{(R)} \geq 0} \left( \sum_{l=1}^{k-1} \left\| \pi_{lk} - \psi_{lk} \sum_{r=1}^R \left(G_l^{(r)} \otimes \bar{G}_k^{(r)}\right) \right\|_2^2 \right.$$
$$+ \sum_{l=k+1}^K \left\| \pi_{kl} - \psi_{kl} \sum_{r=1}^R \left(\bar{G}_k^{(r)} \otimes G_l^{(r)}\right) \right\|_2^2$$
$$\left. + \sum_{(h,l),h<l,h,l\neq k} \left\| \pi_{hl} - \psi_{hl} \sum_{r=1}^R \lambda^{(r)} \left(G_h^{(r)} \otimes G_l^{(r)}\right) \right\|_2^2 \right)$$
$$\text{subject to } \sum_{r=1}^R \bar{G}_k^{(r)}\mathbf{1} = \mathbb{1}, \psi_k \sum_{r=1}^R \bar{G}_k^{(r)} = \pi_k, \text{ for all } i : \bar{G}_k^{(r)}(i\bullet)\mathbf{1} = \lambda^{(r)} \tag{25}$$

If matrices $G_k$ are available from (20), then the second set of constraints can be substituted by $\sum_{r=1}^R G_k^{(r)} = G_k$. In this case, a solution assures that values $J_1^{m,n}(k,k)$ are kept by the resulting MDMAP. The optimization problem is solved for $k = 1, \ldots, K$ and this process is iterated until the objective function becomes 0 for all components or no progress is made any more. Observe that a solution of (25) cannot increase the overall error defined as

$$\sum_{k=1}^K \sum_{l=k+1}^K \left\| \pi_{kl} - \psi_{kl} \left(\sum_{r=1}^R \lambda^{(r)} \left(G_k^{(r)} \otimes G_l^{(r)}\right)\right) \right\|_2^2. \tag{26}$$

If the global error cannot be reduced to 0, then vectors $\pi_{kl}$ have to be computed for the resulting MDMAP from which the joint moments and joint densities can be recomputed.

# 6    Examples

In the following we consider different examples for MDMAPs. First, random vectors where the components of one vector are correlated and subsequent vectors are independent are considered, then random vectors which with correlation between components of one vector and of subsequent vectors are analyzed.

## 6.1    Independent Random Vectors

We begin with random vectors with correlated components. A first simple example are two correlated exponential distributions. We consider exponential distributions with rate 1 and correlation coefficient $R_1(1,2) = R_1(2,1) = 0.5$. To build correlated exponential distributions, the following representation as PHD with $n$ phases is used [7].

$$\pi = \left(n^{-1}, \ldots, n^{-1}\right) \quad D = \begin{pmatrix} -1 & 1 & & \\ & -2 & 2 & \\ & & \ddots & \ddots \\ & & & -n \end{pmatrix} \qquad (27)$$

To obtain a coefficient of correlation of 0.5 at least 5 phases are needed. Observe that in the representation (27) the expected time to absorption is decreasing when entering the distribution at a state with larger index. To obtain a positive correlation if two distributions are coupled, both distributions have to start with a higher probability in the same state. In an MDMAP with two exponential PHDs of order 5, the joint initial vector has 25 entries. Let $\pi(i,j)$ be the probability that the MDMAP starts in phase $i$ of the first and phase $j$ of the second MDMAP. For independent PHDs the probability is $n^{-2}$ for each state, the coefficient of correlation is 0 in this case. By solving (21) we obtain an initial vector with only 7 non-zero entries, namely $\pi(1,1) = \pi(5,5) = 0.1902$, $\pi(2,2) = \pi(3,3) = \pi(4,4) = 0.2$ and $\pi(1,5) = \pi(5,1) = 0.00998$. The resulting MDMAP describes two exponential distribution with rates 1 and correlation coefficient 0.5.

In exactly the same way random vectors with several correlated exponentially distributed components can be generated. It should be mentioned that even in simulation the generation of high dimensional random vectors of correlated exponential distributions is non-trivial. We applied the method from [6,14] which transforms correlated standard normal distributed random vectors into exponential distributions. We consider the case of three correlated exponential distributions all with mean 1 and the following two correlation matrices.

$$R_0 = \begin{pmatrix} 1 & 0.5 & 0.5 \\ 0.5 & 1 & 0.5 \\ 0.5 & 0.5 & 1 \end{pmatrix} \text{ and } R_0' = \begin{pmatrix} 1 & 0.5 & 0.1 \\ 0.5 & 1 & 0.3 \\ 0.1 & 0.3 & 1 \end{pmatrix}$$

Again we use the above representation with 5 states for the exponential distribution. Thus, the joint state space contains 125 states. For the first correlation matrix the algorithm generates an initial vector with 64 non-zero entries that exactly matches the correlation structure. For the second correlation matrix the function *lsqlin* of *octave* or *matlab* generates an MDMAP with the following correlation matrix.

$$\hat{R}_0 = \begin{pmatrix} 1.00000 & 0.4965 & 0.1011 \\ 0.4965 & 1.00000 & 0.2987 \\ 0.1011 & 0.2987 & 1.00000 \end{pmatrix}$$

This is very near to the required correlation but not exactly the same. Interestingly in the resulting initial vector only 14 of the 125 entries are non-zero (if we set values smaller than 1.0e−8 to zero).

If we use a product form approximation, then a rank of 4 is required to approximate matrix $R_0$ with a relative error of less than 1%. The correlation described by $R_0'$ could not be approximated with a small approximation error using a product form representation.

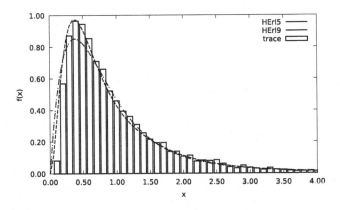

**Fig. 1.** Densities of the log-normal and the fitted Hyper-Erlang distributions.

As a second example we consider log-normal distribution with means and standard deviation 1. In a first step 10,000 samples are drawn from the distribution and are used to fit a Hype-Erlang distribution using the software *gfit* [32]. Figure 1 show the empirical density of the trace and the densities of hyper-Erlang distributions with 5 and 9 states. It can be noticed that both Hyper-Erlang distributions provide a good matching of the empirical density. The 5 state Hype-Erlang distribution consists of 3 Erlang branches, one with 1 phase and the other two with 2 phases. The 9 state Hyper-Erlang distribution contains 4 branches, one with 1 phase, 1 with 2 phases and 1 with 3 phases. Combining two Hyper-Erlang distributions with 5 phases allows us to express a correlation coefficient up to 0.3 whereas 9 phases allow one to model correlation coefficients up to 0.4. If we combine the Hyper-Erlang distribution with 9 phases, modeling the log-normal distribution and the PHD with 5 states representing the exponential distribution, coefficients of correlation between $-0.29$ and $0.34$ can be achieved. To obtain larger coefficient of correlation additional phases have to be added.

If we consider product from representations for the MDMAP with two 9-state hyper-Erlang distributions, then for $R_0^{1,1}(1,2) = 0.1$ and $0.2$ rank 5 representations are computed, whereas for $R_0^{1,1}(1,2) = 0.3$ a rank 3 representation is generated.

## 6.2  Random Processes

We consider again correlated exponential distributions. With the representation of (27) it is not possible to model correlations between subsequent realization because the representation has only a single output state. If we enlarge the number of phases by using a Hyper-Erlang representation where each path starting in phase $i \ (= 1, \ldots, n)$ and ending in phase $n$ is modeled by a single Erlang branch, we obtain a distribution with $n(n+1)/2$ phases, $n$ input and $n$ output states. We analyze an MDMAP with 2 distributions with $n = 5$. The correlation coefficient reachable by these distribution ranges between $-0.46$ and $0.54$.

**Table 1.** Maximal and minimal reachable correlation coefficients.

| $R_0(1,2)$ | max($R_1(1,1)$) max($R_1(1,2)$) | | min($R_1(1,1)$) min($R_1(1,2)$) | | max($R_1(1,1)$) min($R_1(1,2)$) | | min($R_1(1,1)$) max($R_1(1,2)$) | |
|---|---|---|---|---|---|---|---|---|
| | $R_1(1,1)$ | $R_1(1,2)$ | $R_1(1,1)$ | $R_1(1,2)$ | $R_1(1,1)$ | $R_1(1,2)$ | $R_1(1,1)$ | $R_1(1,2)$ |
| $-0.40$ | 0.17 | 0.17 | $-0.17$ | $-0.17$ | 0.53 | $-0.44$ | $-0.44$ | 0.53 |
| $-0.30$ | 0.23 | 0.23 | $-0.24$ | $-0.24$ | 0.50 | $-0.40$ | $-0.40$ | 0.50 |
| $-0.20$ | 0.28 | 0.28 | $-0.28$ | $-0.28$ | 0.46 | $-0.38$ | $-0.38$ | 0.46 |
| $-0.10$ | 0.32 | 0.32 | $-0.31$ | $-0.31$ | 0.43 | $-0.34$ | $-0.34$ | 0.43 |
| $0.00$ | 0.36 | 0.36 | $-0.34$ | $-0.34$ | 0.40 | $-0.30$ | $-0.30$ | 0.40 |
| $0.10$ | 0.40 | 0.40 | $-0.36$ | $-0.36$ | 0.36 | $-0.26$ | $-0.26$ | 0.36 |
| $0.20$ | 0.43 | 0.43 | $-0.38$ | $-0.38$ | 0.32 | $-0.22$ | $-0.22$ | 0.32 |
| $0.30$ | 0.47 | 0.47 | $-0.40$ | $-0.40$ | 0.27 | $-0.17$ | $-0.17$ | 0.27 |
| $0.40$ | 0.49 | 0.49 | $-0.42$ | $-0.42$ | 0.21 | $-0.11$ | $-0.11$ | 0.21 |
| $0.54$ | 0.54 | 0.54 | $-0.46$ | $-0.46$ | 0.05 | 0.04 | 0.04 | 0.05 |

To analyze the flexibility of the representation, we first select some value for $R_0(1,2)$ and compute the corresponding vector $\pi$ by solving (21). Then we try to maximize/minimize $R_1(1,1)$ and $R_1(1,2)$. If we maximize/minimize only one of the two values, then independently of $R_0(1,2)$ the values for $R_1(1,1)$ and $R_1(1,2)$ can range between $-0.45$ and $0.54$, the maximum range reachable by this distribution. If we try to jointly maximize/minimize $R_1(1,1)$ and $R_1(1,2)$, then the range shrinks and depends on $R_0(1,2)$. Results are shown in Table 1 and indicate that there is still a lot flexibility in the representation. The solution of the NNLSPLC problems for this examples requires with *matlab* on a standard PC less than a second.

Finally we consider an MDMAP with three PHDs of the mentioned type. The joint state space of this process contains $15^3$ states, 125 input and 125 output states. We define the following two matrices for the correlation of lag 0 and 1

$$\boldsymbol{R}_0 = \begin{pmatrix} 1 & 0.5 & 0.1 \\ 0.5 & 1 & 0.3 \\ 0.1 & 0.3 & 1 \end{pmatrix} \text{ and } \boldsymbol{R}_1 = \begin{pmatrix} 0.3 & 0.2 & 0.1 \\ 0.2 & 0.3 & 0.2 \\ 0.1 & 0.2 & 0.3 \end{pmatrix}$$

In a first step the initial vector is computed and then matrix $\boldsymbol{G}$ is determined resulting in an MDMAP with the following matrices $\hat{\boldsymbol{R}}_0$ (which is already shown above) and $\hat{\boldsymbol{R}}_1$.

$$\hat{\boldsymbol{R}}_0 = \begin{pmatrix} 1.00000 & 0.4965 & 0.1011 \\ 0.4965 & 1.00000 & 0.2987 \\ 0.1011 & 0.2987 & 1.00000 \end{pmatrix} \text{ and } \hat{\boldsymbol{R}}_1 = \begin{pmatrix} 0.2636 & 0.2341 & 0.0912 \\ 0.2341 & 0.2692 & 0.2072 \\ 0.0912 & 0.2072 & 0.2987 \end{pmatrix}$$

It can be seen that the correlation structure is approximated with small approximation errors. The computation of the initial vector requires negligible time, whereas the solution of the second NNLSPLC problem to determine matrix $\boldsymbol{G}$ requires about an hour of CPU time.

## 7   Conclusion

In this paper we present MDMAPs, a Markov model for random vectors that may be correlated in different dimensions and extends *Phase Type Distributions* and *Markovian Arrival Processes* to the multi-dimensional case. It is shown how MDMAPs can be analyzed and algorithms are presented to fit the parameters of MDMAPs according to joint moments or some values of the conditional distribution function. The proposed model can be applied in input modeling for simulation models where it is an alternative for models that are based on transformed correlated normal distributions. These models usually only use the correlation coefficient to describe dependencies whereas MDMAPs can also use higher order joint moments or values of the conditional distribution function which introduces additional flexibility when real data has to be represented by a stochastic model. Since MDMAPs are Markov models they can be analyzed numerically and can also be used as a stochastic model for correlated failures in dependability models or to represent correlated arrivals and service times in queues with PHD arrivals and services as in [11].

We currently have a first prototype matlab implementation of the algorithms proposed in the paper. This representation will be further improved and then made publically available in the tool ProFiDo [2]. Apart from parameter fitting with respect to moments and joint moments also an EM algorithm for MDMAPs will be considered in future research.

## References

1. Assaf, D., Langberg, N.A., Savits, T.H., Shaked, M.: Multivariate phase-type distributions. Oper. Res. **32**(3), 688–702 (1984). https://doi.org/10.1287/opre.32.3.688

2. Bause, F., Buchholz, P., Kriege, J.: ProFiDo - the processes fitting toolkit dortmund. In: QEST 2010, Seventh International Conference on the Quantitative Evaluation of Systems, Williamsburg, Virginia, USA, 15–18 September 2010, pp. 87–96. IEEE Computer Society (2010). https://doi.org/10.1109/QEST.2010.20

3. Biller, B.: Copula-based multivariate input models for stochastic simulation. Oper. Res. **57**(4), 878–892 (2009). https://doi.org/10.1287/opre.1080.0669

4. Biller, B.: Copula-based multivariate input modeling. Surv. Oper. Res. Manag. Sci. **17**, 69–84 (2012)

5. Biller, B., Ghosh, S.: Multivariate input processes. In: Henderson, S.G., Nelson, B.L. (eds.) Handbook of OR & MS, vol. 13, pp. 123–152. Elsevier, Amsterdam (2006)

6. Biller, B., Nelson, B.L.: Modeling and generating multivariate time-series input processes using a vector autoregressive technique. ACM Trans. Model. Comput. Simul. **13**(3), 211–237 (2003). https://doi.org/10.1145/937332.937333

7. Bladt, M., Nielsen, B.F.: On the construction of bivariate exponential distributions with an arbitrary correlation coefficient. Stoch. Model. **26**(3), 295–308 (2010)

8. Buchholz, P., Felko, I., Kriege, J.: Transformation of acyclic phase type distributions for correlation fitting. In: Dudin, A., De Turck, K. (eds.) ASMTA 2013. LNCS, vol. 7984, pp. 96–111. Springer, Heidelberg (2013). https://doi.org/10.1007/978-3-642-39408-9_8

9. Buchholz, P., Kemper, P., Kriege, J.: Multi-class Markovian arrival processes and their parameter fitting. Perform. Eval. **67**(11), 1092–1106 (2010)

10. Buchholz, P., Kriege, J.: Approximate aggregation of Markovian models using alternating least squares. Perform. Eval. **73**, 73–90 (2014). https://doi.org/10.1016/j.peva.2013.09.001

11. Buchholz, P., Kriege, J.: Fitting correlated arrival and service times and related queueing performance. Queueing Syst. **85**(3–4), 337–359 (2017). https://doi.org/10.1007/s11134-017-9514-5

12. Buchholz, P., Kriege, J., Felko, I.: Input Modeling with Phase-Type Distributions and Markov Models - Theory and Applications. Springer, Cham (2014). https://doi.org/10.1007/978-3-319-06674-5

13. Buchholz, P., Telek, M.: Rational processes related to communicating Markov processes. J. Appl. Probab. **49**(1), 40–59 (2012). https://doi.org/10.1017/S0021900200008858

14. Cario, M.C., Nelson, B.L.: Numerical methods for fitting and simulating autoregressive-to-anything processes. INFORMS J. Comput. **10**(1), 72–81 (1998). https://doi.org/10.1287/ijoc.10.1.72

15. D'Alconzo, A., Drago, I., Morichetta, A., Mellia, M., Casas, P.: A survey on big data for network traffic monitoring and analysis. IEEE Trans. Netw. Serv. Manag. **16**(3), 800–813 (2019). https://doi.org/10.1109/TNSM.2019.2933358

16. Dolgov, S., Savostyanov, D.V.: Alternating minimal energy methods for linear systems in higher dimensions. SIAM J. Sci. Comput. **36**(5) (2014). https://doi.org/10.1137/140953289

17. Holtz, S., Rohwedder, T., Schneider, R.: The alternating linear scheme for tensor optimization in the tensor train format. SIAM J. Sci. Comput. **34**(2) (2012). https://doi.org/10.1137/100818893

18. Hong, Y., Zhang, M., Meeker, W.Q.: Big data and reliability applications: the complexity dimension. J. Qual. Technol. **50**(2), 135–149 (2018)

19. Horváth, G., Telek, M., Buchholz, P.: A MAP fitting approach with independent approximation of the inter-arrival time distribution and the lag-correlation. In: QEST, pp. 124–133. IEEE CS Press (2005)

20. Kriege, J., Buchholz, P.: Traffic modeling with phase-type distributions and VARMA processes. In: Agha, G., Van Houdt, B. (eds.) QEST 2016. LNCS, vol. 9826, pp. 295–310. Springer, Cham (2016). https://doi.org/10.1007/978-3-319-43425-4_20

21. Kroonenberg, P.M., Leeuw, J.D.: Principal component analysis of three-mode data by means of alternating least squares algorithms. Psychometrika **45**, 69–97 (1980). https://doi.org/10.1007/BF02293599

22. Kulkarni, V.G.: A new class of multivariate phase type distributions. Oper. Res. **37**(1), 151–158 (1989). https://doi.org/10.1287/opre.37.1.151

23. Law, A.M.: Simulation Modeling and Analysis, 5th edn. Mc Graw Hill, New York (2013)

24. Lawson, C.L., Hanson, R.J.: Solving Least Squares Problems. Classics in Applied Mathematics. SIAM (1995)

25. Lucantoni, D.M.: New results on the single server queue with a batch Markovian arrival process. Stoch. Model. **7**(1), 1–46 (1991)

26. Mishra, D., Gunasekaran, A., Papadopoulos, T., Childe, S.J.: Big data and supply chain management: a review and bibliometric analysis. Ann. OR **270**(1–2), 313–336 (2018). https://doi.org/10.1007/s10479-016-2236-y

27. Neuts, M.F.: A versatile Markovian point process. J. Appl. Probab. **16**, 764–779 (1979)

28. Neuts, M.F.: Matrix-Geometric Solutions in Stochastic Models - An Algorithmic Approach. Johns Hopkins University Press, Baltimore (1981)

29. Rohwedder, T., Uschmajew, A.: On local convergence of alternating schemes for optimization of convex problems in the tensor train format. SIAM J. Numer. Anal. **51**(2), 1134–1162 (2013). https://doi.org/10.1137/110857520

30. Stewart, W.J.: Probability, Markov Chains, Queues, and Simulation: The Mathematical Basis of Performance Modeling. Princeton University Press, Princeton (2009)

31. Telek, M., Horváth, G.: A minimal representation of Markov arrival processes and a moments matching method. Perform. Eval. **64**(9–12), 1153–1168 (2007)

32. Thümmler, A., Buchholz, P., Telek, M.: A novel approach for phase-type fitting with the EM algorithm. IEEE Trans. Dependable Secur. Comput. **3**(3), 245–258 (2006)

33. Trivedi, K.S.: Probability and Statistics with Reliability, Queuing and Computer Science Applications, 2 edn. Wiley (2016). https://doi.org/10.1002/9781119285441

# Automatic Pre- and Postconditions
# for Partial Differential Equations

Michele Boreale[(✉)]

Dipartimento di Statistica, Informatica, Applicazioni (DiSIA) "G. Parenti",
Università di Firenze, Viale Morgagni 65, 50124 Florence, Italy
michele.boreale@unifi.it
https://local.disia.unifi.it/boreale/

**Abstract.** Based on a simple automata-theoretic and algebraic framework, we study equational reasoning for Initial Value Problems (IVPs) of
polynomial Partial Differential Equations (PDEs). In order to represent
IVPs in their full generality, we introduce *stratified* systems, where function definitions can be decomposed into distinct subsystems, focusing
on different subsets of independent variables. Under a certain coherence
condition, for such stratified systems we prove existence and uniqueness
of formal power series solutions, which conservatively extend the classical
analytic ones. We then give a—in a precise sense, complete—algorithm
to compute weakest preconditions and strongest postconditions for such
systems. To some extent, this result reduces equational reasoning on
PDE initial value (and boundary) problems to algebraic reasoning. We
illustrate some experiments conducted with a proof-of-concept implementation of the method.

## 1 Introduction

Techniques for reasoning on ordinary differential equations (ODEs) are at the
heart of current formal methods and tools for continuous and hybrid systems,
which form an active research area, see e.g. [1–7] and references therein. Although
examples of hybrid systems whose continuous dynamics is described by *partial
differential equations* (PDEs) abound, formal techniques for reasoning on PDEs
have comparably received much less attention. Existing proposals mostly focus
on specific types of equations, such as the Hamilton-Jacobi equations [8,9]. The
present paper, building on [10], is meant as a contribution to developing formal
methods for reasoning on PDEs. Our approach is *formal*, in the sense of being
entirely based on simple coalgebra (automata theory) and algebra (polynomials),
rather than on calculus like most of the previous proposals. Nevertheless, the
resulting notion of PDE solution can be used to reason on the classical analytic
one, in a sense made precise below.

In [10] we have shown that, subject to a certain coherence condition, a system
$\Sigma$ of polynomial PDEs, given an arbitrary initial data specification, admits a
unique solution in the set of commutative formal power series (CFPSs; Sect. 2).

© Springer Nature Switzerland AG 2020
M. Gribaudo et al. (Eds.): QEST 2020, LNCS 12289, pp. 193–210, 2020.
https://doi.org/10.1007/978-3-030-59854-9_15

Most important, this solution can be expressed operationally, in terms of the transition function of a suitable automaton. This lays the basis for mechanical checking of equations: that is, check that a given (polynomial) expression involving the PDE variables becomes identically 0 when the solution is plugged into it. The corresponding procedure is similar in spirit to an on-the-fly bisimulation checking algorithm. Pragmatically, these CFPS solutions conservatively extend classical ones: if an analytic solution of $\Sigma$ in the classical sense exists, then its Taylor expansion from 0, seen as a formal power series, coincides with the unique CFPS solution.

In the present paper, we make two substantial steps forward. First, we introduce *stratified systems*, by which one can represent fairly complicated initial value problems—and, through changes of coordinates, also boundary problems. Second and most crucial, we give a (relatively) complete algorithm to automatically compute *pre-* and *postconditions* of a given system. In particular, this allows one to automatically compute *all* valid polynomial equations that fit a user-specified format (e.g., all conservation laws up to a given degree), rather than just checking the validity of given ones.

More in detail, in a stratified system we have distinct sets of equations (subsystems): in each of them, a distinct subset of the independent variables is fixed to zero. This way, in a system with, say, two independent variables $x$ and $y$, the solution, $f(x,y)$, can be made dependent on constraints involving not only $f(x,y)$ and its derivatives, but also $f(x,0)$ and its $x$-derivatives, and $f(0,y)$ and its $y$-derivatives. This is how initial value problems are formulated in their generality. Under a syntactic acyclicity condition among subsystems, we prove existence and uniqueness of solutions for stratified systems and an automata-theoretic representation of the corresponding Taylor coefficients (Sect. 3).

This result lays the basis of an algorithm to automatically compute both weakest *preconditions* (= sets of initial data specifications) and strongest *postconditions* (= valid polynomial equations). The method is complete, subject to certain assumptions (Sect. 4). This way one can, for example, automatically *discover* all polynomial equations up to a given degree, valid under a given set of initial data specifications. Or vice-versa, compute the largest set of initial data specifications for given equations to be valid. The original IVP is therefore reduced to a purely algebraic system, which can be used for equational reasoning and, in some cases, to find explicit solutions. Concepts from algebraic geometry are used to prove the termination and correctness of this algorithm. Using a proof-of-concept implementation (Sect. 5), we illustrate this algorithm on well-known examples drawn from mathematical physics. Relations with other works, in particular on ODEs [11,12], is discussed in the concluding section (Sect. 6). Proofs and additional technical material omitted here are available in a full version available online [13].

## 2   Background

We review some notation and terminology from the theory of formal power series and from the formal theory of PDEs, including the main result of [10].

*Commutative Formal Power Series and Polynomials.* Assume a finite set $X = \{x_1, ..., x_n\}$ of *independent variables* is given. The set $X$, ranged over by $t, x, ...,$ will be kept fixed for the rest of the paper. Let $X^\otimes$, ranged over by $\tau, \xi, ...,$ be the set of *monomials*[1] that can be formed from the elements of $X$, in other words, the commutative monoid freely generated by $X$. Let us fix any total order $\mathbf{x} = (x_1, ..., x_n)$ of the variables in $X$. Given a vector $\boldsymbol{\alpha} = (\alpha_1, ..., \alpha_n)$ of nonnegative integers (a *multi-index*), we let $\mathbf{x}^\alpha$ denote the monomial $x_1^{\alpha_1} \cdots x_n^{\alpha_n}$. For $\xi = \mathbf{x}^\alpha$ and $\tau = \mathbf{x}^\beta$, we let $\xi \leq \tau$ if for each $i = 1, ..., n$, $\alpha_i \leq \beta_i$. A *commutative formal power series* (CFPS) with indeterminates in $X$ and coefficients in $\mathbb{R}$ is a total function $f : X^\otimes \to \mathbb{R}$. The set of such CFPSs will be denoted by $\mathbb{R}[\![X]\!]$. We will sometimes use the suggestive notation $\sum_{\alpha \in \mathbb{N}^n} f(\mathbf{x}^\alpha) \cdot \mathbf{x}^\alpha$ to denote a CFPS $f$. By slight abuse of notation, for each $\mu \in \mathbb{R}$, we will denote the CFPS that maps $\epsilon$ to $\mu$ and anything else to 0 simply as $\mu$; while $x_i$ will denote the $i$-th identity, the CFPS that maps $x_i$ to 1 and anything else to 0. The definitions the sum $f + g$, (convolution) product $f \cdot g$, inverse $f^{-1}$ (if $f(\epsilon) \neq 0$) and partial derivative $\frac{\partial f}{\partial x}$ operations on CFPS are standard, and enjoy the usual algebraic properties (we also review these operations in [13, App. A]). In particular sum and product make $\mathbb{R}[\![X]\!]$ a ring with 0 and 1 as identities.

If the *support* of $f$, $\text{supp}(f) \stackrel{\triangle}{=} \{\tau : f(\tau) \neq 0\}$, is finite, we will call $f$ a *polynomial*. The set of polynomials, denoted by $\mathbb{R}[X]$, is closed under the above defined operations of sum, product (which make it a ring) and partial derivative, but in general not inverse. It is important to note that, when confining to polynomials, sum, product and partial derivative are well defined even in case the cardinality of the set of indeterminates $X$ is infinite.

*Partial Differential Equations.* The definitions in this paragraph are standard, or slight variations of the standard ones as found in the formal theory of PDEs, cf. [10, 14, 15] and references therein. A finite, nonempty set $U$ of *dependent variables*, disjoint from $X$ and ranged over by $u, v, ...,$ is given. We let $\mathcal{D} \stackrel{\triangle}{=} \{u_\tau : u \in U, \tau \in X^\otimes\}$ be the set of the *derivatives*. Informally, a symbol $u \in U$ represents a function, and $u_\tau$ its partial derivative $\frac{\partial u}{\partial \tau}$; here $u_\epsilon$ will be identified with $u$. We let $\mathcal{P} \stackrel{\triangle}{=} \mathbb{R}[X \cup \mathcal{D}]$, ranged over by $E, F, ...,$ denote the set of *(differential) polynomials* with coefficients in $\mathbb{R}$ and indeterminates in $X \cup \mathcal{D}$. Considered as formal objects, differential polynomials are just finite-support CFPSs, as per previous paragraph. As such, they inherit the operations of sum, product and partial derivative, along with the corresponding properties. Syntactically, we shall write polynomials as expressions of the form $\sum_{\gamma \in M} \lambda_\gamma \cdot \gamma$, for $0 \neq \lambda_\gamma \in \mathbb{R}$ and $M \subseteq_{\text{fin}} (X \cup \mathcal{D})^\otimes$. Note that this notation is consistent with the sum and product operations defined on polynomials. For example, $E = v_z u_{xy} + v_y^2 + u + 5x$ is a polynomial[2]. For an independent variable $x \in X$, the

---

[1] In general, we shall adopt for monomials the same notation we use for strings, as the context is sufficient to disambiguate. In particular, we overload the symbol $\epsilon$ to denote both the empty string and the unit monomial. When $X = \emptyset$, $X^\otimes \stackrel{\triangle}{=} \{\epsilon\}$.

[2] Real arithmetic expressions will be used as a meta-notation for polynomials: e.g. $(u + u_x + 1) \cdot (x + u_y)$ denotes the polynomial $xu + uu_y + xu_x + u_x u_y + x + u_y$.

*total derivative* of $E \in \mathcal{P}$ w.r.t. $x$ is just the derivative of $E$ w.r.t $x$, taking into account that $\frac{\partial u_\tau}{\partial x} = u_{x\tau}$ and the chain rule. Formally, the operator $D_x : \mathcal{P} \to \mathcal{P}$ is defined by (note $\sum$ below has only finitely many nonzero terms)

$$D_x E \triangleq \frac{\partial E}{\partial x} + \sum_{u,\tau} u_{x\tau} \cdot \frac{\partial E}{\partial u_\tau}$$

where $\frac{\partial E}{\partial a}$ denotes the partial derivative of polynomial $E$ along $a \in X \cup \mathcal{D}$. $D_x(\cdot)$ inherits differentiation rules for sum and product that are the analog of those for partial derivatives $\partial(\cdot)/\partial x$. As an example, for the polynomial $E$ above, we have $D_x E = v_{xz} u_{xy} + v_z u_{xxy} + 2v_y v_{xy} + u_x + 5$. In particular, $D_x u_\tau = u_{x\tau}$ and $D_x x^k = k x^{k-1}$. Just as partial derivatives, total derivatives commute with each other, that is $D_x D_y F = D_y D_x F$. This suggests to extend the notation to monomials: for any monomial $\tau = x_1 \cdots x_m$, we let $D_\tau F$ be $D_{x_1} \cdots D_{x_m} F$, where the order of the derivatives is irrelevant. We formally introduce systems of PDEs below, along with the key notions of *parametric* and *principal* derivatives. Informally, parametric derivatives play a role similar to the lower order derivatives in ODEs initial value problems: just like in ODEs, once we fix their values at the origin, the solution of the system should be uniquely determined. On the other hand, equations for principal derivatives depend on the parametric ones, just like higher order derivatives in ODEs depend on the lower order ones.

**Definition 1 (system of PDEs).** *A system of PDEs is a nonempty set $\Sigma$ of equations (pairs) of the form $u_\tau = E$, with $E \in \mathcal{P}$. The set of derivatives $u_\tau$ that appear as left-hand sides of equations in $\Sigma$ is denoted by $\mathrm{dom}(\Sigma)$. Based on $\Sigma$, the set $\mathcal{D}$ is partitioned into the sets of principal and parametric derivatives, defined as follows.*

$$\mathcal{P}r(\Sigma) \triangleq \{ u_{\tau\xi} : u_\tau \in \mathrm{dom}(\Sigma) \text{ and } \xi \in X^{\otimes} \} \qquad \mathcal{P}a(\Sigma) \triangleq \mathcal{D} \setminus \mathcal{P}r(\Sigma).$$

*We let $\mathcal{P}_0(\Sigma) \triangleq \mathbb{R}[X \cup \mathcal{P}a(\Sigma)]$ be the set of $\Sigma$-normal forms.*

*Example 1 (Heat equation).* The Heat equation in one spatial dimension, $u_t(t,x) = u_{xx}(t,x)$, corresponds to $X = \{t, x\}$, $U = \{u\}$ and $\Sigma = \{u_t = u_{xx}\}$. Here we have $\mathcal{P}r(\Sigma) = \{u_{t\tau} : \tau \in X^{\otimes}\}$ and $\mathcal{P}a(\Sigma) = \{u_{x^j} : j \geq 0\}$. See Fig. 1, left.

Note that we do *not* insist that each derivative occurs at most once as left-hand side in $\Sigma$. The *infinite prolongation* of a system $\Sigma$, denoted $\Sigma^{\infty}$, is the system of PDEs of the form $u_{\xi\tau} = D_\xi F$, where $u_\tau = F$ is in $\Sigma$ and $\xi \in X^{\otimes}$. Of course, $\Sigma^{\infty} \supseteq \Sigma$. Moreover, $\Sigma$ and $\Sigma^{\infty}$ induce the *same* sets of principal and parametric derivatives.

We can now introduce the concept of *solution* of PDEs, which is based on a PDE's analog of initial value problems (IVPs). We say a function $\psi : \mathcal{P} \to \mathbb{R}[X]$ is a *homomorphism* if it is a ring homomorphism—preserves sum, product and their identities as expected—and additionally: preserves derivatives, that is

$\psi(D_x E) = \frac{\partial}{\partial x}\psi(E)$, and maps each $x_i \in X$ to the $i$-th identity CFPS. For any function $\psi : U \to \mathbb{R}[\![X]\!]$, its homomorphic extension $\mathcal{P} \to \mathbb{R}[\![X]\!]$ is defined as expected and, by slight abuse of notation, still denoted by "$\psi$". In the definition below, it is useful to bear in mind that, informally, for any $f \in \mathbb{R}[\![X]\!]$, $f(\epsilon)$ is the formal counterpart of $f(0)$, and that for each parametric derivative $u_\tau \in \mathcal{P}a(\Sigma)$, the initial data value $\rho(u_\tau)$ is the formal counterpart of $\frac{\partial u}{\partial \tau}(0)$.

**Definition 2 (initial value problem).** *Let $\Sigma$ be a system of PDEs. An* initial data specification *is a mapping $\rho : \mathcal{P}a(\Sigma) \to \mathbb{R}$. An* initial value problem *(IVP) is a pair* $\mathbf{iP} = (\Sigma, \rho)$.

*A* solution *of* $\mathbf{iP}$ *is a homomorphism $\psi : \mathcal{P} \to \mathbb{R}[\![X]\!]$ such that: (a) the initial value conditions are satisfied, that is $\psi(u_\tau)(\epsilon) = \rho(u_\tau)$ for each $u_\tau \in \mathcal{P}a(\Sigma)$; and (b) all equations are satisfied, that is $\psi(u_\tau) = \psi(F)$ for each $u_\tau = F$ in $\Sigma^\infty$.*

For $\Sigma$ to have a solution, a few syntactic conditions must be imposed, whose purpose is to avoid inconsistencies in the equational theory generated by $\Sigma$. A *ranking* is a total order $\prec$ of $\mathcal{D}$ such that: (a) $u_\tau \prec u_{x\tau}$, and (b) $u_\tau \prec v_\xi$ implies $u_{x\tau} \prec v_{x\xi}$, for each $x \in X$, $\tau, \xi \in X^\otimes$ and $u, v \in U$. Dickson's lemma [16] implies that $\mathcal{D}$ with $\prec$ is a well-order, and in particular that there is no infinite descending chain in it. The system $\Sigma$ is $\prec$-*normal* if, for each equation $u_\tau = E$ in $\Sigma$, $u_\tau \succ v_\xi$, for each $v_\xi$ appearing in $E$. An easy but important consequence of condition (b) above is that if $\Sigma$ is normal then also its prolongation $\Sigma^\infty$ is normal.

Now, consider the equational theory over $\mathcal{P}$ induced by the equations in $\Sigma^\infty$. More precisely, write $E \to_\Sigma F$ if $F$ is the polynomial that is obtained from $E$ by replacing one occurrence of $u_\tau$ with $G$, for some equation $u_\tau = G \in \Sigma^\infty$. Note, in particular, that $E \in \mathcal{P}$ cannot be rewritten if and only if $E \in \mathcal{P}_0(\Sigma)$. We let $=_\Sigma$ denote the reflexive, symmetric and transitive closure of $\to_\Sigma$. The following definition formalizes the key concepts of consistency and coherence of $\Sigma$. Basically, as shown in [10], under the natural requirement of normality, consistency is a necessary and sufficient condition for $\Sigma$ to admit a unique solution under *arbitrary* initial conditions.

**Definition 3 (coherence).** *Let $\Sigma$ be a system of PDEs.*

– *$\Sigma$ is* consistent *if for each $E \in \mathcal{P}$ there is a unique $F \in \mathcal{P}_0(\Sigma)$ such that $E =_\Sigma F$.*
– *Let $\prec$ be a ranking. A system $\Sigma$ is $\prec$-*coherent* if it is $\prec$-normal and consistent.*

As an example, the Heat equation in Example 1 is obviously consistent, as it features just one equation. Moreover, it is $\prec$-coherent w.r.t. the ranking $u_\tau \prec u_\xi$ iff $\tau \prec_{\text{lex}} \xi$, where $\prec_{\text{lex}}$ is the lexicographic monomial order induced by $t > x$. For any consistent system, we can define a *normal form function*

$$S_\Sigma : \mathcal{P} \to \mathcal{P}_0(\Sigma)$$

by letting $S_\Sigma E \overset{\triangle}{=} F$, for the unique $F \in \mathcal{P}_0(\Sigma)$ such that $E \to_\Sigma^* F$. The term $S_\Sigma E$ will be often abbreviated as $SE$, if $\Sigma$ is understood from the context.

Deciding if a (finite) system $\Sigma$ is coherent, for a suitable ranking $\prec$, is of course a nontrivial problem. Since $\prec$ is a well-order, there are no infinite sequences of rewrites $E_1 \to_\Sigma E_2 \to_\Sigma E_3 \to_\Sigma \cdots$: therefore it is possible to rewrite any $E$ into some $F \in \mathcal{P}_0(\Sigma)$ in a finite number of steps. Proving coherence reduces then to proving $\to_\Sigma$ confluent. For our purposes, it is enough to know that completing a given system of equations to make it coherent, or deciding that this is impossible, can be achieved by one of many existing computer algebra algorithms, like those in [14,15]; see the discussion and the references in [10]. In many cases arising from applications, say mathematical physics, transforming the system into a coherent form for an appropriate ranking can be accomplished manually, without much difficulty: see the examples in Sect. 5.

We can now characterize explicitly the solutions of a coherent $\Sigma$. Informally, for any fixed $\rho$, the CFPS associated with $E \in \mathcal{P}$ takes each monomial $\tau \in X^\otimes$ to the real obtained by evaluating the $\tau$-derivative of $E$ under $\rho$, once this derivative is written in normal form. Formally, the characterization is based on a transition function, $\delta_\Sigma : \mathcal{P} \times X \to \mathcal{P}_0(\Sigma)$, defined as

$$\delta_\Sigma(E, x) \triangleq S_\Sigma D_x E. \tag{1}$$

It can be shown (see [10]) that $\delta_\Sigma$ satisfies the following commutation property: $\delta_\Sigma(\delta_\Sigma(E, x), y) = \delta_\Sigma(\delta_\Sigma(E, y), x)$ for all $x, y \in X$. This justifies the notation $\delta_\Sigma(E, \tau)$ for $\tau \in X^\otimes$, with $\delta_\Sigma(E, \epsilon) \triangleq S_\Sigma E$. Next, an initial data specification $\rho : \mathcal{P}a(\Sigma) \to \mathbb{R}$ can be extended homomorphically to a function $\mathcal{P}_0(\Sigma) \to \mathbb{R}$, interpreting $+$ and $\cdot$ as the usual sum and product over $\mathbb{R}$, and letting $\rho(x) \triangleq 0$ for each independent variable $x \in X$. The following theorem of existence and uniqueness of solutions is the main result of [10]. (for the sake of completeness, the proof is also reproduced in [13, App. A]). Below, recall that for $\boldsymbol{\alpha} = (\alpha_1, ..., \alpha_n) \in \mathbb{N}^n$, $\boldsymbol{\alpha}! = \alpha_1! \cdots \alpha_n!$.

**Theorem 1 (existence and uniqueness of solution, [10]).** *Let $\Sigma$ be finite and coherent. For any initial data specification $\rho$, there is a unique solution $\phi_{\mathrm{iP}} : \mathcal{P} \to \mathbb{R}[\![X]\!]$ of the IVP $\mathrm{iP} = (\Sigma, \rho)$. Moreover, $\phi_{\mathrm{iP}}$ satisfies the following formula, for each $E \in \mathcal{P}$ and $\tau = \mathbf{x}^\alpha \in X^\otimes$.*

$$\phi_{\mathrm{iP}}(E)(\tau) = \frac{\rho(\delta_\Sigma(E, \tau))}{\boldsymbol{\alpha}!}. \tag{2}$$

We remark that our concept of solution of a PDE IVP conservatively extends the classical solution concept, in the following sense: if a classical solution exists that is analytic around the origin, then its Taylor expansion, seen as a formal power series, coincides with the CFPS solution [13, App. A].

## 3   Stratified Systems

Consider the Heat equation of Example 1. Suppose we want to specify that the temperature at time $t = 0$ varies along the $x$-line according to, say,

$u(0, x) = \exp(-x) = \sum_{j \geq 0} \frac{(-1)^j}{j!} x^j$. With the pure PDE formalism introduced so far, the only way to specify $u(0, x)$ is by explicitly giving the values of all its derivatives at the origin, via the initial data $\rho$. That is, by specifying the parametric derivatives of $u$: $\rho(u_{x^j}) = (\frac{\partial^j}{\partial x^j} u(0, x))_{|x=0} \triangleq (-1)^j$, for each $j \geq 0$. Such a $\rho$ is an infinite object which does not obviously lend itself to equational and algorithmic manipulations. It would be more natural, instead, to specify $u(0, x)$ simply via a subsystem $\Sigma_0 = \{u_x = -u\}$ (plus the single initial condition $\rho(u) = 1$), somehow prescribing that this equation applies when fixing $t = 0$, so that the resulting function only depends on $x$. More generally, a pure PDE system $\Sigma$ alone cannot express general IVPs, where one wants to specify constraints on the functions obtained by keeping the value of certain independent variables fixed. This limitation is overcome by stratified systems, introduced below.

We first introduce *subsystems*. Let us fix once and for all a nonempty set of dependent variables $U$, and a finite set of independent variables $X$. For $Y \subseteq X$, a $Y$-subsystem defines, informally, functions where variables outside $Y$ have been zeroed. In particular, derivatives can be taken only along variables in $Y$. We need now some standard notation on partial orders. For a partial order $\preceq$ defined over some universe set $A$ and for $B \subseteq A$, we will let $\uparrow_{\preceq} (B) \triangleq \{a \in A : a \succeq b$ for some $b \in B\}$ denote the upward closure of $B$ w.r.t $\preceq$; similarly, we will let $\downarrow_{\preceq} (B)$ denote the downward closure of $B$. Moreover, we will let $\min_{\preceq}(B) \triangleq \{b \in B :$ whenever $b' \in B$ and $b' \preceq b$ then $b' = b\}$ denote the set of $\preceq$-minimal element of $B$. Additionally, we define the following partial order $\leq_Y$ on the set of derivatives $\mathcal{D}$, depending on $Y \subseteq X$: $u_\tau \leq_Y u_{\tau'}$ if and only if $\tau' = \tau \xi$ for some $\xi \in Y^\otimes$. In the definition of subsystem given below, the intuition is that the $\leq_Y$-minimal derivatives, the set $U_\Gamma$, act as the dependent variables of a new system of PDEs with independent variables in $Y$ and derivatives in $\mathcal{D}_\Gamma$.

**Definition 4 (subsystem).** *Let $\Sigma$ a set of equations and $Y \subseteq X$. For $\Gamma = (\Sigma, Y)$, we define the following subsets of $\mathcal{D}$.*

$$U_\Gamma \triangleq \min_{\leq_Y}(\downarrow_{\leq_Y} \{u_\tau : u_\tau \text{ occurs in } \Sigma\}) \qquad \mathcal{D}_\Gamma \triangleq \uparrow_{\leq_Y} (U_\Gamma)$$
$$\mathcal{P}r(\Gamma) \triangleq \uparrow_{\leq_Y} (\text{dom}(\Sigma)) \qquad\qquad\qquad \mathcal{P}a(\Gamma) \triangleq \mathcal{D}_\Gamma \setminus \mathcal{P}r(\Gamma).$$

*We let $\mathcal{P}_\Gamma \triangleq \mathbb{R}[Y \cup \mathcal{D}_\Gamma]$. We say $\Gamma = (\Sigma, Y)$ is a $Y$-subsystem if $U_\Gamma$ is finite, and for each polynomial $E$ appearing in $\Sigma$, $E \in \mathcal{P}_\Gamma$. We call $\Gamma$ a main subsystem if $Y = X$ and $U_\Gamma = U$. Finally, $\Gamma^\infty \triangleq \{u_{\tau\xi} = D_\xi G : u_\tau = G \in \Sigma$ and $\xi \in Y^\otimes\}$.*

Stratified systems can encode initial value problems in their general form. A precedence relation among subsystems, $\Gamma_i \prec \Gamma_j$, formalizes that equations in $\Gamma_j$ depends on parametric variables that are defined (are principal) in $\Gamma_i$.

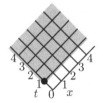
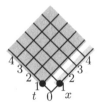

**Fig. 1.** $u$-derivatives arranged according to the partial order: $u_\tau \leq u_\xi$ iff $\tau \leq \xi$. In the Hasse diagrams, derivatives corresponds to line intersections, with elements in some $\mathrm{dom}(\Sigma)$ marked by a black dot. **Left:** system $\Sigma$ of Example 1, where dark-shaded region $= \mathcal{P}r(\Sigma)$, white region $= \mathcal{P}a(\Sigma)$. **Right:** stratified system $H = \{\Gamma_1, \Gamma_2\}$ of Example 2, where dark-shaded region $= \mathcal{P}r(\Gamma_1)$, light-shaded region $= \mathcal{P}r(\Gamma_2)$, white region $= \mathcal{P}a(H)$.

**Definition 5 (stratified system).** *A stratified system is a finite set of subsystems $H = \{\Gamma_1, ..., \Gamma_m\}$ $(m \geq 1, \Gamma_i = (\Sigma_i, X_i), \Sigma_i \neq \emptyset, X_i \subseteq X)$ such that:*

*(a) for some $1 \leq j \leq m$, $\Gamma_j$ is a main subsystem; we will conventionally take $j = 1$;*
*(b) for any $i \neq j$, $\mathcal{P}r(\Gamma_i) \cap \mathcal{P}r(\Gamma_j) = \emptyset$;*
*(c) the binary relation over $\{1, ..., m\}$ defined as $i \prec j$ iff $\mathcal{P}r(\Gamma_i) \cap \mathcal{P}a(\Gamma_j) \neq \emptyset$, is acyclic.*

*The parametric derivatives and normal forms of $H$ are $\mathcal{P}a(H) \triangleq \mathcal{D} \setminus (\cup_{i=1}^m \mathcal{P}r(\Gamma_i))$ and $\mathcal{P}_0(H) \triangleq \mathbb{R}[\mathcal{P}a(H)]$, respectively. $H$ is coherent if all of its subsystems are coherent w.r.t. one and the same ranking on $\mathcal{D}$.*

Note that each $H$ features a unique main subsystem.

*Example 2 (Heat equation with initial temperature).* Consider the Heat equation of Example 1, with an initial temperature exponentially decaying from the origin, $u_x(0, x) = -u(0, x)$. The corresponding stratified system is $H = \{\Gamma_1, \Gamma_2\} = \{(\Sigma_1, X_1), (\Sigma_2, X_2)\}$ with $\Sigma_1 = \{u_t = u_{xx}\}$, $X_1 = X = \{t, x\}$ and $\Sigma_2 = \{u_x = -u\}$, $X_2 = \{x\}$. We have (see Fig. 1, right):

$$U_{\Gamma_1} = \{u\} \quad \mathcal{D}_{\Gamma_1} = \{u_\tau : \tau \in X^\otimes\} \quad \mathcal{P}r(\Gamma_1) = \{u_{t\tau} : \tau \in X^\otimes\} \quad \mathcal{P}a(\Gamma_1) = \{u_{x^j} : j \geq 0\}$$
$$U_{\Gamma_2} = \{u\} \quad \mathcal{D}_{\Gamma_2} = \{u_{x^j} : j \geq 0\} \quad \mathcal{P}r(\Gamma_2) = \{u_{x^j} : j \geq 1\} \quad \mathcal{P}a(\Gamma_2) = \{u\}.$$

Note that $\mathcal{D}_{\Gamma_1} = \mathcal{D}$, so $\Gamma_1$ is the main subsystem, and that $\mathcal{P}a(H) = \{u\}$. Clearly, $2 \prec 1$, as $\mathcal{P}r(\Gamma_2) \cap \mathcal{P}a(\Gamma_1) \neq \emptyset$; on the other hand, $1 \nprec 2$, as $\mathcal{P}r(\Gamma_1) \cap \mathcal{P}a(\Gamma_2) = \emptyset$; so the relation $\prec$ is acyclic. Finally, fixing the lexicographic order induced by $t > x$, $H$ is trivially seen to be coherent.

In order to define solutions of stratified systems, let us introduce some additional notation about CFPSs. For a CFPS $f \in \mathbb{R}[\![X]\!]$ and $Y \subseteq X$, we can consider the CFPS $f_{|Y^\otimes} \in \mathbb{R}[\![Y]\!]$. For an intuitive explanation of this concept, assume e.g. $f$ represents $f(x_1, x_2)$ and $Y = \{x_2\}$: recalling that we take the origin as the expansion point, $f_{|Y^\otimes}$ represents $f(0, x_2)$, that is, $f$ where the variables not in $Y$ have been replaced by 0. Formally, for $\psi : \mathcal{P} \to \mathbb{R}[\![X]\!]$ and a subsystem $\Gamma = (\Sigma, Y)$, we let $\psi_\Gamma : \mathcal{P}_\Gamma \to \mathbb{R}[\![Y]\!]$ be defined as: $\psi_\Gamma(E) \triangleq \psi(E)_{|Y^\otimes}$ for each $E \in \mathcal{P}_\Gamma$.

**Definition 6 (solutions of $H$).** *Let $H$ be a stratified system.*

1. *A solution of $H$ is a homomorphism $\psi : \mathcal{P} \to \mathbb{R}[\![X]\!]$ such that for each $\Gamma_i \in H$, $\psi_{\Gamma_i} : \mathcal{P}_{\Gamma_i} \to \mathbb{R}[\![X_i]\!]$ respects all the equations in $\Gamma_i^\infty$.*

2. *Let $\rho : \mathcal{P}a(H) \to \mathbb{R}$ be an initial data specification and $\Gamma_0 = (\Sigma_0, X_0) \triangleq (\{u_\tau = \rho(u_\tau) : u_\tau \in \mathcal{P}a(H)\}, \emptyset)$. A solution of the initial value problem $\mathbf{iP} = (H, \rho)$ is solution of the stratified system $H \cup \{\Gamma_0\}$.*

We can linearly order the subsystems of $H$ according to a total order compatible with $\prec$ and then lift inductively existence and uniqueness (Theorem 1) to $H$.

**Theorem 2 (existence and uniqueness for $H$).** *Let $H$ be a coherent stratified system. For any initial data specification $\rho$ for $H$, there is a unique solution of $\mathbf{iP} = (H, \rho)$.*

We illustrate the idea behind the proof of Theorem 2 on the Heat equation of Example 2.

*Example 3 (Example 2, cont.).* Let us fix any initial data specification $\rho(u) = u_0 \in \mathbb{R}$ for $H$. As prescribed by Definition 6(2), we consider the extended system $\overline{H} \triangleq H \cup \{\Gamma_0\}$, where $\Gamma_0 = (\{u = u_0\}, \emptyset)$. Note that $U_{\Gamma_0} = \mathcal{D}_{\Gamma_0} = \mathcal{P}r(\Gamma_0) = \{u\}$ and $\mathcal{P}a(\Gamma_0) = \emptyset$. Now we build a sequence of IVPs $\mathbf{iP}_i$, and corresponding solutions $\psi_i : \mathcal{P}_{\Gamma_i} \to \mathbb{R}[\![X_i]\!]$, for the subsystems $\Gamma_i$'s in $\overline{H}$. The construction proceeds inductively on a linear order compatible with $\prec$, that is: $0 \prec 2 \prec 1$. The definition of each initial data specification $\rho_i : \mathcal{P}a(\Gamma_i) \to \mathbb{R}$ relies on the solutions $\psi_j$ for $j \prec i$. The existence of such solutions is guaranteed by Theorem 1. In particular:

- $\mathbf{iP}_0 = (\{u = u_0\}, \rho_0)$, with $\rho_0(u) \triangleq \emptyset$ (empty function), has solution[3] $\psi_0 : \mathcal{P}_{\Gamma_0}(= \mathbb{R}[u]) \to \mathbb{R}[\![\emptyset]\!]$;
- $\mathbf{iP}_2 = (\{u_x = -u\}, \rho_2)$, with $\rho_2(u) \triangleq \psi_0(u)(\epsilon)$, has solution $\psi_2 : \mathcal{P}_{\Gamma_2}(= \mathbb{R}[x, u]) \to \mathbb{R}[\![x]\!]$;
- $\mathbf{iP}_1 = (\{u_t = u_{xx}\}, \rho_1)$, with $\rho_1(u_{x^k}) \triangleq \psi_2(u_{x^k})(\epsilon)$ $(k \geq 0)$, has solution $\psi_1 : \mathcal{P}_{\Gamma_1}(= \mathcal{P}) \to \mathbb{R}[\![t, x]\!]$.

It can be shown—and this is the nontrivial part of Theorem 2—that the solution of the main subsystem, $\psi_1$, is a solution of $\overline{H}$ (Definition 6(1)), and in particular: $(\psi_1)_{\Gamma_i} = \psi_i$ for each $i$. Hence $\psi_1$ is the (unique) solution of $(H, \rho)$.

---

[3] Specifically, $\psi_0(E)(\epsilon) = E(u_0)$ for each $E \in \mathbb{R}[u]$.

In view of the subsequent algorithmic developments, the next step is to obtain a formula for the Taylor coefficients of the solutions of $H$, in analogy with the formula (2) for pure systems. This formula will be based on the transition function of the main subsystem, $\delta_{\Sigma_1}$. However, a pivotal role will now be also played by a reduction function $S_H : \mathcal{P} \to \mathcal{P}_0(H)$, introduced below: it will allow one to rewrite any $E \in \mathcal{P}$ to a normal form in $\mathcal{P}_0(H)$, where it can be evaluated for any given initial data specification $\rho$ for $H$. Below, $\to_{\Sigma_i}$ (resp. $\to_X$) denotes the rewrite relation over $\mathcal{P}$ induced by the equations in $\Gamma_i^\infty$ (resp. $\{x = 0 : x \in X\}$).

**Definition 7 (reduction $S_H$).** *Let $H = \{\Gamma_1, ..., \Gamma_m\}$ be a coherent stratified system. Let $\to_H \subseteq \mathcal{P} \times \mathcal{P}$ be $\to_H \overset{\triangle}{=} \to_{\Sigma_1} \cup \cdots \cup \to_{\Sigma_m} \cup \to_X$. For each $E \in \mathcal{P}$, we let $S_H E$ denote an arbitrarily fixed $F \in \mathcal{P}_0(H)$ such that $E \to_H^* F$.*

Note that $S_H E$ is well defined due to normality[4] of $H$. Let $\phi$ be a solution of an IVP $(H, \rho)$. We remark that in general it is not true that $\phi(E) = \phi(S_H E)$ (trivially, $S_H x = 0$, but $\phi(x) \neq 0$). It *is* true, however, that $\phi(E)(\epsilon) = \phi(S_H E)(\epsilon)$; moreover $\phi(S_H E)(\epsilon) = \rho(S_H E)$. This fact is quite intuitive, recalling the informal interpretation of $f(\epsilon)$ as $f(0)$ for a CFPS $f$. For instance, in the Heat equation system of Example 2, one would have $u_t(0, 0) = u_{xx}(0, 0) = u(0, 0)(= \rho(u))$, where the first and second equality follow from applying $\Sigma_1$ and $\Sigma_2$ (twice), respectively. Formally, we have the following formula, giving the Taylor coefficients of $\phi(E)$. This is also key to the algorithm in the next section.

**Corollary 1 (Taylor coefficients).** *Let $H$ be a coherent stratified system. Denote by $\delta_{\Sigma_1}$ the transition function of the main subsystem of $H$. For any initial data specification $\rho$ for $H$, the unique solution $\phi$ of $(H, \rho)$ enjoys the following, for every $E \in \mathcal{P}$ and $\tau = \mathbf{x}^\alpha \in X^\otimes$.*

$$\phi(E)(\tau) = \frac{\rho(S_H(\delta_{\Sigma_1}(E, \tau)))}{\alpha!}. \tag{3}$$

*Example 4 (Example 2, cont.).* Consider any initial data specification $\rho(u) = u_0 \in \mathbb{R}$ for $H$, let $\psi$ be the solution of $(H, \rho)$ and $f = \psi(u)$. We compute the first few coefficients of $f$ by applying (3) with $E = u$. Let us first compute a few $S_H(\delta_{\Sigma_1}(u, \tau))$ s. Recall that the definition of $=_{\Sigma_i}$ is based on $\Gamma_i^\infty$ $(i = 1, 2)$.

$$S_H(\delta_{\Sigma_1}(u, \epsilon)) = S_H u = u \qquad\qquad S_H(\delta_{\Sigma_1}(u, t)) = S_H u_{xx} = S_H(-u_x) = u$$
$$S_H(\delta_{\Sigma_1}(u, x)) = S_H u_x = -u \qquad\qquad S_H(\delta_{\Sigma_1}(u, tt)) = S_H u_{x^4} = u$$
$$S_H(\delta_{\Sigma_1}(u, tx)) = S_H u_{x^3} = -u \qquad\qquad S_H(\delta_{\Sigma_1}(u, xx)) = S_H u_{xx} = u.$$

In general, one can check that for $\tau = (t, x)^\alpha$, $\alpha = (\alpha_1, \alpha_2) \in \mathbb{N}^2$, $S_H(\delta_{\Sigma_1}(u, \tau)) = (-1)^{\alpha_2} u$. Hence, by (3), we have the CFPS: $f = u_0 + u_0 t - u_0 x + (u_0/2)t^2 - u_0 tx + (u_0/2)x^2 \cdots = \sum_{\tau = \mathbf{x}^\alpha} (-1)^{\alpha_2} (u_0/\alpha!)\tau$.

---

[4] In fact, more is true: $\to_H$ is terminating and confluent, so there is a unique $H$-normal form $F$ s.t. $E \to_H^* F$. See [13, App. A]. Therefore the arbitrariness in Definition 7 is only apparent.

# 4   Algorithms for Pre- and Postconditions

We will first recall some terminology and some basic facts from algebraic geometry, then introduce pre- and postconditions and finally the POST algorithm to compute them.

*Preliminaries.* From now on, we will restrict our attention to the following subclass of systems.

**Definition 8 (FP systems).** *A stratified system $H$ is* finite-parameter *(FP) if $\mathcal{P}a(H)$ is finite.*

For instance, in Example 2 the system $H$ is FP, while $H' \stackrel{\triangle}{=} \{\Gamma_1\}$ is not. In concrete applications, one would expect that most systems are FP. Let us now recall some additional notation and terminology about polynomials. According to (3), the calculation of the Taylor coefficients of a solution of a FP IVP $\mathbf{iP} = (H, \rho)$ involves evaluating expressions in $\mathcal{P}_0(H) = \mathbb{R}[\mathcal{P}a(H)]$. As $k \stackrel{\triangle}{=} |\mathcal{P}a(H)| < +\infty$, elements of $\mathcal{P}_0(H)$ can be treated as usual multivariate polynomials in a *finite* number of indeterminates. In particular, we can identify initial data specifications $\rho$ for $H$ with points in $\mathbb{R}^k$. Accordingly, for polynomials $E \in \mathcal{P}_0(H)$ and initial data specification $\rho \in \mathbb{R}^k$, it is notationally convenient to write $\rho(E)$ as $E(\rho)$, that is the value in $\mathbb{R}$ obtained by evaluating the polynomial $E$ at the point $\rho \in \mathbb{R}^k$.

In what follows, we shall rely on a few basic notions from algebraic geometry, which we quickly review below (a more detailed review can be found in [13, App. A]). See [16, Ch. 2–4] for a comprehensive treatment. An *ideal* $J \subseteq \mathcal{P}_0(H)$ is a nonempty set of polynomials closed under addition, and under multiplication by polynomials in $\mathcal{P}_0(H)$. For $P \subseteq \mathcal{P}_0(H)$, $\langle\, P \,\rangle \stackrel{\triangle}{=} \{\sum_{i=1}^{m} F_i \cdot E_i : m \geq 0, F_i \in \mathcal{P}_0(H), E_i \in P\}$ denotes the smallest ideal which includes $P$, and $\mathbf{V}(P) \subseteq \mathbb{R}^k$ the *(affine) variety* induced by $P$: $\mathbf{V}(P) \stackrel{\triangle}{=} \{\rho \in \mathbb{R}^k : E(\rho) = 0 \text{ for each } E \in P\} \subseteq \mathbb{R}^k$. For $W \subseteq \mathbb{R}^k$, $\mathbf{I}(W) \stackrel{\triangle}{=} \{E \in \mathcal{P}_0(H) : E(\rho) = 0 \text{ for each } \rho \in V\}$ is the ideal induced by $W$. We will use a few basic facts about ideals and varieties: (a) both $\mathbf{I}(\cdot)$ and $\mathbf{V}(\cdot)$ are inclusion reversing: $P_1 \subseteq P_2$ implies $\mathbf{V}(P_1) \supseteq \mathbf{V}(P_2)$ and $W_1 \subseteq W_2$ implies $\mathbf{I}(W_1) \supseteq \mathbf{I}(W_2)$; (b) any ascending chain of ideals $I_0 \subseteq I_1 \subseteq \cdots \subseteq \mathcal{P}_0(H)$ stabilizes in a finite number of steps (Hilbert's basis theorem); (c) for finite $P \subseteq \mathcal{P}_0(H)$, the problem of deciding if $E \in \langle\, P \,\rangle$ is decidable, by computing a Gröbner basis (a set of generators with special properties) of $\langle\, P \,\rangle$.

*Preconditions and Postconditions.* Let $H$ be a coherent, FP system and let $k \stackrel{\triangle}{=} |\mathcal{P}a(H)|$. Informally, computing the *preconditions* of a given set $Q \subseteq \mathcal{P}$ means finding all the initial data specifications $\rho \in \mathbb{R}^k$ under which all the polynomials in $Q$ represent valid equations for the system $H$—that is, they become identically zero when one plugs the solution of $(H, \rho)$ into them. Dually, computing the *postconditions* of a given set of initial data specifications $W \subseteq \mathbb{R}^k$ means finding the set $Q \subseteq \mathcal{P}$ of all polynomial equations that are valid under all initial data

$\rho \in W$. Here, we shall confine ourselves to *algebraic* sets $W$, that is $W = \mathbf{V}(P)$ for some $P \subseteq \mathcal{P}_0(H)$—think of $P$ as a set of constraints on the initial data. Formally, we have the following definition. Recall that for any $\rho \in \mathbb{R}^k$, we let $\phi_{(H,\rho)} : \mathcal{P} \to \mathbb{R}[\![X]\!]$ denote the unique solution of the IVP $(H, \rho)$.

**Definition 9 (pre- and postconditions).** *Let $H$ be coherent and* FP. *Let $P$ and $Q$ be sets of polynomials such that $P \subseteq \mathcal{P}_0(H)$ and $Q \subseteq \mathcal{P}$. We define the sets of weakest preconditions* $\mathrm{wp}_H(Q) \subseteq \mathbb{R}^k$ *and of the strongest postconditions* $\mathrm{sp}_H(P) \subseteq \mathcal{P}$ *as follows.*

$$\mathrm{wp}_H(Q) \overset{\triangle}{=} \{\rho \in \mathbb{R}^k \; : \; \phi_{(H,\rho)}(E) = 0 \; \text{ for each } E \in Q\}$$

$$\mathrm{sp}_H(P) \overset{\triangle}{=} \{E \in \mathcal{P} \; : \; \phi_{(H,\rho)}(E) = 0 \; \text{ for each } \rho \in \mathbf{V}(P)\}\,.$$

Any subset of $\mathrm{wp}_H(Q)$ will be called an (algebraic) precondition for $Q$, and any subset of $\mathrm{sp}_H(P)$ a postcondition for $\mathbf{V}(P)$. We focus here on computing strongest postconditions, which, as we shall see, can be used to compute preconditions as well. Actually, it is computationally convenient to introduce a *relativized* version of this problem.

**Problem 1 (relativized strongest postcondition).** *Let $H$ be coherent,* FP. *Given user-specified sets $P \subseteq_{\mathrm{fin}} \mathcal{P}_0(H)$ and $R \subseteq \mathcal{P}$, find a finite characterization of $\mathrm{sp}_H(P) \cap R$.*

By 'finding a finite characterization', we mean effectively computing a finite set of generators, of an appropriate algebraic type, for the set in question (see next paragraph). Following a well-established tradition in the field of continuous and hybrid system, the set $R$ will be represented by means of a polynomial template.

*The* POST *Algorithm.* We first introduce *polynomial templates* [1], that is, polynomials in $\mathrm{Lin}(\mathbf{a})[X \cup \mathcal{D}]$, where $\mathrm{Lin}(\mathbf{a})$ are (formal) linear combinations of the parameters in $\mathbf{a} = (a_1, ..., a_s)$ (for fixed $s \geq 1$) with real coefficients. For instance, $\ell = 5a_1 + 42a_2 - 3a_3$ is one such expression[5]. In other words, a polynomial template has the form $\pi = \sum_i \ell_i \gamma_i$ for distinct monomials $\gamma_i \in (X \cup \mathcal{D})^{\otimes}$, and $\ell_i$ linear expressions in the parameters $a_i$ s. For example, the following is a template: $\pi = (5a_1 + (3/4)a_3)u_x v^2 x y^2 + (7a_1 + (1/5)a_2)u v_{xy} + (a_2 + 42a_3)$. A *parameter evaluation* is a vector $v = (v_1, ..., v_s) \in \mathbb{R}^s$; we denote by $\pi[v] \in \mathcal{P}$ the polynomial obtained from $\pi$ by replacing each occurrence of $a_i$ with $v_i$ in the linear expressions of $\pi$ and evaluating them. For $V \subseteq \mathbb{R}^s$, $\pi[V] \overset{\triangle}{=} \{\pi[v] \; : \; v \in V\} \subseteq \mathcal{P}$.

For a user specified template $\pi$ with $s$ parametes, our goal is to solve Problem 1 with $R = \pi[\mathbb{R}^s]$. In other words, for a given $P \subseteq \mathcal{P}_0(H)$ that describes an algebraic variety of initial data specifications, we want to compute

$$\mathrm{sp}_H(P) \cap \pi[\mathbb{R}^s]\,. \tag{4}$$

---

[5] Linear expressions with a constant term, such as $2 + 5a_1 + 42a_2 - 3a_3$ are not allowed.

Informally, we will achieve this by building a sequence of vector spaces $\mathbb{R}^s \supseteq V_0 \supseteq V_1 \supseteq \cdots$, such that $\pi[V_i]$ contains polynomials whose derivatives up to order $i$ vanish on all points in $\mathbf{V}(P)$. This sequence converges to a vector space, say $V_m$, such that $\pi[V_m]$ contain polynomials whose derivatives of *every* order vanish on $\mathbf{V}(P)$. On account of Corollary 1, Eq. (3), such polynomials belong to $\mathrm{sp}_H(P)$. A nontrivial point in this scheme is being able to detect convergence of the sequence of vector spaces $V_i$.

Formally, we first extend $\delta_{\Sigma_1}$ and $S_H$ to templates as expected: for $\pi = \sum_i \ell_i \gamma_i$, $\delta_{\Sigma_1}(\pi, x) \triangleq \sum_i \ell_i \delta_{\Sigma_1}(\gamma_i, x)$ and $S_H \pi \triangleq \sum_i \ell_i S_H \gamma_i$, seen as a polynomials in $\mathrm{Lin}(\mathbf{a})[X \cup \mathcal{D}]$ and $\mathrm{Lin}(\mathbf{a})[\mathcal{P}a(H)]$, respectively. We shall make use of the following substitution properties of templates, which hold true in coherent systems (see [13, App. A].). For each $x \in X$ and $v \in \mathbb{R}^s$:

$$\delta_{\Sigma_1}(\pi[v], x) = \delta_{\Sigma_1}(\pi, x)[v] \qquad\qquad S_H(\pi[v]) = (S_H \pi)[v]. \qquad (5)$$

We are now set to introduce the POST algorithm. Given $P \subseteq \mathcal{P}_0(H)$ and a template $\pi$, fix $P_0$ s.t. $I_0 \triangleq \langle\, P_0\, \rangle \subseteq \mathbf{I}(\mathbf{V}(P))$ ($P_0 = P$ is a possible choice). The algorithm consists in generating two sequences of sets, $V_i \subseteq \mathbb{R}^s$ and $J_i \subseteq \mathcal{P}_0(H)$, for $i \geq 0$. The idea is that, at step $i$, $V_i$ collects those $v \in \mathbb{R}^s$ such that $S_H(\pi[v])$, and its derivatives up to order $i$, vanish on $\mathbf{V}(P)$, that is belong to $\mathbf{I}(\mathbf{V}(P))$. As $\mathbf{I}(\mathbf{V}(P))$ may be hard to compute, it is convenient to permit replacing it with some $\langle\, P_0\, \rangle \subseteq \mathbf{I}(\mathbf{V}(P))$. The $J_i$'s are used to detect stabilization. We use $\pi_\tau$ as an abbreviation of $\delta_{\Sigma_1}(\pi, \tau)$.

$$V_i \triangleq \bigcap_{\tau\,:\,|\tau| \leq i} \{v \in \mathbb{R}^s \,:\, (S_H \pi_\tau)[v] \in I_0\} \qquad (6)$$

$$J_i \triangleq \Big\langle\, \bigcup_{\tau\,:\,|\tau| \leq i} (S_H \pi_\tau)[V_i]\, \Big\rangle. \qquad (7)$$

Consider the least $m$ such that *both* $V_m = V_{m+1}$ and $J_m = J_{m+1}$: we let $\mathrm{POST}_H(P_0, \pi) \triangleq (V_m, J_m)$. Note that $m$ is well defined. Indeed, $V_0 \supseteq V_1 \supseteq \cdots$ forms a descending chain of finite-dimensional vector spaces in $\mathbb{R}^s$, which must stabilize at some $m'$; then $J_{m'} \subseteq J_{m'+1} \subseteq \cdots$ forms an ascending chain of ideals in $\mathcal{P}_0(H)$, which must stabilize at some $m \geq m'$. We remark that neither of the two conditions $V_{m+1} = V_m$ or $J_m = J_{m+1}$ taken alone does imply stabilization, in general. The next theorem states correctness and relative completeness of POST. Part (a) says that the set of polynomials $\pi[V_m]$ is a postcondition of $P$ and, in case $\langle\, P_0\, \rangle = \mathbf{I}(\mathbf{V}(P))$, coincides with the strongest postcondition relative to $\pi$, that is (4). Part (b) says that $J_m$ represents the weakest precondition of $\pi[V_m]$: this can be useful to look for preconditions in general, but will not be discussed here.

**Theorem 3 (relative completeness of POST).** *Let $H$ be coherent and* FP. *Let $P \subseteq \mathcal{P}_0(H)$ and $\pi$ be a template. Fix $P_0$ s.t. $I_0 \triangleq \langle\, P_0\, \rangle \subseteq \mathbf{I}(\mathbf{V}(P))$. Let* $\mathrm{POST}_H(P_0, \pi) = (V_m, J_m)$.

(a) $\pi[V_m] \subseteq \mathrm{sp}_H(P) \cap \pi[\mathbb{R}^s]$, *with equality if $I_0 = \mathbf{I}(\mathbf{V}(P))$;*
(b) $\mathbf{V}(J_m) = \mathrm{wp}_H(\pi[V_m])$.

*Proof.* In the proof we shall make use of the following stabilization property of the sequence of the $(V_i, J_i)$s ([13, Lemma A.16]).

$$\mathrm{POST}_H(P_0, \pi) = (V_m, J_m) \text{ implies that for each } j \geq 1, V_m = V_{m+j} \text{ and } J_m = J_{m+j}. \quad (8)$$

Let us consider part (a) of the theorem. Fix any $v \in V_m$, we must prove that $\pi[v] \in \mathrm{sp}_H(P)$, that is $\phi_{(H,\rho)}(\pi[v]) = 0$ for each $\rho \in \mathbf{V}(P)$. By Corollary 1, our task reduces to showing that, for each $\tau$, $(S_H(\pi[v]_\tau))(\rho) = (S_H \pi_\tau)[v](\rho) = 0$ (here we have used (5)), for each $\rho \in \mathbf{V}(P)$. That is, for each $\tau$, $(S_H \pi_\tau)[v] \in \mathbf{I}(\mathbf{V}(P))$. The latter is implied by $(S_H \pi_\tau)[v] \in I_0 \subseteq \mathbf{I}(\mathbf{V}(P))$. By definition (6), this holds for each $\tau$ such that $v \in V_{|\tau|}$. Hence for each $\tau$, as $v \in V_0 \supseteq \cdots \supseteq V_m = V_{m+1} = \cdots$ (by (8)). Assume now that $I_0 = \mathbf{I}(\mathbf{V}(P))$ and consider $v \in \mathbb{R}^s$ such that $\pi[v] \in \mathrm{sp}_H(P)$: we show that $v \in V_m$. Our task is showing that for each $\tau$ with $|\tau| \leq m$, $(S_H \pi_\tau)[v] \in \mathbf{I}(\mathbf{V}(P))$. The latter means precisely that $(S_H \pi_\tau)[v](\rho) = 0$ for each $\rho \in \mathbf{V}(P)$. But this holds by definition of $\pi[v] \in \mathrm{sp}_H(P)$ and Corollary 1: indeed, for each $\tau$, $(S_H(\pi[v]_\tau))(\rho) = (S_H \pi_\tau)[v](\rho) = 0$ (here we have used (5)), for each $\rho \in \mathbf{V}(P)$.

Let us consider part (b). First, consider any $\rho \in \mathrm{wp}_H(\pi[V_m])$. By definition and Corollary 1 (and using (5)), this is equivalent to $(S_H \pi_\tau)[v](\rho) = 0$ for each $v \in V_m$ and $\tau$. By definition of ideal $J_m$, this implies $F(\rho) = 0$ for each $F \in J_m$, that is $\rho \in \mathbf{V}(J_m)$. On the other hand, consider any $\rho \in \mathbf{V}(J_m)$ and any $v \in V_m$. Clearly $\rho \in \mathbb{R}^k$. Then proving that $\rho \in \mathrm{wp}_H(\pi[V_m])$, that is $\phi_{(H,\rho)}(\pi[v]) = 0$, is equivalent, via Corollary 1 (and again (5)), to showing that $(S_H \pi_\tau)[v](\rho) = 0$, for each $\tau$. Consider any such $\tau$: for $k \geq m$ large enough, by definition of $J_k$ and the fact that $V_m = V_k$, we have $J_k \supseteq (S_H \pi_\tau)[V_m]$, hence $J_m = J_k \supseteq (S_H \pi_\tau)[V_m]$ (by (8)), therefore $(S_H \pi_\tau)[v](\rho) = 0$, as required.

The vector spaces $V_i$s in (6) can be effectively represented by the successive linear constraints imposed by (6) on the template parameters $\mathbf{a} = (a_1, ..., a_s)$. In turn, this permits computing finite sets of generators for the ideals $J_i$s in (7). This is illustrated with an example below. For a set of linear expressions $L \subseteq \mathrm{Lin}(\mathbf{a})$, we let $\mathrm{span}(L) \stackrel{\triangle}{=} \{v \in \mathbb{R}^s : \ell[v] = 0 \text{ for each } \ell \in L\} \subseteq \mathbb{R}^s$ be the vector space of parameter evaluations that annihilate all expressions in $L$.

*Example 5 (Example 2, cont.).* Fix $P = P_0 = \emptyset$, hence $\mathbf{V}(P) = \mathbb{R}$ (here $k = |\{u\}| = 1$ and we impose no constraints on the initial data) and $I_0 = \mathbf{I}(\mathbf{V}(P)) = \{0\}$. We seek for linear relations between $u$ and $u_x$, considering the template $\pi \stackrel{\triangle}{=} a_1 u + a_2 u_x$. We compute $\mathrm{POST}_H(P_0, \pi) = (V_m, J_m)$ as follows. Below we reuse the equalities for $S_H(\delta_{\Sigma_1}(u, \tau))$ computed in Example 4.

- $(i = 0)$. $S_H \pi = (a_1 - a_2)u$. Therefore $V_0 = \mathrm{span}(\{a_1 - a_2\}) = \{(\lambda, \lambda) : \lambda \in \mathbb{R}\}$ and $J_0 = \{0\}$.
- $(i = 1)$. $S_H \pi_x = S_H(a_1 u_x + a_2 u_{xx}) = (a_2 - a_1)u$ and $S_H \pi_t = S_H(a_1 u_{xx} + a_2 u_{x^3}) = (a_1 - a_2)u$. Therefore $V_1 = \mathrm{span}(\{a_2 - a_1, a_1 - a_2\}) = V_0$ and similarly $J_1 = J_0$.

Hence the algorithm stabilizes already at $m = 0$, returning $V_0 = \{(\lambda, \lambda) : \lambda \in \mathbb{R}\}$ and $J_0 = \{0\}$. This means that the valid instances of $\pi$ are of the form $\lambda(u + u_x)$, for all $\lambda \in \mathbb{R}$. Or, equivalently, that $u_x = -u$ is a valid equation, under any initial data specification.

Suppose $\mathrm{POST}_H(P_0, \pi) = (V_m, J_m)$. Given a parameter evaluation $v \in \mathbb{R}^s$, checking if $\pi[v] \in \pi[V_m]$ is equivalent to checking if $v \in V_m$: this can be effectively done knowing a basis $B_m$ of the vector space $V_m$. In practice, it is more convenient to succinctly represent the whole set $\pi[V_m]$ returned by $\mathrm{POST}_H$ in terms of a new *result template* $\pi'$ with $s' \leq s$ parameters, such that $\pi'[\mathbb{R}^{s'}] = \pi[V_m]$. In the example above, $\pi' = a_1(u + u_x)$. The result template $\pi'$ can in fact be computed directly from $\pi$, by propagating, via substitutions, the linear constraints on $\mathbf{a}$ arising from (6) as they are generated (further details in [13, App.A]).

# 5   Example: Burgers' Equation

We have put a proof-of-concept implementation of the $\mathrm{POST}$ algorithm of Sect. 4 at work on some IVPs drawn from mathematical physics. We illustrate one case below[6].

We consider the inviscid case of the Burgers' equation [17,18], with a linear initial condition at $t = 0$ (for $b, c$ arbitrary real constants).

$$u_t = -u \cdot u_x \qquad\qquad u(0, x) = bx + c.$$

We fix $X = \{t, x\}$ and $U = \{u, b, c\}$. The above IVP is encoded by the stratified system $H = \{\Gamma_1, \Gamma_2\}$, where

$$\Gamma_1 = (\{u_t = -uu_x\} \cup \Sigma_{aux1}, \{t, x\}) \qquad \Gamma_2 = (\{u_x = b\} \cup \Sigma_{aux2}, \{x\}).$$

$\Sigma_{aux1} = \{b_t = 0, c_t = 0, c_x = 0\}$ and $\Sigma_{aux2} = \{b_x = 0\}$ just encode that $b, c$ are constants. As $\mathcal{P}a(H) = \{u, b, c\}$, the system is FP. Moreover, $H$, with the lexicographic order induced by $u > b > c$ and $t > x$, is coherent. We fix the set of possible initial data specifications to $\mathbf{V}(P)$ where $P = \{u - c\}$: this just ensures that $u(0, 0) = c$. In order to discover interesting postconditions of $P$, we consider a complete polynomial template of total degree 3 over the indeterminates $Z \triangleq \{t, x\} \cup \mathcal{P}a(H)$, $\pi = \sum_{\gamma_i \in Z^\otimes, |\gamma_i| \leq 3} a_i \gamma_i$, which consists of $s = 56$ terms. Letting $P_0 = P$, we run $\mathrm{POST}_H(P, \pi)$, which halts at the iteration $m = 5$, returning $(V_5, J_5)$. This took about 6s in our experiment. The algorithm returns $V_5$ in the form of a 1-parameter result template $\pi'$, such that $\pi'[\mathbb{R}] = \pi[V_5]$: the set of all instances of $\pi'$ forms a valid postcondition of $P$. In this case Theorem 3(a)

---

[6] Additional examples, concerning conservation laws and boundary problems, are reported in [13]. Code and examples are available at https://github.com/micheleatunifi/PDEPY/blob/master/PDE.py. Execution times reported here are for a Python Anaconda distribution running under Windows 10 on a Surface Pro laptop.

implies that $\pi'[\mathbb{R}] = \mathrm{sp}_H(P) \cap \pi[\mathbb{R}^s]$. Specifically, we find, for $a_1$ a template parameter:

$$\pi' = a_1 \cdot (ctu + u - b - cx).$$

In other words, up to the multiplicative constant $a_1$, $ctu + u = b + cx$ is the only equation of degree $\leq 3$ satisfied by the solutions of $H$, for initial data specifications $\rho \in \mathbf{V}(P)$. This equation can be easily solved algebraically for $u$—note that we are actually manipulating CFPSs—and yields the unique solution of the IVP:

$$u = \frac{cx + b}{ct + 1}.$$

## 6   Related Work

The present paper is broadly related to recent and ongoing work in the field of formal tools for ODEs, such us the theory of differential equivalences by Cardelli et al., see [7] and references therein. More specifically, our development here conceptually parallels and extends our previous work on polynomial ODEs, in particular [11,12]. The POST algorithm has a similar structure to the algorithm by the same name in [12]. Technically, though, the case of PDEs is remarkably more challenging, for the following reasons. (a) In PDEs, both the existence of solutions and the transition structure itself depend on coherence. In ODEs, (analytic) solutions always exist in the polynomial case, coherence is trivial and the resulting transition structure is quite simple. (b) In PDE IVPs and the related stratified systems, a prominent role is played by the acyclicity of their structure, which is again trivial in ODEs. (c) In PDEs, differential polynomials live in the infinite-indeterminates space $\mathcal{P}$, which requires reduction to $\mathcal{P}_0(H)$ via $S_H$, and, for the POST algorithm, a finiteness assumption on parametric derivatives; in ODEs, $\mathcal{P} = \mathcal{P}_0(\Sigma)$ has always finitely many indeterminates.

Our work is related to the field of Differential Algebra (DA), see [14,15,19–21] and references therein. In particular, Boulier et al.'s RosenfeldGröbner algorithm [19], computes the ideal of the differential and polynomial consequences of a system $\Sigma$. This ideal, for pure systems and no constraints on the initial data, is related to our strongest postcondition; however, how to encode general IVPs, pre- and postconditions in their format is far from trivial, if possible at all. More generally, while DA techniques can be used to reduce systems to a coherent form, which is required by our approach, they do not seem to be concerned with IVPs or boundary problems as such. The only exceptions we are aware of are [22,23], which focus on linear ODEs.

## References

1. Sankaranarayanan, S., Sipma, H., Manna, Z.: Non-linear loop invariant generation using Gröbner bases. In: POPL 2004. ACM (2004)

2. Sankaranarayanan, S.: Automatic invariant generation for hybrid systems using ideal fixed points. In: HSCC 2010, pp. 221–230. ACM (2010)
3. Platzer, A.: Logics of dynamical systems. In: LICS 2012, pp. 13–24. IEEE (2012)
4. Ghorbal, K., Platzer, A.: Characterizing algebraic invariants by differential radical invariants. In: Ábrahám, E., Havelund, K. (eds.) TACAS 2014. LNCS, vol. 8413, pp. 279–294. Springer, Heidelberg (2014). https://doi.org/10.1007/978-3-642-54862-8_19. http://reports-archive.adm.cs cmu.edu/anon/2013/CMU-CS-13-129.pdf
5. Kong, H., Bogomolov, S., Schilling, C., Jiang, Y., Henzinger, T.: Safety verification of nonlinear hybrid systems based on invariant clusters. In: HSCC 2017, pp. 163–172. ACM (2017)
6. Boreale, M.: Algorithms for exact and approximate linear abstractions of polynomial continuous systems. In: HSCC 2018, pp. 207–216. ACM (2018)
7. Cardelli, L., Tribastone, M., Tschaikowski, M., Vandin, A.: Symbolic computation of differential equivalences. Theor. Comput. Sci. **777**, 132–154 (2019)
8. Claudel, C.G., Bayen, A.M.: Solutions to switched Hamilton-Jacobi equations and conservation laws using hybrid components. In: Egerstedt, M., Mishra, B. (eds.) HSCC 2008. LNCS, vol. 4981, pp. 101–115. Springer, Heidelberg (2008). https://doi.org/10.1007/978-3-540-78929-1_8
9. Platzer, A.: Differential hybrid games. ACM Trans. Comput. Log **18**(3), 19–44 (2017)
10. Boreale, M.: On the coalgebra of partial differential equations. In: MFCS 2019, LIPIcs, vol. 138, pp. 24:1–24:13. Schloss Dagstuhl - Leibniz-Zentrum für Informatik (2019). https://drops.dagstuhl.de/opus/volltexte/2019/10968/pdf/LIPIcs-MFCS-2019-24.pdf
11. Boreale, M.: Algebra, coalgebra, and minimization in polynomial differential equations. In: Esparza, J., Murawski, A.S. (eds.) FoSSaCS 2017. LNCS, vol. 10203, pp. 71–87. Springer, Heidelberg (2017). https://doi.org/10.1007/978-3-662-54458-7_5. arXiv.org:1710.08350. Full version in Logical Methods in Computer Science 15(1)
12. Boreale, M.: Complete algorithms for algebraic strongest postconditions and weakest preconditions in polynomial ODE'S. In: Tjoa, A.M., Bellatreche, L., Biffl, S., van Leeuwen, J., Wiedermann, J. (eds.) SOFSEM 2018. LNCS, vol. 10706, pp. 442–455. Springer, Cham (2018). https://doi.org/10.1007/978-3-319-73117-9_31. Full version in Sci. Comput. Program. 193
13. Boreale, M.: Automatic pre- and postconditions for partial differential equations (2020). https://github.com/micheleatunifi/PDEPY/blob/master/FullPDEprepost.pdf. Full version of the present paper
14. Reid, G., Wittkopf, A., Boulton, A.: Reduction of systems of nonlinear partial differential equations to simplified involutive forms. Eur. J. Appl. Math. **7**(6), 635–666 (1996)
15. Marvan, M.: Sufficient set of integrability conditions of an orthonomic system. Found. Comput. Math. **6**(9), 651–674 (2009)
16. Cox, D.A., Little, J., O'Shea, D.: Ideals, Varieties, and Algorithms An Introduction to Computational Algebraic Geometry and Commutative Algebra. UTM. Springer, Cham (2015). https://doi.org/10.1007/978-3-319-16721-3
17. Bateman, H.: Some recent researches on the motion of fluids. Mon. Weather. Rev. **43**(4), 163–170 (1915)
18. Burgers, J.M.: A mathematical model illustrating the theory of turbulence. In: Advances in Applied Mechanics, vol. 1, pp. 171–199. Elsevier (1948)
19. Boulier, F., Lazard, D., Ollivier, F., Petitot, M.: Computing representations for radicals of finitely generated differential ideals. Appl. Algebra Eng. Commun. Comput. **20**(1), 73–121 (2009)

20. Robertz, D.: Formal Algorithmic Elimination for PDEs. LNM, vol. 2121. Springer, Cham (2014). https://doi.org/10.1007/978-3-319-11445-3
21. Rust, C.J., Reid, G.J., Wittkopf, A.D.: Existence and uniqueness theorems for formal power series solutions of analytic differential systems. In: ISSAC, vol. 1999, pp. 105–112 (1999)
22. Rosenkranz, M., Regensburger, G.: Solving and factoring boundary problems for linear ordinary differential equations in differential algebras. J. Symb. Comput. **43**(8), 515–544 (2008)
23. Rosenkranz, M., Regensburger, G., Tec, L., Buchberger, B.: Symbolic analysis for boundary problems: from rewriting to parametrized Gröbner bases (2012). coRR abs/1210.2950

# Importance of Interaction Structure and Stochasticity for Epidemic Spreading: A COVID-19 Case Study

Gerrit Großmann[(✉)], Michael Backenköhler, and Verena Wolf

Saarland Informatics Campus, Saarland University, 66123 Saarbrücken, Germany
{gerrit.grossmann,michael.backenkoehler,verena.wolf}@uni-saarland.de

**Abstract.** In the recent COVID-19 pandemic, computer simulations are used to predict the evolution of the virus propagation and to evaluate the prospective effectiveness of non-pharmaceutical interventions. As such, the corresponding mathematical models and their simulations are central tools to guide political decision-making. Typically, ODE-based models are considered, in which fractions of infected and healthy individuals change deterministically and continuously over time.

In this work, we translate an ODE-based COVID-19 spreading model from literature to a stochastic multi-agent system and use a contact network to mimic complex interaction structures. We observe a large dependency of the epidemic's dynamics on the structure of the underlying contact graph, which is not adequately captured by existing ODE-models. For instance, existence of super-spreaders leads to a higher infection peak but a lower death toll compared to interaction structures without super-spreaders. Overall, we observe that the interaction structure has a crucial impact on the spreading dynamics, which exceeds the effects of other parameters such as the basic reproduction number $R_0$.

We conclude that deterministic models fitted to COVID-19 outbreak data have limited predictive power or may even lead to wrong conclusions while stochastic models taking interaction structure into account offer different and probably more realistic epidemiological insights.

**Keywords:** COVID-19 · Epidemic stochastic simulation · SEIR model · SARS-CoV-2 · 2019–2020 coronavirus pandemic

## 1 Introduction

On March 11th, 2020, the World Health Organization (WHO) officially declared the outbreak of the *coronavirus disease 2019* (COVID-19) to be a pandemic. By this date at the latest, curbing the spread of the virus became a major worldwide concern. Given the lack of a vaccine, the international community relied on non-pharmaceutical interventions (NPIs) such as social distancing, mandatory quarantines, or border closures. Such intervention strategies, however, inflict high costs on society. Hence, for political decision-making it is crucial to forecast the spreading dynamics and to estimate the effectiveness of different interventions.

© Springer Nature Switzerland AG 2020
M. Gribaudo et al. (Eds.): QEST 2020, LNCS 12289, pp. 211–229, 2020.
https://doi.org/10.1007/978-3-030-59854-9_16

Mathematical and computational modeling of epidemics is a long-established research field with the goal of predicting and controlling epidemics. It has developed epidemic spreading models of many different types: data-driven and mechanistic as well as deterministic and stochastic approaches, ranging over many different temporal and spatial scales (see [15,50] for an overview).

Computational models have been calibrated to predict the spreading dynamics of the COVID-19 pandemic and influenced public discourse. Most models and in particular those with high impact are based on ordinary differential equations (ODEs). In these equations, the fractions of individuals in certain compartments (e.g., infected and healthy) change continuously and deterministically over time, and interventions can be modeled by adjusting parameters.

In this paper, we compare the results of COVID-19 spreading models that are based on ODEs to results obtained from a different class of models: stochastic spreading processes on contact networks. We argue that virus spreading models taking into account the interaction structure of individuals and reflecting the stochasticity of the spreading process yield a more realistic view on the epidemic's dynamics.

If an underlying interaction structure is considered, not all individuals of a population meet equally likely as assumed for ODE-based models. A well-established way to model such structures is to simulate the spreading on a network structure that represents the individuals of a population and their social contacts. Effects of the network structure are largely related to the *epidemic threshold* which describes the minimal *infection rate* needed for a pathogen to be able to spread over a network [38]. In the network-free paradigm the basic reproduction number ($R_0$), which describes the (mean) number of susceptible individuals infected by patient zero, determines the evolution of the spreading process. The value $R_0$ depends on both, the connectivity of the society *and* the infectiousness of the pathogen. In contrast, in the network-based paradigm the interaction structure (given by the network) and the infectiousness (given by the infection rate) are decoupled.

Here, we focus on contact networks as they provide a universal way of encoding real-world interaction characteristics like super-spreaders, grouping of different parts of the population (e.g. senior citizens or children with different contact patterns), as well as restrictions due to spatial conditions and mobility, and household structures. Moreover, models based on contact networks can be used to predict the efficiency of interventions [5,35,39].

Here, we analyze in detail a network-based stochastic model for the spreading of COVID-19 with respect to its differences from existing ODE-based models and the sensitivity of the spreading dynamics on particular network features. We calibrate both, ODE-models and stochastic models with interaction structure to the same basic reproduction number $R_0$ or to the same infection peak and compare the corresponding results. In particular, we analyze the changes in the effective reproduction number over time. For instance, early exposure of super-spreaders leads to a sharp increase of the reproduction number, which results in a strong increase of infected individuals. We compare the times at which the number of infected individuals is maximal for different network structures as well as the death toll. Our results show that the interaction structure has a major impact on the spreading dynamics and, in particular, important characteristic values deviate strongly from those of the ODE model.

## 2    Related Work

In the last decade, research focused largely on epidemic spreading, where inter-actions were constrained by contact networks, i.e., a graph representing the indi-viduals (as nodes) and their connectivity (as edges). Many generalizations, e.g. to weighted, adaptive, temporal, and multi-layer networks exist [32,45]. Here, we focus on simple contact networks without such extensions.

Spreading characteristics on various contact networks based on the well-known Susceptible-Infected-Susceptible (SIS) or Susceptible-Infected-Recovered (SIR) compartment model have been investigated intensively. In such models, each individual (node) successively passes through the individual stages (com-partments). For an overview, we refer the reader to [36]. Qualitative and quan-titative differences between network structures and network-free models have been investigated in [2,23]. In contrast, this work considers a specific COVID-19 spreading model and focuses on those characteristics that are most relevant for COVID-19 and which have, to the best of our knowledge, not been analyzed in previous work.

SIS-type models require knowledge of the spreading parameters (infection strength, recovery rate, etc.) and the contact network, which can partially be inferred from real-world observations. Currently, inferred data for COVID-19 seems to be of very poor quality [25]. However, while the spreading parame-ters are subject to a broad scientific discussion. Publicly available data, which could be used for inferring a realistic contact network, practically does not exist. Therefore real-world data on contact networks is rare [24,31,33,44,46] and not available for large-scale populations. A reasonable approach is to generate the data synthetically, for instance by using mobility and population data based on geographical diffusion [3,18,37,47]. For instance, this has been applied to the influenza virus [34]. Due to the major challenge of inferring a realistic contact network, most of these works, however, focus on how specific network features shape the spreading dynamics.

### 2.1    COVID-19 Spreading Models

Literature abounds with proposed models of the COVID-19 spreading dynamics. Very influential is the work of Neil Ferguson and his research group that regularly publishes reports on the outbreak (e.g. [11]). They study the effects of different interventions on the outbreak dynamics. The computational modeling is based on a model of influenza outbreaks [12,20]. They present a very high-resolution spatial analysis based on movement-data, air-traffic networks etc. and perform sensitivity analysis on the spreading parameters, but to the best of our knowledge not on the interaction data. Interaction data were also inferred locally at the beginning of the outbreak in Wuhan [4] or in Singapore [41] and Chicago [13]. Models based on community structures, however, consider isolated (parts of) cities and are of limited significance for large-scale model-based analysis of the outbreak dynamic.

Another work focusing on interaction structure is the modeling of outbreak dynamics in Germany and Poland done by Bock et al. [6]. The interaction structure within households is modeled based on census data. Inter-household interactions are expressed as a single variable and are inferred from data. They then generated "representative households" by re-sampling but remain vague on many details of the method.

A more rigorous model of stochastic propagation of the virus is proposed by Arenas et al. [1]. They take the interaction structure and heterogeneity of the population into account by using demographic and mobility data. They analyze the model by deriving a mean-field equation. Mean-field equations are more suitable to express the mean of a stochastic process than other ODE-based methods but tend to be inaccurate for complex interaction structures. Moreover, the relationship between networked-constrained interactions and mobility data remains unclear to us.

Other notable approaches use SIR-type methods, but cluster individuals into age-groups [29,40], which increases the model's accuracy. Rader et al. [42] combined spatial-, urbanization-, and census-data and observed that the crowding structure of densely populated cities strongly shaped the epidemics intensity and duration. In a similar way, a meta-population model for a more realistic interaction structure has been developed [8] without considering an explicit network structure.

The majority of research, however, is based on deterministic, network-free SIR-based ODE-models. For instance, the work of José Lourenço et al. [30] infers epidemiological parameters based on a standard SIR model. Similarly, Dehning et al. [9] use an SIR-based ODE-model, where the infection rate may change over time. They use their model to predict a suitable time point to loosen NPIs in Germany. Khailaie et al. analyze how changes in the reproduction number ("mimicking NPIs") affect the epidemic dynamics [26], where a variant of the deterministic, network-free SIR-model is used and modified to include states (compartments) for hospitalized, deceased, and asymptomatic patients. Otherwise, the method is conceptually very similar to [9,30] and the authors argue against a relaxation of NPIs in Germany. Another popular work is the online simulator *covidsim*[1]. The underlying method is also based on a network-free SIR-approach [51,52]. However, the role of an interaction structure is not discussed and the authors explicitly state that they believe that the stochastic effects are only relevant in the early stages of the outbreak. A very similar method has been developed at the German Robert-Koch-Institut (RKI) [7]. Jianxi Luo et al. proposed an ODE-based SIR-model to predict the end of the COVID-19 pandemic[2],

---

[1] Available at `covidsim.eu`.

[2] Available at `ddi.sutd.edu.sg`.

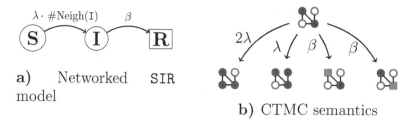

**a)    Networked    SIR model**

**b) CTMC semantics**

Fig. 1. Networked SIR model. (a) Compartments with instantaneous transition rates. Each node successively passes through the three compartments/states: susceptible (S), infected (I), and recovered/removed (R). (b) Four possible transitions on a 4-node contact network based on the CTMC semantics.

which is regressed with daily updated data. ODE-models have also been used to project the epidemic dynamics into the "postpandemic" future by Kissler et al. [28]. Some groups also resort to branching processes, which are inherently stochastic but not based on a complex interaction structure [22,43].

## 3    Translating SIR-Type Models for Epidemic Spreading

A very popular class of epidemic models is based on the assumption that during an epidemic individuals are either susceptible (S), infected (I), or recovered/removed (R). The mean number of individuals in each compartment evolves according to the following system of ordinary differential equations

$$
\begin{aligned}
\frac{d}{dt}s(t) &= -\frac{\lambda_{ODE}}{N}s(t)i(t) \\
\frac{d}{dt}i(t) &= \frac{\lambda_{ODE}}{N}s(t)i(t) - \beta i(t) \\
\frac{d}{dt}r(t) &= \beta i(t),
\end{aligned}
\tag{1}
$$

where $N$ denotes the total population size, $\lambda_{ODE}$ and $\beta$ are the infection and recovery rates. Typically, one assumes that $N = 1$ in which case the equation refers to fractions of the population, leading to the invariance $s(t)+i(t)+r(t) = 1$ for all $t$. It is trivial to extend the compartments and transitions.

### 3.1    Network-Based Spreading Model

A stochastic network-based spreading model is a continuous-time stochastic process on a discrete state space. The underlying structure is given by a graph, where each node represents one individual (or any other entity of interest). At each point in time, each node occupies a compartment, for instance: S, I, or R. Moreover, nodes can only receive or transmit infections from neighboring nodes (according to the edges of the graph). For the general case with $m$ possible compartments, this yields a state space of size $m^n$, where $n$ is the number of nodes.

The jump times until events happen are typically assumed to follow an exponential distribution. Note that in the ODE model, residual residence times in the compartments are not tracked, which naturally corresponds to the exponential distribution in the network model. Hence, the underlying stochastic process is a continuous-time Markov Chain (CTMC) [27]. The extension to non-Markovian semantics is trivial. We illustrate the three-compartment case in Fig. 1. The transition rates of the CTMC are such that an infected node transmits infections at rate $\lambda$. Hence, the rate at which a susceptible node is infected is $\lambda \cdot \#\text{Neigh}(\text{I})$, where $\#\text{Neigh}(\text{I})$ is the number of its infected direct neighbors. Spontaneous recovery of a node occurs at rate $\beta$. The size of the state space renders a full solution of the model infeasible and approximations of the mean-field [14] or Monte-Carlo simulations are common ways to analyze the process.

**General Differences to the ODE Model.** The aforementioned formalism yields some fundamental differences from network-free ODE-based approaches. The most distinct difference is the decoupling of infectiousness and interaction structure. The infectiousness $\lambda$ (i.e., the infection rate) is assumed to be a parameter expressing how contagious a pathogen inherently is. It encodes the probability of a virus transmission *if* two people meet. That is, it is independent from the social interactions of individuals (it might however depend on hygiene, masks, etc.). The influence of social contacts is expressed in the (potentially time-varying) connectivity of the graph. Loosely speaking, it encodes the possibility *that* two individuals meet. In the ODE-approach both are combined in the basic reproduction number. Note that, throughout this manuscript, we use $\lambda$ to denote the infectiousness of COVID-19 (as an instantaneous transmission rate).

Another important difference is that ODE-models consider fractions of individuals in each compartment. In the network-based paradigm, we model absolute numbers of entities in each compartment and extinction of the epidemic may happen with positive probability. While ODE-models are agnostic to the actual population size, in network-based models, increasing the population by adding more nodes inevitably changes the dynamics.

A key link between the two paradigms is that if the network topology is a *complete graph* (resp. clique) then the ODE-model gives an accurate approximation of the expected fractions of the network-based model. In systems biology this assumption is often referred to as *well-stirredness*. In the limit of an infinite graph size, the approximation approaches the true mean.

## 3.2   From ODE-Models to Networks

To transform an ODE-model to a network-based model, one can simply keep rates relating to spontaneous transitions between compartments as these transitions do not depend on interactions (e.g., recovery at rate $\beta$). Translating the infection rate is more complicated. In ODE-models, one typically has given an infection rate and assumes that each infected individual can infect all susceptible ones. To make the model invariant to the actual number of individuals, one typically divides the rate by the population size (or assumes the population size is one and the ODEs express fractions). Naturally, in a contact network, we do not work with fractions but each node relates to one entity.

Here, we propose to choose an infection rate such that the network-based model yields the same basic reproduction number $R_0$ as the ODE-model. The basic reproduction number describes the (expected) number of individuals that an infected person infects in a completely susceptible population. We calibrate our model to this starting point of the spreading process, where there is a single infected node (*patient zero*). We assume that $R_0$ is either explicitly given or can implicitly be derived from an ODE-based model specification. Hence, when we pick a random node as patient zero, we want it to infect on average $R_0$ susceptible neighbors (all neighbors are susceptible at that point in time) before it recovers or dies.

Let us assume that, like in the aforementioned SIR-model, infectious node infect their susceptible neighbors with rate $\lambda$ and that an infectious node loses its infectiousness (by dying, recovering, or quarantining) with rate $\beta$. According to the underlying CTMC semantics of the network model, each susceptible neighbor gets infected with probability $\frac{\lambda}{\beta+\lambda}$ [27]. Note that we only take direct infections from patient zero into account and, for simplicity, assume all neighbors are only infected by patient zero. Hence, when patient zero has $k$ neighbors, the expected number of neighbors it infects is $k\frac{\lambda}{\beta+\lambda}$. Since the mean degree of the network is $k_{\mathrm{mean}}$, the expected number of nodes infected by patient zero is

$$R_0 = k_{\mathrm{mean}} \frac{\lambda}{\beta + \lambda}. \tag{2}$$

Now we can calibrate $\lambda$ to relate to any desired $R_0$. That is

$$\lambda = \frac{\beta R_0}{k_{\mathrm{mean}} - R_0}. \tag{3}$$

Note that $R_0$ will always be smaller than $k_{\mathrm{mean}}$ which follows from Eq. (2), considering that $k_{\mathrm{mean}} \geq 1$ (by construction of the network) and $\beta > 0$. In contrast, in the deterministic paradigm this relationship is given by the equation (cf. [9,30]):

$$\lambda_{\mathrm{ODE}} = R_0 \beta. \tag{4}$$

Note that the recovery rate $\beta$ is identical in the ODE- and network-model. We can translate the infection rate of an ODE-model to a corresponding network-based stochastic model with the equation

$$\lambda = \frac{\lambda_{\mathrm{ODE}}}{k_{\mathrm{mean}} - R_0}, \tag{5}$$

while keeping $R_0$ fixed. In the limit of an infinite complete network, this yields $\lim_{n\to\infty} \lambda = \frac{\lambda_{\mathrm{ODE}}}{n}$, which is equivalent to the effective infection rate in the ODE-model $\frac{\lambda_{\mathrm{ODE}}}{N}$ for population size $N$ (cf. Eq. (1)).

**Example.** Consider a network where each node has exactly 5 neighbors (a 5-regular graph) and let $R_0 = 2$. We also assume that the recovery rate is $\beta = 1$, which then yields $\lambda_{\text{ODE}} = 2$. The probability that a random neighbor of patient zero becomes infected is $\frac{2}{5} = \frac{\lambda}{(\beta+\lambda)}$, which gives $\lambda = \frac{2}{3}$.

**Extensions of** SIR. It is trivial to extent the compartments and transitions, for instance by including an *exposed* compartment for the time-period where an individual is infected but not yet infectious. The derivation of $R_0$ remains the same. The only requirement is the existence of a distinct infection and recovery rate, respectively. In the next section, we discuss a more complex case.

## 4   A Network-Based COVID-19 Spreading Model

We consider a network-based model that is strongly inspired by the ODE-model used in [26] and document it in Fig. 2. We use the same compartments and transition-types but simplify the notation compared to [26] to make the intuitive meaning of the variables clearer[3].

We denote the compartments by $\mathcal{C} = \{S, E, C, I, H, U, R, D\}$, where each node can be *susceptible* (S), *exposed* (E), a *carrier* (C), *infected* (I), *hospitalized* (H), in the *intensive care unit* (U), *dead* (D), or *recovered* (R). Exposed agents are already infected but symptom-free and not infectious. Carriers are also symptom-free but already infectious. Infected nodes show symptoms and are infectious. Therefore, we assume that their infectiousness is reduced by a factor of $\gamma$ ($\gamma \leq 1$, sick people will reduce their social activity). Individuals that are hospitalized (or in the ICU) are assumed to be properly quarantined and cannot infect others.

Accurate spreading parameters are very difficult to infer in general and the high number of undetected cases complicates the problem further in the current pandemic. Here, we choose values that are within the ranges listed in [26], where the ranges are rigorously discussed and justified. We document them in Table 1. We remark that there is a high amount of uncertainty in the spreading parameters. However, our goal is not a rigorous fit to data but rather a comparison of network-free ODE-models to stochastic models with an underlying network structure.

Note that the mean number of days in a compartment is the inverse of the cumulative instantaneous rate to leave that compartment. For instance, the mean residence time in compartment H is $\frac{1}{(1-r_h)\mu_h + r_h\mu_h} = \frac{1}{\mu_h}$. As a consequence of the race condition of the exponential distribution [48], $r_h$ modulates the probability of entering the successor compartment. That is, with probability $r_h$, the successor compartment will be R and not U.

Inferring the infection rate $\lambda$ for a fixed $R_0$ is somewhat more complex than in the previous section because this model admits two compartments for infectious agents. We first consider the expected number of nodes that a randomly

---

[3] At the time of finalizing this manuscript, the model of Khailaie et al. seems to be updated in a similar way. However, it also became more complex (see gitlab.com/simm/covid19/secir/-/wikis/Report).

chosen patient zero infects, while being in state C. We denote the corresponding basic reproduction number by $\widehat{R_0}$. We calibrate the only unknown parameter $\lambda$ accordingly (the relationships from the previous section remain valid). We explain the relation to $R_0$ when taking C and I into account in the Appendix (available in [16]). Substituting $\beta$ by $\mu_c$ gives

$$\lambda = \frac{\lambda_{\mathrm{ODE}}}{k_{\mathrm{mean}} - \widehat{R_0}} = \frac{\lambda_{\mathrm{ODE}}}{k_{\mathrm{mean}} - \frac{\lambda_{\mathrm{ODE}}}{\mu_c}}. \tag{6}$$

### 4.1   Human-to-Human Contact Networks

Naturally, it is extremely challenging to reconstruct large-scale contact-networks based on data. Here, we test different types of contact networks with different features, which are likely to resemble important real-world characteristics. The contact networks are specific realizations (i.e., variates) of random graph models. Different graph models highlight different (potential) features of the real-world interaction structure. The number of nodes ranges from 100 to $10^5$. We only use strongly connected networks (where each node is reachable from all other nodes). We refer to [10] or the NetworkX [19] documentation for further information about the network models discussed in the sequel. We provide a schematic visualization in Fig. 3.

We consider **Erdős–Rényi** (ER) random graphs as a baseline, where each pair of nodes is connected with a certain (fixed) probability. We also compute results for **Watts–Strogatz** (WS) random networks. They are based on a ring topology with random re-wiring. The re-wiring yields to a small-world property of the network. Colloquially, this means that one can reach each node from each other node with a small number of steps (even when the number of nodes increases). We further consider **Geometric Random Networks** (GN), where nodes are randomly sampled in an Euclidean space and randomly connected such that nodes closer to each other have a higher connection probability. We also consider **Barabási–Albert** (BA) random graphs, that are generated using a preferential attachment mechanism among nodes, as well as graphs generated using the **Configuration Model** (CM-PL) which are—except from being constrained on having power-law degree distribution—completely random. The latter two models contain a very small number of nodes with very high degree, which act as super-spreaders. We also test a synthetically generated **Household** (HH) network that was loosely inspired by [2]. Each household is a clique, the edges between households represent connections stemming from work, education, shopping, leisure, etc. We use a configuration model to generate the global inter-household structure that follows a power-law distribution. We also use a **complete graph** (CG) as a sanity check. It allows the extinction of the epidemic, but otherwise similar results to those of the ODE are expected.

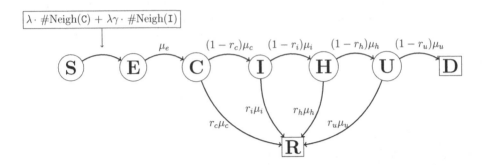

**Fig. 2.** Multi-agent compartment model for COVID-19 with instantaneous transition rates. The infection rate is $\lambda$. Exposed (E) nodes are newly infected. Carriers (C) are already infectious but still symptomless, infected nodes (I) develop symptoms and reduce their social activity (modulated by $\gamma$). Nodes that are hospitalized (H) and in the ICU (U) are properly quarantined. Recovered (R) nodes remain recovered.

**Table 1.** Model parameters

| Parameter | Value | Meaning |
|---|---|---|
| $\lambda$ | — | Infection rate, set w.r.t. Eq. (6) (using $\widehat{R_0}$) |
| $\gamma$ | 0.2 | How much social contact is maintained when sick |
| $\#\text{Neigh}(x)$ | $\in \mathbb{Z}_{\geq 0}$ | Current number of neighbors in compartment $x$ |
| $\lambda_{\text{ODE}}$ | 0.29 | Infection rate in ODE-model (denoted by $R_1$ in [26]) |
| $\widehat{R_0}$ | 1.8 | $R_0$ assuming $\gamma = 0$ |
| $R_0$ | $\approx 2.05$ | $R_0$, when assuming $\gamma = 0.2$ (cf. Appendix in [16]) |
| $x(t)$ | — | Number of (expected) nodes in $x \in \mathcal{C}$ at time $t$ |
| $\text{I}_{total}(t)$ | — | $e(t) + c(t) + i(t) + h(t) + u(t)$ |
| $r_x$ | $\in [0,1]$ | Recovery probability when node is in compartment $x$ |
| $\mu_x$ | $> 0$ | Instantaneous rate of leaving $x$ |
| $\mu_e$ | $\frac{1}{5.2}$ | Rate of transitioning from E to C |
| $r_c$ | 0.08 | Recovery probability when node is a carrier |
| $\mu_c$ | $\frac{1}{5.2}$ | Rate of leaving C |
| $r_i$ | 0.8 | Recovery probability when node is infected |
| $\mu_i$ | $\frac{1}{5}$ | Rate of leaving I |
| $r_h$ | 0.74 | Recovery probability when node is hospitalized |
| $\mu_h$ | $\frac{1}{10}$ | Rate of leaving H |
| $r_u$ | 0.46 | Recovery probability when node is in the ICU |
| $\mu_u$ | $\frac{1}{8}$ | Rate of leaving U |

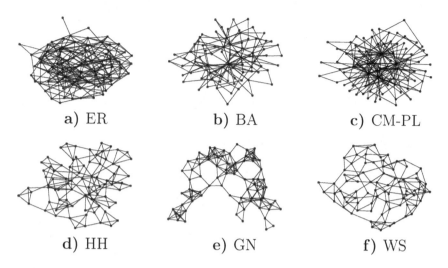

**Fig. 3.** Schematic visualizations of the random graph models with 80 nodes.

## 4.2   Parameter Calibration

We are interested in the relationship between the contact network structure, $R_0$, the height and time point of the infection-peak, and the number of individuals ultimately affected by the epidemic. Therefore, we run different network models with different $\widehat{R_0}$. For one series of experiments, we fix $\widehat{R_0} = 1.8$ and derive the corresponding infection rate $\lambda$ and the value for $\lambda_{\mathrm{ODE}}$ in the ODE model. In the second experiments, calibrate $\lambda$ and $\lambda_{\mathrm{ODE}}$ such that all infection peaks lie on the same level.

## 4.3   Interventions

In the sequel, we do not explicitly model NPIs. However, we note that the network-based paradigm makes it intuitive to distinguish between NPIs related to the probability that people meet (by changing the contact network) and NPIs related to the probability of a transmission happening when two people meet (by changing the infection rate $\lambda$). Political decision-making is faced with the challenge of transforming a network structure which inherently supports COVID-19 spreading to one which tends to suppress it. Here, we investigate how changes in $\lambda$ affect the dynamics of the epidemic in Sect. 5 (Experiment 3).

## 5   Numerical Results

We compare the solution of the ODE model (using numerical integration) with the solution of the corresponding stochastic network-based model (using Monte-Carlo simulations). Code will be made available[4]. We investigate the evolution

---

[4] github.com/gerritgr/StochasticNetworkedCovid19.

of mean fractions in each compartment over time, the evolution of the so-called *effective reproduction number*, and the influence of the infectiousness $\lambda$.

**Setup.** We used contact networks with $n = 1000$ nodes (except for the complete graph where we used 100 nodes). To generate samples of the stochastic spreading process, we used event-driven simulation (similar to the rejection-free version in [17]). Specifically, we utilized a simulation scheme, where all future events (i.e., transitions of nodes) were sorted in a priority queue (according to their application time). In each simulation step, the next event is drawn from the queue, the event is applied to the network, and the time is updated accordingly. Depending on the type of transition (a) new event(s) is (are) generated and pushed to the queue. Some events already in the queue might become irrelevant and are removed. The event queue is initialized by generating one event for each node.

The simulation started with three random seed nodes in compartment C (and with an initial fraction of 3/1000 for the ODE model). One thousand simulation runs were performed on a fixed variate of a random graph. We remark that results for other variates were very similar. Hence, for better comparability, we refrained from taking an average over the random graphs. The parameters to generate a graph are: ER: $k_{mean} = 6$, WS: $k = 4$ (numbers of neighbors), $p = 0.2$ (re-wire probability), GN: $r = 0.1$ (radius), BA: $m = 2$ (number of nodes for attachment), CM-PL: $\gamma = 2.0$ (power-law parameter), $k_{min} = 2$, HH: household size is 4, global network is CM-PL with $\gamma = 2.0$, $k_{min} = 3$. The CPU time for a single simulation on a standard desktop computer was in the range of a few hours.

**Experiment 1: Results with Homogeneous $\widehat{R_0}$.** In our first experiment, we compare the epidemic's evolution (cf. Fig. 4) while $\lambda$ is calibrated such that all networks admit an $\widehat{R_0}$ of 1.8. and $\lambda$ is set (w.r.t. the mean degree) according to Eq. (6). Thereby, we analyze how well certain network structures generally support the spread of COVID-19. The evolution of the mean fraction of nodes in each compartment is illustrated in Fig. 4 and Fig. 5.

Based on the Monte-Carlo simulations, we analyzed how $R_t$, the *effective reproduction number*, changes over time. The number $R_t$ denotes the average number of neighbors that an infectious node (who got exposed at day $t$) infects over time (cf. Fig. 6). For $t = 0$, the estimated effective reproduction number always starts around the same value and matched the theoretical prediction. Independent of the network, $\widehat{R_0} = 1.8$ yields $R_0 \approx 2.05$ (cf. Appendix [16]).

In Fig. 6 we see that the evolution of $R_t$ differs tremendously for different contact networks. Unsurprisingly, $R_t$ decreases on the complete graph (CG), as nodes, that become infectious later, will not infect more of their neighbors. This also happens for GN- and WS-networks, but they cause a much slower decline of $R_t$ which is around 1 in most parts (the sharp decrease in the end stems from the end of the simulation being reached). This indicates that the epidemic slowly "burns" through the network.

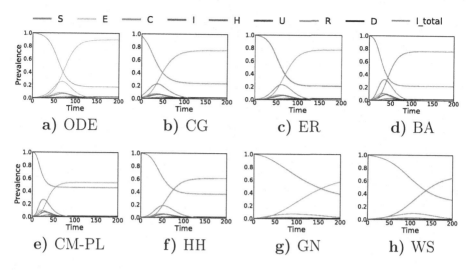

**Fig. 4. Exp. 1**: Evolution of the mean fractions in each compartment over time with 95% confidence intervals (barely visible).

In contrast, in networks that admit super-spreaders (CM-PL, HH, and also BA), it is principally possible for $R_t$ to increase. For the CM-PL network, we have a very early and intense peak of the infection while the number of individuals ultimately affected by the virus (and consequently the death toll[5]) remains comparably small (when we remove the super-spreaders from the network while keeping the same $R_0$, the death toll and the time point of the peak increase, plot not shown). Note that the high value of $R_t$ in Fig. 6c in the first days results from the fact that super-spreaders become exposed, which later infect a large number of individuals. As there are very few super-spreaders, they are unlikely to be part of the seeds. However, due to their high centrality, they are likely to be one of the first exposed nodes, leading to an "explosion" of the epidemic. In HH-networks this effect is way more subtle but follows the same principle.

**Experiment 2: Calibrating $\widehat{R_0}$ to a Fixed Peak.** Next, we calibrate $\lambda$ such that each network admits an infection peak (regarding $I_{total}$) of the same height (0.2). Results are shown in Fig. 7. They emphasize that there is no direct relationship between the number of individuals affected by the epidemic and the height of the infection peak, which is particularly relevant in the light of limited ICU capacities. It also shows that vastly different infection rates and basic reproduction numbers are acceptable when aiming at keeping the peak below a certain threshold.

---

[5] The number of fatalities in the figures is difficult to see, but it is (in the time limit) proportional to the number of recovered nodes.

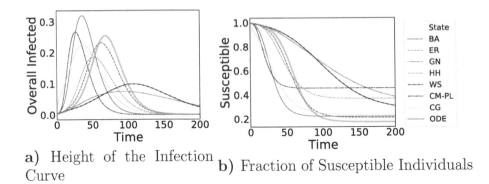

**a)** Height of the Infection Curve

**b)** Fraction of Susceptible Individuals

**Fig. 5. Exp. 1:** Same data as in Fig. 4 but only the evolution $I_{total}$ and $S$ are shown to highlight differences between networks.

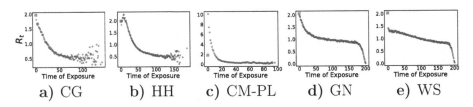

**a)** CG     **b)** HH     **c)** CM-PL     **d)** GN     **e)** WS

**Fig. 6. Exp. 1, Effective Reproduction Number**: Evolution of the (mean) effective reproduction number, $R_t$, over time, empirically evaluated. $x$-axis: Day at which a node becomes exposed, $y$-axis: (mean) number of neighbors this node infects while being a carrier or infected. Note that at later time points results are more noisy as the number of samples decreases. The first data-point is the simulation-based estimation of $R_0$ and is shown as a blue square.

**Experiment 3: Sensitivity Regarding $\lambda$.** Assume we have an estimate of the infectiousness, $\lambda$, of COVID-19. How do changes of $\lambda$ (e.g., by better hygiene) influence epidemic's properties and what is the impact of uncertainty about the value? Here, we investigate how the height of the infection-peak and $\widehat{R_0}$ scale with $\lambda$ for different topologies. Our results are illustrated in Fig. 8.

Noticeably, the relationship is concave for most network models but almost linear for the ODE model. This indicates that the networks models are more sensitive to small changes of $\lambda$ (and $R_0$). This suggests that the use of ODE models might lead to a misleading sense of confidence because, roughly speaking, it will tend to yield similar results when adding some noise to $\lambda$. That makes them seemingly robust to uncertainty in the parameters, while in reality the process is much less robust. Assuming that BA-networks resemble some important features of real social networks, the non-linear relationship between infection peak and infectiousness indicates that small changes of $\lambda$ (which could be achieved through proper hand-washing, wearing masks, keeping distance, etc.) can significantly "flatten the curve".

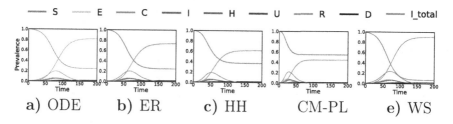

**Fig. 7. Exp. 2**: Evolution of mean-fractions in each compartment over time with infectiousness calibrated such that the peak has the same height.

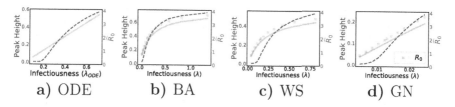

**Fig. 8. Exp. 3**: $\widehat{R}_0$ and maximal expected height of $\mathtt{I}_{\text{total}}$ (*expected* refers to samples, *maximal* refers to time) w.r.t. $\lambda$. For the network-models, $R_0$ (cf. Eq. (7) in [16]) is drawn as a scatter plot. Note the different scales on $x$- and $y$-axis

## 5.1   Discussion

In the series of experiments, we tested how various network types influence an epidemic's dynamics. The network types highlight different potential features of real-world social networks. Most results do not contradict with real-world observations. For instance, we found that better hygiene and the truncation of super-spreaders will likely reduce the peak of an epidemic by a large amount. We also observed that, even when $R_0$ is fixed, the evolution of $R_t$ largely depends on the network structure. For certain networks, in particular those admitting super-spreaders, it can even increase. An increasing reproduction number can be seen in many countries, for instance in Germany [21]. How much of this can be attributed to super-spreaders is still being researched. Note that super-spreaders do not necessarily have to correspond to certain individuals. It can also, on a more abstract level, refer to a type of events. We also observed that CM-PL networks have a very early and very intense infection peak. However, the number of people ultimately affected (and therefore also the death toll) remain comparably small. This is somewhat surprising and requires further research. We speculate that the fragmentation in the network makes it difficult for the virus to "reach every corner" of the graph while it "burns out" relatively quickly in region of the more central high-degree nodes.

# 6    Conclusions and Future Work

We presented results for a COVID-19 case study that is based on the translation of an ODE model to a stochastic network-based setting. We compared several interaction structures using contact graphs where one was (a finite version of) the implicit underlying structure of the ODE model, the complete graph. We found that inhomogeneity in the interaction structure significantly shapes the epidemic's dynamic. This indicates that fitting deterministic ODE models to real-world data might lead to qualitatively and quantitatively wrong results. The interaction structure should be included into computational models and should undergo the same rigorous scientific discussion as other model parameters.

Contact graphs have the advantage of encoding various types of interaction structures (spatial, social, etc.) and they decouple the infectiousness from the connectivity. We found that the choice of the network structure has a significant and counterintuitive impact and it is very likely that this is also the case for the inhomogeneous interaction structure among humans. Specifically, networks containing super-spreaders consistently lead to the emergence of an earlier and higher peak of the infection. Moreover, the almost linear relationship between $R_0$, $\lambda_{ODE}$, and the peak intensity in ODE-models might also lead to misplaced confidence in the results. Regarding the network structure in general, we find that super-spreaders can lead to a very early "explosion" of the epidemic. Small-worldness, by itself, does not admit this property. Generally, it seems that—unsurprisingly—a geometric network is best at containing a pandemic. This is an indication for the effectiveness of corresponding mobility restrictions. Surprisingly, we found a trade-off between the height of the infection peak and the fraction of individuals affected by the epidemic in total.

For future work, it would be interesting to investigate the influence of non-Markovian dynamics. ODE-models naturally correspond to an exponentially distributed residence times in each compartment [17,49]. Moreover, it would be interesting to reconstruct more realistic contact networks. They would allow to investigate the effect of NPIs in the network-based paradigm and to have a well-founded scientific discussion about their efficiency. From a risk-assessment perspective, it would also be interesting to focus more explicitly on worst-case trajectories (taking the model's inherent stochasticity into account). This is especially relevant because the costs to society do not scale linearly with the characteristic values of an epidemic. For instance, when ICU capacities are reached, a small additional number of severe cases might lead to dramatic consequences.

**Acknowledgements.** We thank Luca Bortolussi and Thilo Krüger for helpful comments regarding the manuscript. This work was partially funded by the DFG project MULTIMODE.

# References

1. Arenas, A., et al.: Derivation of the effective reproduction number R for COVID-19 in relation to mobility restrictions and confinement. medRxiv (2020)

2. Ball, F., Sirl, D., Trapman, P.: Analysis of a stochastic sir epidemic on a random network incorporating household structure. Math. Biosci. **224**(2), 53–73 (2010)
3. Barrett, C.L., et al.: Generation and analysis of large synthetic social contact networks. In: Proceedings of the 2009 Winter Simulation Conference (WSC), pp. 1003–1014. IEEE (2009)
4. Bi, Q., et al.: Epidemiology and transmission of COVID-19 in Shenzhen China: analysis of 391 cases and 1,286 of their close contacts. MedRxiv (2020)
5. Bistritz, I., Bambos, N., Kahana, D., Ben-Gal, I., Yamin, D.: Controlling contact network topology to prevent measles outbreaks. In: 2019 IEEE Global Communications Conference (GLOBECOM), pp. 1–6. IEEE (2019)
6. Bock, W., et al.: Mitigation and herd immunity strategy for COVID-19 is likely to fail. medRxiv (2020)
7. Buchholz, U., et al.: Modellierung von beispielszenarien der sars-cov-2-ausbreitung und schwere in deutschland (2020). (only available in German)
8. Chinazzi, M., et al.: The effect of travel restrictions on the spread of the 2019 novel coronavirus (COVID-19) outbreak. Science **368**, 395–400 (2020)
9. Dehning, J., et al.: Inferring COVID-19 spreading rates and potential change points for case number forecasts. arXiv preprint arXiv:2004.01105 (2020)
10. Estrada, E., Knight, P.A.: A First Course in Network Theory. Oxford University Press, Oxford (2015)
11. Ferguson, N., et al.: Report 9: impact of non-pharmaceutical interventions (NPIs) to reduce COVID19 mortality and healthcare demand (2020)
12. Ferguson, N.M., Cummings, D.A., Fraser, C., Cajka, J.C., Cooley, P.C., Burke, D.S.: Strategies for mitigating an influenza pandemic. Nature **442**(7101), 448–452 (2006)
13. Ghinai, I., et al.: Community transmission of SARS-CoV-2 at two family gatherings-Chicago, Illinois, February–March 2020 (2020)
14. Gleeson, J.P.: Binary-state dynamics on complex networks: pair approximation and beyond. Phys. Rev. X **3**(2), 021004 (2013)
15. Grassly, N.C., Fraser, C.: Mathematical models of infectious disease transmission. Nat. Rev. Microbiol. **6**(6), 477–487 (2008)
16. Grossmann, G., Backenkoehler, M., Wolf, V.: Importance of interaction structure and stochasticity for epidemic spreading: a COVID-19 case study. ResearchGate (2020). https://www.researchgate.net/publication/341119247_Importance_of_Interaction_Structure_and_Stochasticity_for_Epidemic_Spreading_A_COVID-19_Case_Study
17. Großmann, G., Bortolussi, L., Wolf, V.: Rejection-based simulation of non-Markovian agents on complex networks. In: Cherifi, H., Gaito, S., Mendes, J.F., Moro, E., Rocha, L.M. (eds.) COMPLEX NETWORKS 2019. SCI, vol. 881, pp. 349–361. Springer, Cham (2020). https://doi.org/10.1007/978-3-030-36687-2_29
18. Hackl, J., Dubernet, T.: Epidemic spreading in urban areas using agent-based transportation models. Future Internet **11**(4), 92 (2019)
19. Hagberg, A., Swart, P., S Chult, D.: Exploring network structure, dynamics, and function using NetworkX. Technical report, Los Alamos National Laboratory (LANL), Los Alamos, NM (United States) (2008)
20. Halloran, M.E., et al.: Modeling targeted layered containment of an influenza pandemic in the United States. Proc. Nat. Acad. Sci. **105**(12), 4639–4644 (2008)
21. Hamouda, O., et al.: Schätzung der aktuellen entwicklung der sars-cov-2-epidemie in deutschland-nowcasting (2020)
22. Hellewell, J., et al.: Feasibility of controlling COVID-19 outbreaks by isolation of cases and contacts. Lancet Glob. Health **8**, 488–496 (2020)

23. Holme, P.: Representations of human contact patterns and outbreak diversity in sir epidemics. IFAC-PapersOnLine **48**(18), 127–131 (2015)

24. Huang, C., et al.: Insights into the transmission of respiratory infectious diseases through empirical human contact networks. Sci. Rep. **6**, 31484 (2016)

25. Ioannidis, J.P.: Coronavirus disease 2019: the harms of exaggerated information and non-evidence-based measures. Eur. J. Clin. Invest. **50**(4), e13222 (2020)

26. Khailaie, S., et al.: Estimate of the development of the epidemic reproduction number RT from coronavirus SARS-CoV-2 case data and implications for political measures based on prognostics. medRxiv (2020)

27. Kiss, I.Z., Miller, J.C., Simon, P.L.: Mathematics of Epidemics on Networks. IAM, vol. 46. Springer, Cham (2017). https://doi.org/10.1007/978-3-319-50806-1

28. Kissler, S., Tedijanto, C., Goldstein, E., Grad, Y., Lipsitch, M.: Projecting the transmission dynamics of SARS-CoV-2 through the post-pandemic period (2020)

29. Klepac, P., et al.: Contacts in context: large-scale setting-specific social mixing matrices from the BBC pandemic project. medRxiv (2020). https://doi.org/10.1101/2020.02.16.20023754, https://www.medrxiv.org/content/early/2020/03/05/2020.02.16.20023754

30. Lourenço, J., et al.: Fundamental principles of epidemic spread highlight the immediate need for large-scale serological surveys to assess the stage of the SARS-CoV-2 epidemic. medRxiv (2020)

31. Machens, A., Gesualdo, F., Rizzo, C., Tozzi, A.E., Barrat, A., Cattuto, C.: An infectious disease model on empirical networks of human contact: bridging the gap between dynamic network data and contact matrices. BMC Infect. Dis. **13**(1), 185 (2013). https://doi.org/10.1186/1471-2334-13-185

32. Masuda, N., Holme, P. (eds.): Temporal Network Epidemiology. TB. Springer, Singapore (2017). https://doi.org/10.1007/978-981-10-5287-3

33. McCaw, J.M., et al.: Comparison of three methods for ascertainment of contact information relevant to respiratory pathogen transmission in encounter networks. BMC Infect. Dis. **10**(1), 166 (2010)

34. Milne, G.J., Kelso, J.K., Kelly, H.A., Huband, S.T., McVernon, J.: A small community model for the transmission of infectious diseases: comparison of school closure as an intervention in individual-based models of an influenza pandemic. PloS One **3**(12), e4005 (2008)

35. Nowzari, C., Preciado, V.M., Pappas, G.J.: Analysis and control of epidemics: a survey of spreading processes on complex networks. IEEE Control Syst. Mag. **36**(1), 26–46 (2016)

36. Pastor-Satorras, R., Castellano, C., Van Mieghem, P., Vespignani, A.: Epidemic processes in complex networks. Rev. Mod. Phys. **87**(3), 925 (2015)

37. Perez, L., Dragicevic, S.: An agent-based approach for modeling dynamics of contagious disease spread. Int. J. Health Geogr. **8**(1), 50 (2009). https://doi.org/10.1186/1476-072X-8-50

38. Prakash, B.A., Chakrabarti, D., Valler, N.C., Faloutsos, M., Faloutsos, C.: Threshold conditions for arbitrary cascade models on arbitrary networks. Knowl. Inf. Syst. **33**(3), 549–575 (2012). https://doi.org/10.1007/s10115-012-0520-y

39. Preciado, V.M., Zargham, M., Enyioha, C., Jadbabaie, A., Pappas, G.: Optimal vaccine allocation to control epidemic outbreaks in arbitrary networks. In: 52nd IEEE Conference on Decision and Control, pp. 7486–7491. IEEE (2013)

40. Prem, K., et al.: The effect of control strategies to reduce social mixing on outcomes of the COVID-19 epidemic in Wuhan, China: a modelling study. Lancet Publ. Health **5**, e261–e270 (2020)

41. Pung, R., et al.: Investigation of three clusters of COVID-19 in Singapore: implications for surveillance and response measures. Lancet **395**, 1039–1046 (2020)
42. Rader, B., et al.: Crowding and the epidemic intensity of COVID-19 transmission. medRxiv (2020). https://doi.org/10.1101/2020.04.15.20064980, https://www.medrxiv.org/content/early/2020/04/20/2020.04.15.20064980
43. Riou, J., Althaus, C.L.: Pattern of early human-to-human transmission of wuhan2019 novel coronavirus (2019-nCoV), December 2019 to January 2020. Eurosurveillance **25**(4), 2000058 (2020)
44. Salathé, M., Kazandjieva, M., Lee, J.W., Levis, P., Feldman, M.W., Jones, J.H.: A high-resolution human contact network for infectious disease transmission. Proc. Nat. Acad. Sci. **107**(51), 22020–22025 (2010)
45. Salehi, M., Sharma, R., Marzolla, M., Magnani, M., Siyari, P., Montesi, D.: Spreading processes in multilayer networks. IEEE Trans. Netw. Sci. Eng. **2**(2), 65–83 (2015)
46. Sapiezynski, P., Stopczynski, A., Lassen, D.D., Lehmann, S.: Interaction data from the Copenhagen Networks Study. Sci. Data **6**(1), 1–10 (2019)
47. Soriano-Panos, D., Ghoshal, G., Arenas, A., Gómez-Gardenes, J.: Impact of temporal scales and recurrent mobility patterns on the unfolding of epidemics. J. Stat. Mech. Theory Exp. **2020**(2), 024006 (2020)
48. Stewart, W.J.: Probability, Markov Chains, Queues, and Simulation: The Mathematical Basis of Performance Modeling. Princeton university Press, Princeton (2009)
49. Van Mieghem, P., Van de Bovenkamp, R.: Non-Markovian infection spread dramatically alters the susceptible-infected-susceptible epidemic threshold in networks. Phys. Rev. Lett. **110**(10), 108701 (2013)
50. Vynnycky, E., White, R.: An Introduction to Infectious Disease Modelling. OUP Oxford, Oxford (2010)
51. Wilson, N., Barnard, L.T., Kvalsig, A., Verrall, A., Baker, M.G., Schwehm, M.: Modelling the potential health impact of the COVID-19 pandemic on a hypothetical European country. medRxiv (2020)
52. Wilson, N., Barnard, L.T., Kvalsvig, A., Baker, M.: Potential health impacts from the COVID-19 pandemic for New Zealand if eradication fails: report to the NZ ministry of health (2020)

# Applications

# The Dynamic Fault Tree Rare Event Simulator

Carlos E. Budde[1]([✉]) [iD], Enno Ruijters[2] [iD], and Mariëlle Stoelinga[1,3] [iD]

[1] Formal Methods and Tools, University of Twente, Enschede, The Netherlands
{c.e.budde,m.i.a.stoelinga}@utwente.nl
[2] BetterBe, Enschede, The Netherlands
mail@ennoruijters.nl
[3] Department of Software Science, Radboud University, Nijmegen, The Netherlands

**Abstract.** The dynamic-fault-tree rare event simulator, DFTRES, is a statistical model checker for dynamic fault trees (DFTs), supporting the analysis of highly dependable systems, e.g. with unavailability or unreliability under $10^{-30}$. To efficiently estimate such low probabilities, we apply the Path-ZVA algorithm to implement Importance Sampling with minimal user input. Calculation speed is further improved by selective automata composition and bisimulation reduction. DFTRES reads DFTs in the Galileo or JANI textual formats. The tool is written in Java 11 with multi-platform support, and it is released under the GPLv3. In this paper we describe the architecture, setup, and input language of DFTRES, and showcase its accurate estimation of dependability metrics of (resilient) repairable DFTs from the FFORT benchmark suite.

## 1 Introduction

Our modern societies depend heavily on complex electro-mechanical systems, making it essential to ensure that such systems are reliable. An industry-standard technique to assess reliability is fault tree analysis. However, an unavoidable bottleneck of this technique is that exact analysis becomes too memory-intensive for complex *dynamic fault trees* (DFTs [6]). Alternatively, Monte Carlo simulation can be used to statistically estimate the likelihood of undesired events such as system failure. Although constant in memory usage, this approach takes unacceptably long times to converge when a system failure is rare, i.e. highly unlikely. An effective solution then is to use *rare event simulation* (RES [15]).

This paper presents **DFTRES**[1]: a statistical analysis tool for DFTs that applies *Importance Sampling* (IS [10]). IS is one of the most efficient approaches to perform RES analyses, and allows **DFTRES** to drastically speed up accurate estimations of rare failures in repairable DFTs. Whereas most RES techniques rely on expert input, **DFTRES** allows a fully automatic application of IS [18].

---

[1] Available at https://github.com/utwente-fmt/DFTRES.

This work was partially funded by NWO project 15474 (*SEQUOIA*).

---

The original version of this chapter was revised: Reference 5 has been corrected. The correction to this chapter is available at https://doi.org/10.1007/978-3-030-59854-9_21

M. Gribaudo et al. (Eds.): QEST 2020, LNCS 12289, pp. 233–238, 2020.
https://doi.org/10.1007/978-3-030-59854-9_17

**Related Work.** Various tools exist to analyse DFTs, see [19]. The model checker Storm [11] offers a DFT front-end. Storm produces exact results through model checking, requiring the full state-space, and does not support repairs. Other tools for rare event simulation of automata include Plasma Lab [12], where the user must manually parameterise the model, and **FIG** [2] and **modes** [3], which implement a RES method other than IS, less suited to analyse DFTs.

Previous versions of **DFTRES** were experimentally evaluated in [18] and [9], where it was called "FTRES." In Sect. 3 we mention new features that have been implemented ever since, most prominently weak-bisimulation reduction during initial automata composition, and so-called forcing for time-bounded properties.

**Organization of the Paper.** After some background in Sect. 2, we explain the operation and structure of **DFTRES** in Sect. 3, and show its performance in Sect. 4.

## 2   Rare Event Simulation for Fault Trees

Fault trees are an industry-standard graphical formal-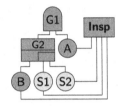
ism for reliability analysis [19]. A (dynamic) fault tree
models possible failures of a system by decomposing
it into *basic events*, denoted by circles and representing elemental failure causes of components, and *gates*,
denoted by various symbols and representing how failures interact and which combinations of smaller fail-
ures lead to system failure. Figure 1 shows an exam-     **Fig. 1.** A repairable DFT
ple: the top *AND*-gate (G1) means that both G2 and
A must fail for the system to fail. G2 is a *SPARE*-gate, meaning that B and its spares S1 and S2 must fail; but the spares cannot not fail before they are used. Insp denotes a periodic simultaneous inspection and repair of all basic events.

When basic events are decorated with failure probabilities or rates, it is possible to compute numerical resilience metrics of the system. These include *reliability*, the probability that the system remains functional until some given

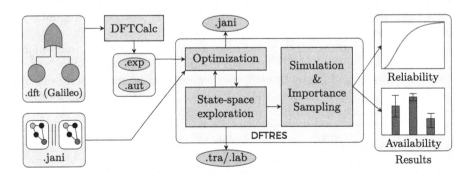

**Fig. 2.** The overall structure of **DFTRES**

"mission time," and also (for systems with repairable components) *availability*, the average fraction of time that the system is functional.

For large fault trees, particularly with complex dynamic gates describing time-dependent failure effects or with complex repair policies, exact numerical analysis becomes infeasible due to time and memory exhaustion. Such systems may still be analyzed using Monte Carlo simulation, at the expense of requiring many simulation runs for high accuracy, particularly when the event of interest (system failure) is highly unlikely.

**DFTRES** addresses this problem using Importance Sampling with the Path-ZVA algorithm [14]. This IS scheme effectively adjusts the failure rates to make system failures more likely, performs simulation runs, then corrects for the adjusted failure rates to estimate the original failure probability. This allows for high-accuracy estimations in relatively few simulation samples [18].

## 3    DFTRES

The architecture of **DFTRES** is depicted in Fig. 2: a fault tree in the widely-used Galileo format [9,11,17,20] (e.g. Fig. 3) is translated into a network of automata by DFTCalc [1], and input into **DFTRES**. Alternatively, a network of automata in JANI format [4]

```
1  toplevel "G1";
2  "G1" and "G2" "A";
3  "G2" wsp "B" "S1" "S2";
4  "A"  lambda=1.7e-5 dorm=1 phases=2 interval=1;
5  "B"  lambda=1.1e-3 dorm=1 phases=3 interval=2;
6  "S1" lambda=0.0021 dorm=0 phases=1 interval=1;
7  "S2" lambda=0.0021 dorm=0 phases=1 interval=1;
8  "Insp" 2insp4 "A" "B" "S1" "S2";
```

**Fig. 3.** (Extended) Galileo for Fig. 1

can be input directly. **DFTRES** then performs several optimizations to reduce the state-space, generates (a part of) the composed state-space, and performs (IS) simulations to estimate numeric metrics such as system reliability. The automata and composed system can also be output for analysis by other tools.

**DFTRES** begins its analysis with an optimization stage (new since [9]): transitions of the automata that cannot synchronize are removed and all automata are reduced modulo weak bisimulation. Further, so-called *don't care* optimization is performed by collapsing and discarding groups of states without observable behavior. Pairs of automata with a small composed state-space (by default at most 256 states) are composed and reduced again, and this process is repeated until no more compositions can be made.

Finally, to compute relevant metrics, simulation is performed using IS, namely the Path-ZVA algorithm [14] and, (new since [9]) for time-bounded properties, forcing [13]. Supported metrics are reliability (time-bounded or -unbounded reachability) and availability (steady-state probability). Mean time to failure (expected reward) can also be estimated, but not using IS. Simulation runs are sampled, in parallel on multi-core systems, until a specified time bound or simulation number is reached, or a desired relative or absolute estimated error is reached. Results are presented as (by default) 95% *confidence intervals* (CIs)[2].

---

[2] While every effort is made to provide accurate confidence intervals, their coverage can fall considerably below 95% due to the extreme probability distributions involved [8].

**DFTRES** is released under the GPLv3, and is cross-platform due to its implementation in Java, without run-time dependencies. It requires only a Java compiler and Make [7] to build. Galileo input is provided by DFTCalc, which is supported on Linux and Mac. **DFTRES** is designed to be easily extensible to additional input formats and IS schemes. **DFTRES**'s command-line interface provides many options, but typically requires only the model file, property, and desired accuracy. For instance, "java -jar DFTRES.jar -a --relErr 0.05 model.dft" estimates availability (-a) to a relative error of 0.05. More examples can be found in an artifact prepared for experimental reproduction [5].

## 4    Experimental Evaluation

We estimated the (un)reliability and (un)availability of four repairable DFTs from the FFORT benchmark [17]: Cabinets-2-2, FTPP-2-2-repair, HECS-2-2-repair, and RBC. All experiments ran in an 8-core Intel® i7-6700 with 24 GB RAM. The results are shown in Fig. 4.

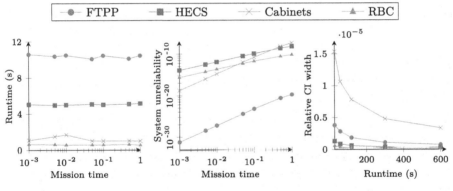

(a) Unreliability: runtime    (b) Unreliability: estimate    (c) Unavailability: rel. error

**Fig. 4.** Experimental results

We estimated the system unreliability (i.e. the probability that the system fails before) mission times 1.0, 0.5, 0.1, 0.05, 0.01, 0.005, and 0.001. We built 95% CIs for 5% relative error: Fig. 4b shows how the unreliability decreases exponentially—from right to left—as a function of the mission time. Figure 4a plots the runtime needed for CI with 5% accuracy. Unlike traditional simulation, runtime is almost independent of the value being estimated. Instead, the model structure and complexity is the primary factor affecting analysis time, mainly governed by the length of the shortest path(s) to a rare event.

Figure 4c shows unavailability analyses. We let estimations run for 0.5, 1, 2, 5, and 10 min, and measured the relative width of the resulting CI. With longer runtime **DFTRES** builds more accurate, narrower intervals: the precision improves approximately as the square root of time, which can be explained by observing that the standard error of the mean decreases as the square root of the number of samples.

In [5] we provide an artifact to easily reproduce our experiments. It runs in Debian-based Linux distributions, such as the virtual machine available at https://figshare.com/articles/tacas20ae_ova/9699839.

# References

1. Arnold, F., Belinfante, A., Van der Berg, F., Guck, D., Stoelinga, M.: DFTCALC: a tool for efficient fault tree analysis. In: Bitsch, F., Guiochet, J., Kaâniche, M. (eds.) SAFECOMP 2013. LNCS, vol. 8153, pp. 293–301. Springer, Heidelberg (2013). https://doi.org/10.1007/978-3-642-40793-2_27

2. Budde, C.E.: FIG: the finite improbability generator. TACAS 2020. LNCS, vol. 12078, pp. 483–491. Springer, Cham (2020). https://doi.org/10.1007/978-3-030-45190-5_27

3. Budde, C.E., D'Argenio, P.R., Hartmanns, A., Sedwards, S.: An efficient statistical model checker for nondeterminism and rare events. Int. J. Softw. Tools Technol. Transf. 1–22 (2020). https://doi.org/10.1007/s10009-020-00563-2

4. Budde, C.E., Dehnert, C., Hahn, E.M., Hartmanns, A., Junges, S., Turrini, A.: JANI: quantitative model and tool interaction. In: Legay, A., Margaria, T. (eds.) TACAS 2017. LNCS, vol. 10206, pp. 151–168. Springer, Heidelberg (2017). https://doi.org/10.1007/978-3-662-54580-5_9

5. Budde, C.E., Ruijters, E., Stoelinga, M.: The dynamic fault tree rare event simulator: experimental replication package (2020). https://figshare.com/articles/software/The_Dynamic_Fault_Tree_Rare_Event_Simulator/12235889, https://doi.org/10.6084/m9.figshare.12235889.v2

6. Dugan, J., Boyd, S.B.M.: Fault trees and sequence dependencies. In: Annual Proceedings on Reliability and Maintainability Symposium, pp. 286–293 (1990). https://doi.org/10.1109/ARMS.1990.67971

7. Feldman, S.I.: Make - a program for maintaining computer programs. Softw. Pract. Exp. 9(4), 255–265 (1979). https://doi.org/10.1002/spe.4380090402

8. Glynn, P.W., Rubino, G., Tuffin, B.: Robustness properties and confidence interval reliability issues. In: Rubino and Tuffin [16], pp. 63–84. https://doi.org/10.1002/9780470745403.ch4

9. Hartmanns, A., et al.: The 2019 comparison of tools for the analysis of quantitative formal models. In: TACAS. LNCS, vol. 11429, pp. 69-92. Springer (2019). https://doi.org/10.1007/978-3-030-17502-3_5

10. Heidelberger, P.: Fast simulation of rare events in queueing and reliability models. ACM Trans. Model. Comput. Simul. 5(1), 43–85 (1995). https://doi.org/10.1145/203091.203094

11. Hensel, C., Junges, S., Katoen, J.P., Quatmann, T., Volk, M.: The probabilistic model checker storm. arXiv e-prints arXiv:2002.07080 (2020). https://arxiv.org/abs/2002.07080

12. Jégourel, C., Legay, A., Sedwards, S.: Command-based importance sampling for statistical model checking. Theor. Comput. Sci. 649, 1–24 (2016). https://doi.org/10.1016/j.tcs.2016.08.009

13. Nicola, V.F., Shahabuddin, P., Nakayama, M.: Techniques for fast simulation of models of highly dependable systems. IEEE Trans. Reliab. **50**(3), 246–264 (2001). https://doi.org/10.1109/24.974122

14. Reijsbergen, D., de Boer, P.T., Scheinhardt, W., Juneja, S.: Path-ZVA: general, efficient and automated importance sampling for highly reliable Markovian systems. ACM TOMACS **28**(3), 22:1–22:25 (2018). https://doi.org/10.1145/3161569

15. Rubino, G., Tuffin, B.: Introduction to rare event simulation. In: Rubino and Tuffin [16], pp. 1–13. https://doi.org/10.1002/9780470745403.ch1

16. Rubino, G., Tuffin, B. (eds.): Rare Event Simulation Using Monte Carlo Methods. Wiley, Hoboken (2009). https://doi.org/10.1002/9780470745403

17. Ruijters, E., et al.: FFORT: a benchmark suite for fault tree analysis. In: ESREL, pp. 878–885 (2019). https://doi.org/10.3850/978-981-11-2724-3_0641-cd

18. Ruijters, E., Reijsbergen, D., de Boer, P.T., Stoelinga, M.: Rare event simulation for dynamic fault trees. Reliab. Eng. Syst. Safety **186**, 220–231 (2019). https://doi.org/10.1016/j.ress.2019.02.004

19. Ruijters, E., Stoelinga, M.: Fault tree analysis: a survey of the state-of-the-art in modeling, analysis and tools. Comput. Sci. Rev. **15–16**, 29–62 (2015). https://doi.org/10.1016/j.cosrev.2015.03.001

20. Sullivan, K.J., Dugan, J.B.: Galileo user's manual & design overview, v2.1-alpha (1998). www.cse.msu.edu/~cse870/Materials/FaultTolerant/manual-galileo.htm

# Entropy Measurement of Concurrent Disorder

Victor Cook(ID), Christina Peterson(✉)(ID), Zachary Painter(ID),
and Damian Dechev(ID)

University of Central Florida, Orlando, FL, USA
{victor.cook,clp8199,zacharypainter}@knights.ucf.edu, dechev@cs.ucf.edu

**Abstract.** There is an imminent demand to understand the relationship between correctness and performance to deliver highly scalable multiprocessor programs. The motivation for this relationship is that relaxed correctness conditions provide performance benefits by reducing contention on data structure hot spots. Previous approaches propose metrics for characterizing relaxed correctness conditions that measure the number of method calls or state transitions to be shifted to arrive at a legal sequential history. The reason the existing metrics cannot measure the performance effects of a correctness condition is that they ignore delays in method calls since delayed responses from method calls yield correct behavior in even the strictest correctness conditions. We observe that method call delays can be captured by measuring the disorder in method call ordering using a metric from information theory known as *entropy*. We propose entropy as the first metric for multiprocessor programs that evaluates the trade-offs between correctness and performance. We measure entropy for a variety of concurrent stacks, queues, and sets from the Synchrobench micro-benchmark suite and correlate entropy, correctness, and performance. Our main insight is that lower entropy corresponds to better performance for strict correctness conditions and higher entropy corresponds to better performance for relaxed correctness conditions.

**Keywords:** Concurrency · Scalability · Entropy

## 1 Introduction

Highly scalable multiprocessor programs are essential to fully utilize the processing power available in multicore architectures. Strict correctness guarantees expected for multiprocessor programs can limit scalability since concurrent processes must be synchronized to provide the illusion of a sequential execution. A *correctness condition* is a definition of correct behavior for a multiprocessor program. For example, *linearizability* [22] is a correctness condition such that a concurrent history of method calls is equivalent to a sequential history, and each method call appears to take effect at some instant between its invocation and response. If the effects of a method call occur between its invocation and

© Springer Nature Switzerland AG 2020
M. Gribaudo et al. (Eds.): QEST 2020, LNCS 12289, pp. 239–257, 2020.
https://doi.org/10.1007/978-3-030-59854-9_18

response, the method call takes effect in *real-time order*. A correctness condition that allows method calls to deviate from real-time order is referred to as a *relaxed correctness condition*. A data structure designed for a relaxed correctness condition implies non-linearizability.

A fundamental metric of concurrent data structures is *throughput* – the number of method calls completed per time unit. Other hardware metrics such as instructions executed, cache-misses, or branch mispredictions influence throughput. We use the term *performance* to cover all such metrics that affect data structure throughput. Many concurrent data structures exploit relaxed correctness conditions to achieve significant performance improvements [1,8,14,24,36,38]. These data structures employ contention reducing techniques to achieve performance gains such as allowing method calls to be performed in parallel in different segments of the data structure by relaxing the real-time ordering requirement [1,8,24] or relaxing the semantics of inherently sequential operations such as the DELETEMIN of a priority queue [36,38]. Understanding the relationship between correctness and performance is paramount in the development of multiprocessor programs at scale.

Previous works on relaxed correctness conditions [1,20] have proposed a metric for the number of method calls or state transitions to be shifted such that the history of method calls is in real-time order. However, these metrics do not characterize the effects of the applied correctness condition on performance. The reason that the existing metrics are not capable of measuring the performance effects of a correctness condition is that they neglect the delays in method calls since delayed method calls are considered correct behavior in all correctness conditions. Delays in method calls are caused by events such as process scheduling, hardware interrupts, software interrupts, or contention on frequently accessed memory locations.

Consider $N$ processes where each process enqueues an element into a linearizable First-In-First-Out (FIFO) queue. The queue must endure a large volume of contention as processes battle to update the tail of the queue to point to their element using a read-modify-write synchronization primitive such as *Compare-And-Swap* (CAS)[1]. The *method call order* is the order in which the method calls take effect. Since the ordering between overlapping method calls is indistinguishable until the outcome is observed, there are $N!$ ways in which these elements may be ordered in the queue. We account for method call delays in the observed method call order by measuring the deviations in the order that the method calls are invoked, referred to as the *invocation order*. Method call order that deviates from invocation order is referred to as *disorder*.

In information systems, *Shannon entropy* [33] measures the disorder in a probability distribution [12]. A sequential program whose outcome is deterministic is *predictable* because there is only one possible ordering of method calls,

---

[1] CAS accepts a memory location, expected value, and new value as parameters. If the dereferenced value of the memory location is equivalent to the expected value, then the memory location value is updated to the new value and true is returned. Otherwise, no change is made and false is returned.

yielding zero entropy. When non-determinism is introduced by executing multiple processes, the ordering of method calls is *unpredictable* due to the $N!$ possible orderings for $N$ overlapping methods, yielding increased entropy.

When entropy is *normalized* such that the range of entropy falls between zero and one, it represents the efficiency of the communication channels [39]. For a concurrent data structure, a communication channel is a unique memory location that a method call touches to access the concurrent data structure, referred to as an *entry point*. The number of entry points that are permitted in the data structure is affected by the correctness condition. For example, if elements in a queue are allowed to be enqueued out of order up to a distance of $k$, the queue can be divided into segments of length $k$ to reduce contention on the head and tail while maintaining the k-FIFO property [24]. The number of entry points utilized, which is computed from normalized entropy, determines the number of method calls that can be performed in parallel. Our main result is a theoretical analysis that expresses the relationship between normalized entropy and speedup of a concurrent program over a sequential program.

In this paper, we propose entropy as the first metric for multiprocessor programs that characterizes the relationship between correctness and performance. Entropy brings new insights into the relationship between correctness and performance by providing a measurement for the efficiency of the concurrent data structure entry points. We apply this knowledge in the optimization of concurrent data structures by finding opportunities for performance gains through relaxed correctness conditions. A key observation in our work is that when analyzing linearizable data structures, lower entropy implies better performance. We attribute this observation to increased predictability in method call ordering due to a design strategy that minimizes failed CAS attempts or delayed lock acquisitions during periods of high contention. We also show that the larger the unpredictability of the outcome of a multiprocessor program (higher entropy), the larger the opportunity for improved scalability by relaxing the correctness condition to reduce contention on the entry points of the data structure.

Entropy has broad applicability to domains in which the reordering of instructions/operations can achieve performance gains. For example, entropy can be applied to the C++ memory model [4] to gain insights on how unpredictability of instruction ordering due to relaxed semantics affects performance. We use the term *correlate* to describe the identification of a relationship between two or more properties of concurrent programs.

The contributions of this work include:

- We propose entropy as the first metric for multiprocessor programs that correlates correctness with performance.
- We propose a model that characterizes the relationship between entropy, correctness, and speedup that determines how the correctness condition affects scalability.
- We measure entropy for the Treiber stack [35], Elimination Backoff Stack [19], QStack [8], Michael–Scott queue [28], CC-queue [9], LCR-queue [29], and four set structures from the Synchrobench micro-benchmark suite [13].

– We provide a case study that uses the entropy metric to obtain performance gains by relaxing the correctness condition and evaluate the trade-offs between correctness and performance.

## 2    Related Work

Several works acknowledge the trade-off between scalability and correctness and have presented formal models for relaxed behaviors of concurrent data structures. Afek et al. [1] present quasi-linearizability, a correctness condition that allows a concurrent history to deviate from a legal sequential history by some bounded distance $k$, referred to as the *quasi-linearization factor*. Henzinger et al. [20] propose a framework for formally describing and quantifying relaxed semantics of concurrent data structures. A *sequential specification* of an object is a set of sequential histories for the object [22]. A sequential specification $S$ is described using a particular labeled transition system whose states are sets of sequences in $S$ and the transitions are labeled by method calls. A local transition cost function assigns a penalty to each wrong transition and a global path cost function accumulates the local costs to compute the overall distance of a sequence, referred to as the *sequence distance*.

The quasi-linearization factor [1] and sequence distance [20] measure how out of order a concurrent execution is with respect to real-time order. Since the frame of reference is a sequential specification for linearizable objects, these metrics do not suffice for the measurement of the disorder/unpredictability in executions where 1) the designed correctness condition is not linearizability, or 2) the history of method calls is correct but admits unpredictability (i.e., if two threads concurrently push a unique element onto a stack, it is unpredictable which push occurs first until the result is observed by a pop). Shannon entropy [33] is a suitable metric for quantifying the disorder/unpredictability in multiprocessor programs because it accounts for unpredictability due to non-determinism in concurrent executions that is applicable to any correctness condition.

Relaxed memory models [4] provide memory operations with weaker semantics than sequential consistency [27] with the benefit of improved performance through compiler optimizations. The weak semantics of relaxed memory models allow atomic operations to be reordered based on the specification. The `memory_order_consume` specification does not allow reads or writes in the current thread dependent on the value currently loaded to be reordered before this load. The `memory_order_acquire` specification does not allow reads or writes in the current thread to be reordered before this load. The `memory_order_release` specification does not allow reads or writes in the current thread to be reordered after this store. The `memory_order_relaxed` specification does not place any ordering constraint between other reads and writes.

A correctness model [30] is proposed that requires justified behaviors to be defined that express the portion of the concurrent history that justifies the non-deterministic behavior observed by the method calls that use relaxed semantics. To prevent undesirable non-deterministic behavior, Bender et al. [2] present a declarative fence insertion approach for specifying the expected orders of memory operations. Since justified behaviors and declarative fence insertion are focused on acceptable method call behavior resulting from a reordering of low-level atomic operations rather than a reordering of the high-level method calls, these formalisms are unsuitable for the measurement of disorder/unpredictability of the method calls in a multiprocessor program.

Entropy, having wide applicability due to its prevalence in information theory, is used as a metric for software engineering. Bianchi [3] measures entropy as the degree of disorder in a software system traceability to assess its degradation. Hassan et al. [16] propose entropy to evaluate the complexity of software development, which is measured according to source code change history. Hassan [15] extends the work of Hassan et al. [16] to predict the incidence of faults in a software system. Confora et al. [5] empirically investigates the relationship of entropy with factors such as changes occurring to software systems, design patterns in the source code, and the number of contributors that modified the source code files. Singh et al. [34] propose an entropy-based bug prediction approach using kernel based support vector regression.

Entropy is also adopted as a metric for software testing. The cross entropy method [31] solves both continuous multiextremal and discrete optimization problems by solving a sequence of simple auxiliary smooth optimization problems based on importance sampling, Markov chain, Boltzmann distribution, and Kullback–Leibler cross-entropy [23]. Rubinstein et al. [32] adapt the cross-entropy method for rare event simulation. Chockler et al. [6] present a software testing approach based on the cross-entropy methods that defines a performance function that is higher for the error or pattern of interest. Chockler et al. [7] develop an approach for replay in concurrent programs based on the cross-entropy method. The approach by Chockler et al. [6] is adapted for approximate replay by defining a performance function that reaches its global maximum on executions that are as close to the recorded execution as possible.

Our proposed entropy metric for multiprocessor programs differs from the use of entropy in software engineering because the disorder/unpredictability is measured for a concurrent execution rather than source code complexity. More similarities exist between our use of entropy and the usage of entropy in software testing. The main difference between our approach and the cross-entropy method [6,7] for software testing in concurrent programs is that software testing is concerned with bug-finding, while our approach is concerned with measuring disorder/unpredictability for concurrent executions to assess the relationship between correctness and performance.

# 3   Shannon Entropy Applied to Concurrent Data Structures

Information theory accounts for the idea that the occurrence of a low probability event conveys more information content than the occurrence of a high probability event because it reveals information regarding an unexpected event.

The information content, referred to as the *surprisal*, is presented in Eq. 1 for an event $e$ with probability $p(e)$.

$$I(e) = -\log p(e) \tag{1}$$

Shannon entropy [33] quantifies the average surprisal of a probability distribution. Equation 2 presents the Shannon entropy for a discrete random variable $X$ with possible values $\{x_1, \cdots, x_n\}$ and probability mass function $P(X)$.

$$H(X) = -\sum_{i=1}^{n} P(x_i)\log P(x_i) \tag{2}$$

The computation of the probability mass function $P(X)$ varies for each of the abstract data types for concurrent data structures. We now present our proposed technique for measuring entropy in concurrent data structures.

### 3.1   Queues and Stacks

A queue incorporates a FIFO ordering for its elements, while a stack incorporates a Last-In-First-Out (LIFO) ordering for its elements. An unpredictable outcome occurs when the elements of a queue or stack are ordered in a way that deviates from FIFO (queue) or LIFO (stack) semantics. The element order in a queue or stack represents the method call order. The expected order of method calls is determined according to invocation order. The deviations in the expected ordering of method calls is measured by counting the observed inversions. Given a queue or stack, let each term $a_j$ in list $a_1, a_2, ..., a_n$ represent the invocation order of element $j$ in the queue or stack. For example, if the elements of the data structure follow FIFO semantics, $a_1 = 1$, $a_2 = 2$, ..., $a_n = n$. The inversion count $x(j)$ for queues is defined for each list term $a_j$ in Eq. 3 [25], where each position of $x(j)$ is initialized to zero. The value of $x(j)$ represents the number of swaps required to sort element $j$ into its expected order in the data structure.

```
for (j = 1; j <= n; j++)
  for (i = 1; i < j; i++)
    if (a_i > a_j)
      x(j) = x(j) + 1
```
(3)

```
for (j = 1; j <= n; j++)
  for (i = 1; i < j; i++)
    if (a_i < a_j)
      x(j) = x(j) + 1
```
(4)

```
for (i = 0; i < n; i++)
  for (j = 1; j <= n; j++)
    if (x(j) == i)
      k_i = k_i + 1
```
(5)

If the elements of a data structure follow LIFO semantics, $a_1 = n$, $a_2 = n-1$, ..., $a_n = 1$. The inversion count $x(j)$ for stacks is defined for each list term $a_j$ in Eq. 4 [25], where each position of $x(j)$ is initialized to zero.

For a queue or stack with $n$ elements, the inversion count for an element is between 0 to $n - 1$. The discrete random variable $X$ for an element of a queue or stack is the inversion count. To compute the probability mass function, we measure the probability that the discrete random variable is equal to some value in an execution. The possible values for discrete random variable $X$ (i.e. number of inversions) are $\{0, \cdots, n-1\}$. Let $k_i$ be a count of the number of elements in the queue or stack that observe inversion count $i$. The value of $k_i$ is computed by Eq. 5, where $k_i$ is initialized to zero for each $i$.

The probability mass function for inversion count $i$ is given by Eq. 6. The Shannon entropy for queues and stacks is presented in Eq. 7, derived by replacing the probability mass function in Eq. 2 with the probability mass function presented in Eq. 6.

$$P(i) = k_i/n \qquad (6) \qquad H(X) = -\sum_{i=0}^{n-1} \frac{k_i}{n} \log \frac{k_i}{n} \qquad (7)$$

### 3.2 Sets and Maps

Sets and maps represent a collection of elements. A set labels each of its elements with a unique key. A map extends the set abstract data type to also include a value for each key. An element's order in the data structure implementing the set or map is referred to as *concrete order*. Since the concrete order does not reflect the order that the method calls take effect, we use the invocation order of the set/map method calls rather than the concrete order. The deviations in the expected ordering of method calls is measured by counting the shortest distance, i.e. number of method call swaps, with respect to the invocation order to produce a legal sequential history. Given a set or map, let list $m_1, m_2, ..., m_n$ be a list of method calls where each term $m_j$ is the method call with invocation order $j$ in the set or map. Each field $m_j.op$ is the operation insert (ins) or remove (rem). Each field $m_j.key$ is the key of the operation. The distance $x(j)$ for sets and maps is defined in Eq. 8, where each position of $x(j)$ is initialized to zero. The *SkipSet* prevents double counting the required swaps. For a set or map with $n$ method calls, the possible values for discrete random variable $X$ (i.e. distance) are $\{0, \cdots, n-1\}$. Let $k_i$ be the the number of method calls that observe distance $i$. The value of $k_i$ is computed by Eq. 5, where $k_i$ is initialized to zero for each $i$. The probability mass function is computed using Eq. 6 and Shannon entropy is computed using Eq. 7.

```
for (j = 1; j <= n; j++) {                    else  if (m_j.op == rem && m_j ∉ SkipSet)
  if (m_j.op == ins && m_j ∉ SkipSet)          if (m_j.key ∈ S)  S = S \ m_j.key
  if (m_j.key ∉ S)  S = S ∪ m_j.key            else  {
  else {                                         x(j) = dist(m_i) such that  i > j,      (8)
    x(j) = dist(m_i) such that  i > j,            m_i.op = ins, m_i.key = m_j.key
    m_i.op = rem, m_i.key = m_j.key               SkipSet = SkipSet ∪ m_i
    SkipSet = SkipSet ∪ m_i                      }
  }                                           } end for − loop
}
```

## 4  Theoretical Analysis of the Correlation Between Entropy, Correctness, and Speedup

We now derive an equation that correlates entropy, correctness, and speedup. Our theoretical analysis is based on the notion of an *entry point* which denotes a unique memory location that a method call must touch to access the concurrent data structure. For example, a queue has two entry points – the head and the tail The correctness condition affects the number of entry points that can be supported by the data structure such that any history of method calls invoked on the data structure is correct.

Entropy is computed from the probability of the inversion events for a data structure. If the probability distribution for the inversion events is uniform, then entropy approaches its maximum value. If the probability distribution for the inversion events is non-uniform, then entropy decreases. If any particular inversion event has a 100% chance of occurrence, then entropy is 0 because there is no uncertainty and therefore no new information is conveyed. When all entry points of a data structure are equally utilized the probability of an inversion event exhibits a uniform distribution. The effective use of the communication channels (i.e. the entry points of a data structure) is quantified by *efficiency*, which is directly related to entropy [39]. Let $n$ be the maximum inversion count observed for a data structure. Efficiency is computed by Eq. 9, also referred to as *normalized entropy* [26].

$$\eta(X) = -\sum_{i=0}^{n-1} \frac{P(x_i)\log P(x_i)}{\log n} = \frac{H(X)}{\log n} \tag{9}$$

We now deduce the relationship between normalized entropy and the number of entry points utilized, then use this relationship to provide a theoretical analysis of the correlation between normalized entropy and theoretical speedup. For the purposes of computing the theoretical speedup, we assume that the theoretical maximum number of entry points is equal to the maximum inversion count. However, due to the design constraints of the data structure, only a fraction of the available entry points will be utilized. We assume that the entry points utilized, denoted by $u$, are accessed with a uniform distribution. The probability of accessing one of the utilized entry points is $P(x_i) = \frac{n/u}{n} = \frac{1}{u}$. Since $P(x_i) = \frac{1}{u}$ for $u$ entry points and $P(x_i) = 0$ for all other entry points, normalized entropy in terms of entry points utilized is expressed by Eq. 10.

$$\eta(X) = -\frac{u \cdot \frac{1}{u} \cdot \log \frac{1}{u}}{\log n} = \frac{\log \frac{1}{u}}{\log n} = -\frac{(\log 1 - \log u)}{\log n} = \frac{\log u}{\log n} \tag{10}$$

We now solve Eq. 10 for the number of entry points utilized, expressed by Eq. 11 where $b$ is the log base.

$$u = b^{\eta(X) \cdot \log n} = b^{\log n^{\eta(X)}} = n^{\eta(X)} \tag{11}$$

The general equation for speedup of a multiprocessor program is provided in Eq. 12.

$$\text{speedup} = \frac{\textit{sequential execution time}}{\textit{concurrent execution time}} \tag{12}$$

Let $m$ be the execution time of a method call. Let $op$ be the total number of method calls to be invoked on the concurrent data structure. Let $p$ be the number of processes. The sequential execution time is $op \cdot m$. Assume that the number of utilized entry points is less than or equal to the number of processes. Since method calls that access different entry points run in parallel in the optimal scenario, the minimum achievable concurrent execution time is $\frac{1}{u} \cdot op \cdot m$. The theoretical speedup is expressed by Eq. 13.

$$\text{theoretical speedup} = \frac{op \cdot m}{\frac{1}{u} \cdot op \cdot m} = u \ (u \leq p) \tag{13}$$

The term $u$ in Eq. 13 is replaced by Eq. 11 to quantify the theoretical speedup in relation to normalized entropy when the number of entry points is less than or equal to the number of processes, expressed by Eq. 14.

$$\text{theoretical speedup} = n^{\eta(X)} \ (n \leq p) \tag{14}$$

If normalized entropy is 1, then the theoretical speedup equals the theoretical maximum number of entry points, which is the maximum inversion count. If normalized entropy is 0, then the theoretical speedup equals 1. The theoretical speedup is often not achieved in practice due to the overhead of concurrency synchronization and interactions of caches and pipelines, which we demonstrate in Sect. 6. However, computing theoretical speedup is still useful for choosing a correctness condition in the design and optimization of a concurrent data structure.

Now let's assume that the number of entry points exceeds that number of processes. Since the total number of entry points will be accessed $\frac{p}{n}$ fraction of the time, the maximum number of entry points utilized is $n \cdot \frac{p}{n} = p$. The theoretical speedup in relation to normalized entropy when the number of entry points exceeds the number of processes is expressed by Eq. 15.

$$\text{theoretical speedup} = p^{\eta(X)} \ (n > p) \tag{15}$$

The correlations between entropy, correctness, and observed performance as derived from our empirical analysis is provided in Sect. 5.5.

## 5   Experimental Evaluation

We measure the entropy for concurrent stacks, queues, and sets presented in literature and correlate the entropy metric with correctness and performance. Experiments were run on an AMD® EPYC® server of 2 GHz clock speed and 128 GB memory, with 32 cores delivering a maximum of 64 simultaneous multi-threads. The operating system is Ubuntu 18.04 LTS and code is compiled with gcc 7.3.0 using -O3 optimizations.

### 5.1   Experiment Design

Memory is pre-allocated to separate execution overhead from the entropy measurement. The thread counts vary from 1 to 32. The experiments comprise various configurations of producer/consumer interleavings. Each testing configuration includes 1000 iterations of 100 producer invocations followed by 100 consumer invocations. We do not use a global barrier to guarantee that all threads have completed their producer invocations prior to starting their consumer invocations. This would have the potential for creating cache misses and would introduce thread idle time into the experiment.

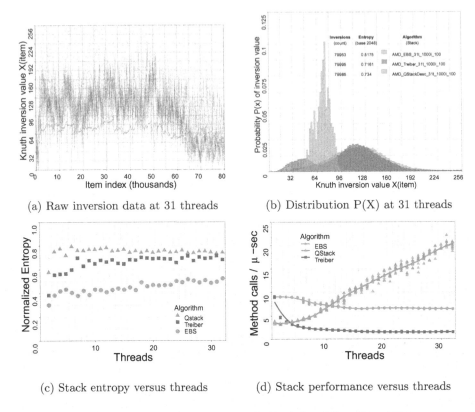

(a) Raw inversion data at 31 threads

(b) Distribution P(X) at 31 threads

(c) Stack entropy versus threads

(d) Stack performance versus threads

**Fig. 1.** Stack results, 1000 iterations of 100 pushes followed by 100 pops.

## 5.2 Stack Results

The stacks evaluated in the experiment include the Treiber stack [35], Elimination Backoff Stack (EBS) [19], and QStack [8]. The raw inversion data at 31 threads for the stacks is shown in Fig. 1a. The probability distribution of the inversion events at 31 threads with 1000 iterations of 100 pushes, 100 pops is shown in Fig. 1b. The EBS and Treiber stack are linearizable, while the QStack is designed for the quantifiability correctness condition [8]. Quantifiability allows method calls to be out of order with respect to real-time order.

Despite the allowable reordering of method calls permitted by quantifiability, the Treiber stack and EBS observe approximately the same number of inversion totals as the QStack. The inversion total for the Treiber stack is attributed to high contention on the stack top. The inversion total for EBS is due to the collision array that allows pairs of pushes and pops to meet in a separate location and eliminate each other's effects to reduce contention on the top of the stack. Method calls may wait in the collision array for some period of time prior to being eliminated which leads to inversion events due to deviations from invocation order. Since the inversion totals for the EBS are due to contention avoidance rather than unpredictable CAS failures, the observed inversion events for EBS are concentrated between 48 and 96. As a result, the EBS has the lowest entropy measurement of 0.5175 since the inversion events are more predictable than the QStack and Treiber stack. The QStack has the most uniform probability distribution, yielding the highest entropy measurement of 0.734. The Treiber stack has a probability distribution similar to the QStack, yielding the second highest entropy measurement of 0.7161.

The entropy measured for thread counts varying from 1 to 32 at 1000 iterations of 100 pushes, 100 pops is shown in Fig. 1c. As expected, the entropy increases as the thread count increases since larger thread counts introduce more unpredictability in the method call ordering. The general trend over all thread counts is that the QStack has the highest entropy, the Treiber stack has the second highest entropy, and the EBS has the lowest entropy.

The performance of the stack data structures is shown in Fig. 1d. The EBS outperforms the Treiber stack at all thread counts. The QStack has linear speedup with respect to the thread count. The QStack surpasses the Treiber stack at 3 threads and surpasses the EBS at 10 threads. Since the EBS entropy is lower than the Treiber entropy, it is clear that lower entropy indicates better performance for linearizable data structures due to lower contention causing fewer operation delays. However, the QStack has the highest entropy and also the best performance. This is caused by the QStack creating new branches for the top of the stack when contention is experienced. The QStack is leveraging the unpredictability in method call ordering due to high contention by diverting threads to other data structure access points.

## 5.3   Queue Results

The queues evaluated in the experiment include the Michael–Scott queue [28], CC-queue [9], and LCR-queue [29]. Entropy instrumentations are built on top of an ACM verified open source queue benchmark [37]. The probability distribution of the inversion events at 31 threads with 1000 iterations of 100 enqueues, 100 dequeues is shown in Figure 2b and the raw inversion in Figure 2a. The Michael–Scott queue, CC-queue, and LCR-queue are linearizable.

The LCR-queue has the highest throughput and also maintains an entropy comparable to the CC-queue. The inversion total for the CC-queue is less than the inversion total for the Michael–Scott queue because the CC-queue uses a flat combining technique [18] that enables the thread that acquires a lock to perform all pending requests that require access to this lock. Since the requests are performed in the same order in which the requests were atomically swapped into the request list, the majority of the observed inversions events are 5 or less. The Michael–Scott queue has a uniform probability distribution over inversion events of 0 to 32 due to unpredictable method call ordering caused by high contention at the head and tail.

The entropy measured for thread counts varying from 1 to 32 at 1000 iterations of 100 enqueues, 100 dequeues is shown in Fig. 2c. The CC-queue experiences the fewest inversions and the inversion values are predictable due to the

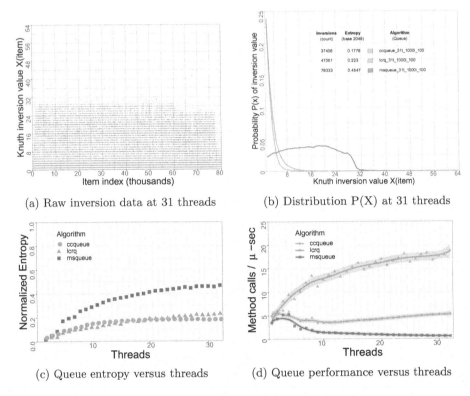

(a) Raw inversion data at 31 threads

(b) Distribution P(X) at 31 threads

(c) Queue entropy versus threads

(d) Queue performance versus threads

**Fig. 2.** Queue results, 1000 iterations of 100 pushes followed by 100 pops.

flat combining technique. As a result, the entropy for the CC-queue is the lowest of the three studied here. Figure 2d shows the throughput versus thread count for the queues. The CC-queue outperforms the Michael–Scott queue, and the LCR-queue outperforms both the CC-queue and Michael–Scott queue. This corresponds with our observation that lower entropy can yield higher performance for linearizable data structures.

## 5.4   Set Results

The set evaluated in the experiment include the Fraser skiplist [11], hand-over-hand list (hoh-list) [21], a lock-free list [10], and a lazy list [17]. Since set operations on different keys are commutative, the total number of inversions observed for sets is much lower than the total number of inversions observed for stacks or queues. The probability for the 0 inversion event is not shown in the plots for the set data because it diminishes the visibility of the trends of the probabilities for the remaining inversion events.

Figure 3a shows the probability distribution of the inversion events at 32 keys, 80% read operations, and various thread counts for the hoh-list. The normalized entropy is the lowest for the 8 thread configuration at 0.0491. The normalized

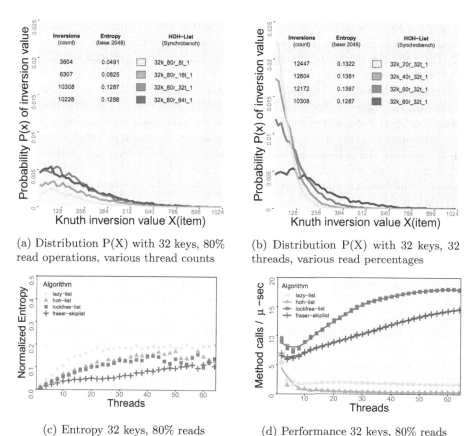

(a) Distribution P(X) with 32 keys, 80% read operations, various thread counts

(b) Distribution P(X) with 32 keys, 32 threads, various read percentages

(c) Entropy 32 keys, 80% reads

(d) Performance 32 keys, 80% reads

**Fig. 3.** Synchrobench set data structure results.

entropy is the highest for the 64 thread configuration at 0.1288. The majority of the inversion events for all thread configurations is 0 inversions. The 80% read configuration causes very little synchronization overhead, resulting in very few inversions. Figure 3b shows the probability distribution of the inversion events at 32 keys, 32 threads, and various read percentages for the hoh-list. The 80% read configuration obtains the lowest entropy at 0.1287 since a larger percentages of reads leads to very few inversions. The 60% read configuration obtains the highest entropy at 0.1397.

Figure 3c shows the entropy versus thread count at 32 keys, 80% reads for the lazy list, hoh-list, lock-free list, and Fraser skiplist. The Fraser skiplist has the lowest entropy since operations quickly traverse the list through shortcuts, minimizing method call delays. The lock-free list entropy is higher than the Fraser skiplist because the backlink traversal varies between each operation depending on the number of nodes to be traversed to arrive at a node not flagged for deletion. The hoh-list entropy is slightly higher than the lock-free list because the hoh-list locks both the predecessor and current node, eliminating the need to retraverse the list. The lazy list has the highest entropy because it optimistically traverses the list without acquiring locks and if a conflict exists with lock acquisition it will retraverse the list, leading to long method call delays.

Figure 3d shows the throughput versus thread count at 32 keys, 80% reads for the lazy list, hoh-list, lock-free skiplist, and Fraser skiplist. The lock-free list has the best performance because the backlinks enable quick recovery on a failed CAS. Although the lazy list has the highest entropy, it performs better than the hoh-list because the hoh-list incurs high overhead due to locking the predecessor and current node for each traversed node. Since all method calls endure this overhead, the hoh-list method call order is closer to the invocation order than the lazy list.

### 5.5   Correlations Between Entropy, Correctness, and Performance

The entropy metric provides different insights for different correctness conditions. For linearizable data structures, lower entropy corresponds to better performance. This occurs because high contention causes unpredictable method call ordering which yields high entropy. High inversion totals do not necessarily correspond to unpredictable method call ordering. For example, the EBS has a collision array that provides multiple entry points to reduce contention on the stack top when contention is high. The EBS has high inversion totals but low entropy because the collision array buffers pending operations during periods of high contention, making the inversion events predictable. Although low inversion totals are an indication of low contention in linearizable data structures, low inversion totals also imply that method calls are being filtered through a sequential bottleneck to achieve method call orderings that are close to invocation order. However, a method call ordering that is close to the invocation order can be an indication of efficient resource management (e.g. the LCR queue) since method calls that wait to acquire resources will cause deviations from invocation order.

For non-linearizable data structures, high entropy to corresponds to better performance. Since correctness conditions that allow methods to be called out of order with respect to real-time order can support multiple entry points into the data structure to reduce contention, unpredictable method call ordering is an indication of full utilization of the entry points. Data structure performance can be maximized by adding additional entry points as permitted by the constraints of the correctness condition.

## 6   Case Study: k-FIFO Queue

Our case study demonstrates how to use the entropy metric to reveal insights on the relationship between correctness and performance at various values of $k$ for the unbounded k-FIFO queue [24]. The unbounded k-FIFO queue maintains an

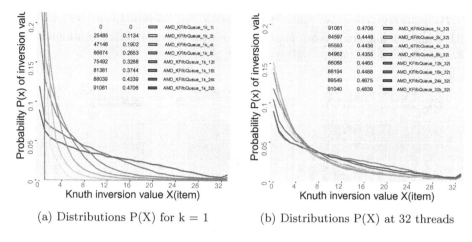

(a) Distributions P(X) for k = 1          (b) Distributions P(X) at 32 threads

**Fig. 4.** Distributions P(X) and entropies for kFIFO queue.

unbounded list of segments of size $k$, allowing up to $k$ concurrent enqueue and $k$ dequeue operations.

Figure 4a shows the probability distribution of the inversion events at the $k = 1$ configuration, which is equivalent to a linearizable FIFO queue, with 1000 iterations of 100 enqueues, 100 dequeues. The $k = 1$ configuration at 1 thread is equivalent to a sequential queue, so the entropy is 0. The entropy increases as the number of threads increases due to high contention on the head and tail. The 32 thread configuration obtain the highest entropy at 0.4706. Applying Eq. 14 with two entry points (the head and the tail), the theoretical speedup is $2^{0.4706} = 1.39$. Regardless of the number of threads operating on the queue at $k = 1$, the maximum theoretical speedup is capped at 2. This indicates that a relaxed correctness condition is required to achieve a higher speedup.

Figure 4b shows the probability distribution of the inversion events at 32 threads with 1000 iterations of 100 enqueues, 100 dequeues. When the number of threads is increased to 32, the entropy is approximately equivalent for all configurations of $k$. This occurs because the contention at the head and tail causes the inversions to be nearly as high as 32 for all values of $k$ due to the possibility for 32 overlapping method calls. Figure 5a maps the entropy over the full range of $K$ and $N$. Consistent with the probability distributions in Fig. 4a and Fig. 4b, the entropy is high where either $K$ or $N$ approach the upper limit of 32. The case study shows how entropy is a useful metric for multi-core programmers and system architects. Equation 15 for theoretical speedup is applicable to the k-FIFO because the number of entry points is greater than the number of threads. However, Fig. 5b shows that the actual speedup is much lower than the theoretical speedup. Selecting the optimal thread count $N$ and relaxation of semantics $K$ is a trade-off between the desired speedup and required correctness guarantees.

The methodology we propose is to use entropy and performance data together with application specific requirements to find the optimal performance entropy mix, which we define as the data structure *efficacy*.

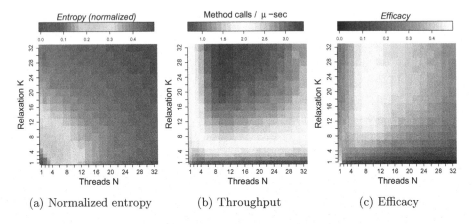

(a) Normalized entropy          (b) Throughput          (c) Efficacy

**Fig. 5.** Heatmaps over K, N for kFIFO queue case study.

For example, application "A" may require a FIFO queue with low entropy, say less than 0.25. Searching the performance map where entropy is less than 0.25, we find the highest value (1.85 ops per μs) is found at $N = 4, K = 6$. For application "A" the k-FIFO queue should use these parameters.

Application "B" might be more flexible and seek the highest performance while being able to tolerate relaxed semantics that result in high entropy. The formula used in Fig. 5c is $Efficacy = Throughput \cdot (1 - Entropy^{0.5})$, representing a trade-off between maximum throughput and minimum entropy. The efficacy map shows a "sweet spot" at $K = 20, N = 10$ with the optimal efficacy of 1.2 yielding 3.12 ops per μs. These parameters acheive significant speedup with an acceptable level of disorder, given the requirements for application "B".

## 7    Conclusion

We have presented entropy as the first metric for multiprocessor programs that correlates correctness and performance. Entropy brings new insights into the analysis of concurrent data structures by providing a measurement for the efficiency of the data structure entry points. To assist designers with determining the trade-offs between correctness and performance, we derive an equation that characterizes the relationship between entropy, correctness, and theoretical speedup. This equation enables a quantitative approach for comparing the theoretical speedup for different correctness conditions.

We show experimental results measuring entropy for concurrent stacks, queues, and sets and show that there is a relationship between entropy, correctness, and performance. Our observation is that lower entropy corresponds to better performance for strict correctness conditions because method calls that efficiently access a data structure experience few delays and observe a method call ordering close to invocation order. Higher entropy corresponds to better performance for relaxed correctness conditions because the allowable out of order method call execution enables the processes to uniformly distribute their work among the memory location access points of the data structure.

Our case study demonstrates how to use the entropy metric to select the optimal thread count and $k$ value for the k-FIFO queue by plotting efficacy, which finds a balance between maximum throughput and minimum entropy. The experimental evaluation and case study motivate the adoption of relaxed correctness conditions that are demonstrated to be scalable using entropy.

# References

1. Afek, Y., Korland, G., Yanovsky, E.: Quasi-linearizability: relaxed consistency for improved concurrency. In: Lu, C., Masuzawa, T., Mosbah, M. (eds.) OPODIS 2010. LNCS, vol. 6490, pp. 395–410. Springer, Heidelberg (2010). https://doi.org/10.1007/978-3-642-17653-1_29
2. Bender, J., Lesani, M., Palsberg, J.: Declarative fence insertion. In: Proceedings of the 2015 ACM SIGPLAN International Conference on Object-Oriented Programming, Systems, Languages, and Applications, pp. 367–385. ACM (2015)
3. Bianchi, A., Caivano, D., Lanubile, F., Visaggio, G.: Evaluating software degradation through entropy. In: Proceedings Seventh International Software Metrics Symposium, pp. 210–219. IEEE (2001)
4. Boehm, H.J., Adve, S.V.: Foundations of the c++ concurrency memory model. In: Proceedings of the 29th ACM SIGPLAN Conference on Programming Language Design and Implementation, pp. 68–78. ACM (2008)
5. Canfora, G., Cerulo, L., Di Penta, M., Pacilio, F.: An exploratory study of factors influencing change entropy. In: 2010 IEEE 18th International Conference on Program Comprehension, pp. 134–143. IEEE (2010)
6. Chockler, H., Farchi, E., Godlin, B., Novikov, S.: Cross-entropy based testing. In: Formal Methods in Computer Aided Design (FMCAD 2007), pp. 101–108. IEEE (2007)
7. Chockler, H., Farchi, E., Godlin, B., Novikov, S.: Cross-entropy-based replay of concurrent programs. In: Chechik, M., Wirsing, M. (eds.) FASE 2009. LNCS, vol. 5503, pp. 201–215. Springer, Heidelberg (2009). https://doi.org/10.1007/978-3-642-00593-0_14
8. Cook, V., Peterson, C., Painter, Z., Dechev, D.: Quantifiability: concurrent correctness from first principles. arXiv preprint arXiv:1905.06421 (2019)
9. Fatourou, P., Kallimanis, N.D.: Revisiting the combining synchronization technique. In: Proceedings of the 17th ACM SIGPLAN Symposium on Principles and Practice of Parallel Programming, pp. 257–266. ACM (2012)
10. Fomitchev, M., Ruppert, E.: Lock-free linked lists and skip lists. In: Proceedings of the Twenty-Third Annual ACM Symposium on Principles of Distributed Computing, pp. 50–59 (2004)
11. Fraser, K.: Practical lock-freedom. Technical report, University of Cambridge, Computer Laboratory (2004)
12. Goodfellow, I., Bengio, Y., Courville, A.: Deep Learning. MIT Press, Cambridge (2016). http://www.deeplearningbook.org

13. Gramoli, V.: More than you ever wanted to know about synchronization: synchrobench, measuring the impact of the synchronization on concurrent algorithms. In: Proceedings of the 20th ACM SIGPLAN Symposium on Principles and Practice of Parallel Programming (PPoPP 2015), pp. 1–10. ACM (2015)

14. Haas, A., et al.: Distributed queues in shared memory: multicore performance and scalability through quantitative relaxation. In: Proceedings of the ACM International Conference on Computing Frontiers, pp. 1–9 (2013)

15. Hassan, A.E.: Predicting faults using the complexity of code changes. In: Proceedings of the 31st International Conference on Software Engineering, pp. 78–88. IEEE Computer Society (2009)

16. Hassan, A.E., Holt, R.C.: The chaos of software development. In: Sixth International Workshop on Principles of Software Evolution, 2003. Proceedings, pp. 84–94. IEEE (2003)

17. Heller, S., Herlihy, M., Luchangco, V., Moir, M., Scherer, W.N., Shavit, N.: A lazy concurrent list-based set algorithm. In: Anderson, J.H., Prencipe, G., Wattenhofer, R. (eds.) OPODIS 2005. LNCS, vol. 3974, pp. 3–16. Springer, Heidelberg (2006). https://doi.org/10.1007/11795490_3

18. Hendler, D., Incze, I., Shavit, N., Tzafrir, M.: Flat combining and the synchronization-parallelism tradeoff. In: Proceedings of the Twenty-Second Annual ACM Symposium on Parallelism in Algorithms and Architectures, pp. 355–364. ACM (2010)

19. Hendler, D., Shavit, N., Yerushalmi, L.: A scalable lock-free stack algorithm. In: Proceedings of the Sixteenth Annual ACM Symposium on Parallelism in Algorithms and Architectures, pp. 206–215. ACM (2004)

20. Henzinger, T.A., Kirsch, C.M., Payer, H., Sezgin, A., Sokolova, A.: Quantitative relaxation of concurrent data structures. In: Proceedings of the 40th Annual ACM SIGPLAN-SIGACT Symposium on Principles of programming languages (POPL 2013), pp. 317–328. ACM (2013)

21. Herlihy, M., Shavit, N.: The Art of Multiprocessor Programming. Morgan Kaufmann, Burlington (2012)

22. Herlihy, M.P., Wing, J.M.: Linearizability: a correctness condition for concurrent objects. ACM Trans. Program. Lang. Syst. (TOPLAS) **12**(3), 463–492 (1990)

23. Kapur, J.N., Kesavan, H.K.: Entropy optimization principles and their applications. In: Singh, V.P., Fiorentino, M. (eds.) Entropy and Energy Dissipation in Water Resources. Water Science and Technology Library, vol. 9, pp. 3–20. Springer, Dordrecht (1992). https://doi.org/10.1007/978-94-011-2430-0_1

24. Kirsch, Christoph M., Lippautz, Michael, Payer, Hannes: Fast and scalable, lock-free k-FIFO queues. In: Malyshkin, Victor (ed.) PaCT 2013. LNCS, vol. 7979, pp. 208–223. Springer, Heidelberg (2013). https://doi.org/10.1007/978-3-642-39958-9_18

25. Knuth, D.E.: The Art of Computer Programming: Volume 3: Sorting and Searching. Addison-Wesley Professional, Boston (1998)

26. Kumar, U., Kumar, V., Kapur, J.N.: Normalized measures of entropy. Int. J. Gener. Syst. **12**(1), 55–69 (1986)

27. Lamport, L.: How to make a multiprocessor computer that correctly executes multiprocess program. IEEE Trans. Comput. **28**(9), 690–691 (1979)

28. Michael, M.M., Scott, M.L.: Simple, fast, and practical Non-Blocking and blocking concurrent queue algorithms. Tech. Rep. **600**, 267–275 (1995)

29. Morrison, A., Afek, Y.: Fast concurrent queues for x86 processors. ACM SIGPLAN Not. **48**(8), 103–112 (2013). https://doi.org/10.1145/2517327.2442527

30. Ou, P., Demsky, B.: Checking concurrent data structures under the c/c++11 memory model. In: Proceedings of the 22nd ACM SIGPLAN Symposium on Principles and Practice of Parallel Programming (PPoPP 2017), January 2017

31. Rubinstein, R.Y.: Optimization of computer simulation models with rare events. Eur. J. Oper. Res. **99**(1), 89–112 (1997)

32. Rubinstein, R.Y., Kroese, D.P.: The Cross-entropy Method: A Unified Approach to Combinatorial Optimization, Monte-Carlo Simulation and Machine Learning. Springer, Heidelberg (2013)

33. Shannon, C.E.: A mathematical theory of communication. Bell Syst. Tech. J. **27**(3), 379–423 (1948)

34. Singh, V., Chaturvedi, K.: Entropy based bug prediction using support vector regression. In: 2012 12th International Conference on Intelligent Systems Design and Applications (ISDA), pp. 746–751. IEEE (2012)

35. Treiber, R.K.: Systems programming: coping with parallelism. Technical Report RJ 5118, IBM Almaden Research Center, April 1986. San Jose, CA (1986)

36. Wimmer, M., Gruber, J., Träff, J.L., Tsigas, P.: The lock-free k-LSM relaxed priority queue. In: Proceedings of the 20th ACM SIGPLAN Symposium on Principles and Practice of Parallel Programming (PPoPP 2015), vol. 50, no. 8, pp. 277–278 (2015)

37. Yang, C., Mellor-Crummey, J.: A wait-free queue as fast as fetch-and-add. In: Proceedings of the 21st ACM SIGPLAN Symposium on Principles and Practice of Parallel Programming - PPoPP 2016, pp. 1–13. ACM Press, New York (2016)

38. Zhang, D., Dechev, D.: A lock-free priority queue design based on multidimensional linked lists. IEEE Trans. Parallel Distrib. Syst. **27**(3), 613–626 (2015)

39. Zunino, L., Zanin, M., Tabak, B.M., Pérez, D.G., Rosso, O.A.: Forbidden patterns, permutation entropy and stock market inefficiency. Physica A **388**(14), 2854–2864 (2009)

# Hardening Critical Infrastructure Networks Against Attacker Reconnaissance

Kartik Palani[(✉)] and David M. Nicol

Information Trust Institute, University of Illinois at Urbana-Champaign,
Urbana, IL, USA
{palani2,dmnicol}@illinois.edu

**Abstract.** The knowledge an attacker gathers about the critical infrastructure network they infiltrate allows them to customize the payload and remain undetected while causing maximum impact. This knowledge is a consequence of internal reconnaissance in the cyber network by lateral movement and is enabled by exploiting discovered vulnerabilities. This stage of the attack is also the longest, thereby giving a defender the biggest opportunity to detect and react to the attacker.

This paper helps a defender minimize the information an attacker might gain once in the network. This can be done by curbing lateral movement, misdirecting the attacker or inhibiting reachability to a critical device. We use a linear threshold models of attack propagation to analyze potential attack loss and use this to find actions that a defender might invest in while staying within their budgetary constraints. We show that while finding the best solution subject to these constraints is computationally intractable, the objective function is supermodular, allowing for a tractable technique with a known approximation bound.

## 1 Introduction

An attack on a critical infrastructure is most effective when the attacker understands the cyber network as well as the physical process under control. The attacker primarily gains this knowledge in the reconnaissance stage. The goal during this phase is to learn more about the environment by lateral movement which is enabled by exploiting vulnerabilities in devices and thereby increasing network penetration. This stage is also the longest phase of the attack, thereby giving the best opportunity to discover and possibly respond to the presence of an attacker.

This paper provides the defender with a tool to minimize the information an attacker might gain during the reconnaissance stage. The defender response can be some combination of curbing lateral movement, misdirecting the attacker or inhibiting reachability to a critical device. We model the propagation of attack through the network, and use this to find actions that a defender might invest in while staying within their budgetary constraints.

© Springer Nature Switzerland AG 2020
M. Gribaudo et al. (Eds.): QEST 2020, LNCS 12289, pp. 258–275, 2020.
https://doi.org/10.1007/978-3-030-59854-9_19

Attack graphs have been used to map possible paths an attacker might take within a network to reach their goal. Given the scale of the network and the possible states it can be in, the size of an attack graph can be in the order of thousands of nodes. And, while there has been work in the past that looks at generating such graphs and verifying them [23,24], very little work has been done on using such a graph to harden the network [8]. We argue that at the scale of current attack graphs, it is very hard for an analyst to make useful decisions without exploiting properties of the graph and the metric in question.

We pose the question of network hardening in terms of modifying the attack graph (deleting edges) and show that despite the problem being computationally complex, there exist properties that can be exploited to get guarantees on the analysis.

**Our Contributions**

In this paper we formally define the network hardening problem against a model of attacker reconnaissance. The paper makes the following contributions:

1. We describe the attack propagation model under an attacker whose goal is to maximize knowledge of the network.
2. We show that the network hardening problem is NP-hard.
3. Using special properties of the defender metric, we show that a greedy algorithm performs with a provable approximation bound of $1 - \frac{1}{e}$.
4. We improve the complexity of the greedy algorithm from quadratic to linear.

We will first present some background on the cyber kill chain in critical infrastructures, diffusion processes and supermodularity as it relates to the network hardening problem (Sect. 2). We then proceed to formally define the network hardening under reconnaissance problem by describing models of attack propagation through attack graphs and of defender actions (Sect. 3). We then use the notion of live-edge graphs to inform our proofs of supermodularity and monotonicity of the objective function under defender actions (Sect. 4). Finally, we present the algorithm for efficiently solving the supermodular optimization problem we describe (Sect. 5) before presenting a discussion (Sect. 7) and concluding.

## 2   Background

### 2.1   Critical Infrastructure Attacks

Our analysis of recent attacks, including the power outage in Ukraine in 2015 and 2016 [16] as well as the 2017 attack on a Saudi Arabian oil and gas facility [15], shows that in order to remain inconspicuous, attackers avoid using malware and techniques that can be associated with adversarial behavior. This type of attack is known as *living off the land*. The success of these attacks comes from the ability of the attacker to move in the network by building a knowledge base in and about the network. It is important to note that while most defensive counter measures dwell on monitoring and protecting against the final impact of an intrusion, the biggest opportunity lies in the steps that lead up to the final step. To this end, the Industrial Control System (ICS) Cyber Kill Chain which was developed to help the defender characterize an attack, can be used as a tool to understand the stages of an intrusion leading up to an attack. A detailed description of the kill chain can be found at [16], but our focus will be on the internal reconnaissance stage.

Internal reconnaissance is the phase where the attacker attempts to find potentially interesting targets, as well as tries to acquire passwords or other credentials in order to attain an increased access to the system. Using these acquired credentials, attackers move laterally and repeat while maintaining stealthy presence. The goal is to be able to understand the network and the process being controlled well enough to be able to develop and deliver an attack that has the intended impact. An example of this can be observed in the second attack on Ukraine's power grid where the attacker understood the ability of the safety control system to thwart their intended impact and used this knowledge to develop a device specific DoS attack for the safety controls. Gaining this information and formulating usable knowledge from it takes attackers a long time, in certain cases up to a year (Ukraine 2015). Thus, while the delivery of the final attack might be instantaneous thereby requiring detection to be high precision and to act in real-time, a greater opportunity for detecting/preventing attack is to monitor/block the reconnaissance activity of an attacker.

Implementing a defense-in-depth strategy improves security by raising the cost of an intrusion for an adversary while simultaneously improving the probability of detection by the defender. The end goal is to reduce the opportunities for an adversary to take advantage of the ability to move laterally through a critical infrastructure network. The use of multiple layers not only helps prevent direct attacks against critical systems but also greatly increases the difficulty of reconnaissance activities on ICS networks. Our goal in this paper is to develop a systematic methodology for implementing such hardening techniques.

## 2.2 Diffusion Networks

Diffusion networks are used to model the spreading behavior of disease, influence or information through large networks. In its basic form a diffusion network is a set of nodes with directed edges that indicate the potential transmission of information, disease etc. A transmission matrix is used to indicate pairwise transmission rates between the nodes. Various diffusion models [7,10] proposed in the literature mostly differ in the how the transmission of infection is modeled.

**Linear Threshold Model.** A commonly used diffusion process (and of special interest to us) that models influence spread is the linear threshold model [9]. At its core, it is a weighted directed graph $G = (V, E, w)$ called the influence graph, where $V$ is the set of nodes, $E$ is a set of directed edges and $w : V \times V \to [0, 1]$ is the weighting function on the edges. For any edge $(u, v) \notin E$, $w(u, v)$ is not defined. Further, for every node $v \in V$, it is required that $\sum_{u:(u,v)\in E} w(u, v) \leq 1$

i.e. the sum of weights from all incoming neighbors is at most 1. Given such an influence graph and a source node $S_0 = \{a\}$, the cascade diffusion process proceeds in discrete time steps $t = 0, 1, 2, \ldots$ as follows:

1. at the initial time step $t = 0$, every node $v \in V$ independently selects a threshold $\theta_v \in [0, 1]$ uniformly at random. This captures our uncertainty in nodes' true thresholds against influence;
2. in every subsequent step $(t + 1)$, an inactive node, $v$, becomes active if the sum of incoming influence exceeds the threshold

$$\sum_{u:u \in S_t, (u,v) \in E} w(u, v) \geq \theta_v$$

where $S_t$ is the set of nodes activated until the previous timestep $t$;
3. the process terminates when no more nodes can be activated.

### 2.3 Supermodular Set Functions

In this work we will reduce the question of finding defensive interventions into an optimization problem which minimizes the knowledge gained by an adversary subject to budgetary constraints. We will see that while the objective function will turn out to be NP-hard to optimize, it possesses the special property of *supermodularity*.

A set function $f : 2^S \to \mathbb{R}$ defined over the power set $2^S$ of a set $S$ is called supermodular iff $\forall A \subseteq B \subset S, \forall s \in S \setminus B$:

$$f(A \cup \{s\}) - f(A) \leq f(B \cup \{s\}) - f(B)$$

The property essentially states that for a non-decreasing supermodular set function $f$, the marginal utility obtained by adding a new element to a larger set is greater than the utility of adding the element to any subset of the larger set. This property is referred to as the *increasing differences* property, as opposed to the more commonly seen diminishing returns property of submodular functions [12].

Most optimization literature focuses on *submodular* objective functions, which we will use to inform our analysis of our supermodular objective. It has been shown that submodular maximization is NP-hard [4] as can be intuited by the combinatorial explosion of possible subsets. However, a proof by Nemhauser et al. [21] shows that a greedy algorithm for maximizing a monotone submodular function (or in our case minimizing a monotone supermodular function) while subject to cardinality constraints can provide a constant factor approximation of $1 - \frac{1}{e} \approx 63\%$.

## 3  Problem Statement

In order to find a set of actions that the defender can implement to harden the network, we need a model of the services in the network (knowledge about these services is what the attacker is after), a model for how the attacker gains this knowledge through internal reconnaissance, and a measure of how good the defender strategy is, thereby allowing us to quantitatively harden the system. In this section, we describe each of these requirements and formalize the network hardening problem. At a high level, we need to find a hardening policy $\pi$ that minimizes knowledge an attacker gains $\mathcal{Q}(\pi)$ while making sure that the cost of the policy stays within a given security budget $\mathcal{B}$. We explore each of the elements of the following optimization later in the paper.

$$\underset{\pi}{\text{minimize}} \quad \mathcal{Q}(\pi) \text{ subject to } \quad cost(\pi) \leq \mathcal{B}$$

## 3.1   Attack Graph

An attack graph is a graphical model that represents the defenders' knowledge about network components, services, their vulnerabilities and their interactions, showing the different paths an attacker can follow to reach a given goal by exploiting a set of vulnerabilities. Along each attack path, vulnerabilities are exploited in sequence, so that each successful exploit gives the attacker more knowledge thereby leading to an increased foothold in the network.

*Uncertain attack graphs* [22], extend the notion of an attack graph by allowing for uncertainty in edge existence. Formally, an uncertain attack graph $\mathcal{A} = (V, E, p)$ is a directed graph, where the nodes in $V$ represent states of an attack and a directed edge in $E$ represents the transition between states. Each edge $(i, j) \in E$ is associated with a transition probability $p_{ij}$ i.e. the likelihood that an attacker can compromise state $j$ given the knowledge from the current compromised state $i$.

A state in the model represents an atomic unit of knowledge, gaining which allows the attacker to make better decisions about the subsequent stages of reconnaissance. Think of the attacker starting blind (or with partial visibility) and each node that they get a foothold on adds to their visibility into the system and its processes. One simple way of depicting a state is as a tuple of host identity, service and privilege level on the service. Thus the knowledge gained from a state is the knowledge available on that host service when attacker achieves the privilege level needed. A transition corresponds to an atomic attacker action that leads to an increased gain in knowledge. Examples of transition include password guessing to escalate privilege, vulnerability exploitation to gain foothold on previously uncompromised host and active scans to detect new services.

A successful reconnaissance campaign is a sequence of transitions or paths in the graph that leads to knowledge states from where an attacker can affect the process being controlled in the critical infrastructure. One metric to measure success of attacker reconnaissance is to count the number of nodes that are infected at the end of the attack propagation phase, but we will dive deeper on these issues in the upcoming sections.

## 3.2    Attack Propagation Model

We define the attack propagation model as follows, based on the linear threshold diffusion model in [9].

1. The attacker starts with an initial knowledge $S_0$, where $S_0 \in V$ is a subset of nodes in the uncertain attack graph.
2. Each node $v$ has a threshold $\theta_v \in [0, 1]$. This represents the weighted fraction of neighbors that must be compromised in order for $v$ to be compromised. The value for the threshold is drawn at random. This threshold is the reason why the attack propagation process is stochastic.

Let us explore the threshold $\theta_v$ a little more. By definition, it can be interpreted as the likelihood that a node is compromised given that it's neighbors have been compromised. In the case of complete knowledge, about the services running on a node or attacker capabilities, the threshold is no longer a random variable. However, we more commonly encounter situations of uncertainty regarding $\theta_v$. One scenario being the lack of detailed knowledge of services, their versions and the vulnerabilities in the network. Note that sometimes, despite perfect knowledge of the network, uncertainties arise from unknown unknowns such as zero-day vulnerabilities. Another scenario is limited understanding of attacker capabilities, for example an adversary may possess information that is gained external to the reconnaissance activity, thereby allowing them to compromise a node without having to compromise all its neighbors.

Given this uncertainty in the node threshold $\theta_v$, the subsequent question is how to model it. Traditionally in the Linear Threshold model, it is assumed that $\theta_v$ is sampled from a standard uniform distribution ($U[0, 1]$). However, this does not capture any knowledge that the defender might possess. We suggest using a Beta distribution to sample the node threshold. The parameters $\alpha$ $and$ $\beta$ of a $Beta(\alpha, \beta)$ allow the defender to capture their prior knowledge. In the base case of $\alpha = \beta = 1$, we end up with a $U[0, 1]$ which captures complete lack of knowledge.

Thus, given the initial knowledge of the attacker (an initial compromised set) and the thresholds of all the uncompromised nodes, the attack propagation process unfolds in discrete time steps: at time $t$ all nodes that were compromised in timestep $t - 1$ remain compromised and any uncompromised node $v$ is activated as:

$$\sum_{w:(v,w)\in E} p_{vw} \geq \theta_v$$

## 3.3    Modeling Prior Knowledge of Attacker

Attackers can often gain auxiliary knowledge about the control system devices and processes from sources external to the network. They might know device manufacturer names and model numbers from reading public documents such as public presentations and requests for tender (similar to the attack on the Kudankulam Nuclear Power Plant [1]). This has in the past, allowed attackers to

successfully conduct watering hole attacks by adding malware to vendor websites
[20]. Adversaries can also acquire attacks for known vulnerabilities on the dark
web, thereby making the lateral movement process faster. They might acquire
stolen passwords for system operators of one system from an attack that was
not targeted for it, like in the case of the Ukraine attack where some passwords
were gained as a consequence of the ransomware notPetya (that did not target
ICS specifically) [16]. Thus, an attacker starts in the network with some prior
knowledge. We model this by designating a subset of the nodes as compromised
at the start of the attack propagation process. $S_0 \subseteq V$ is the knowledge an
attacker possesses at time $t_0$. In order to simplify computation we add a dummy
node $s$ to the set of nodes $V$ with edges to the nodes $S_0$ each with probability
1. This allows us to designate $s$ as the source node for attacks while allowing us
to maintain $S_0$ as the initial foothold of the attacker.

### 3.4   Defender Actions

Knowledge reduction is equivalent to reducing the coverage of nodes in the uncertain attack graph. We define two actions a defender can perform to harden the
network.

1. Add a security rule. We consider scenarios where a firewall or intrusion prevention system is present in the network and the defender adds a fixed rule or
   signature that the security appliance must match against to decide if a flow
   is permitted or not. Adding a rule is equivalent to reducing the edge traversal
   probability of an edge in the attack graph. If the defender believes that a rule
   prevents the attacker from reaching the next knowledge state then the edge
   is deleted i.e. edge traversal probability goes to zero. In this work we only
   consider the latter scenario where an edge is completely removed from the
   attack graph. We discuss the limitation of this in Sect. 7.
2. Add a security appliance or apply a patch. When a new security appliance
   is added to the network with the correct configuration or when a known
   vulnerability is patched, multiple edge traversal probabilities are modified by
   a single action. However, such an action has a higher cost associated with
   it. In the case of a new security appliance this might be the capital cost of
   the technology and its deployment as well as the operational cost of hiring
   analysts to dig through false positives. In the case of patching the cost is
   mostly that of testing the patch in an identical environment before deeming
   it fit to be reproduced in the operational network.

The set of actions that a defender undertakes is known as the *policy* $\pi$. Each
policy $\pi$ is a set of deleted edges corresponding to defender actions. We denote
the implementation of a policy as a modification to the uncertain attack graph $\mathcal{A}$:

$$\mathcal{A}_\pi = \mathcal{A}(V, E \setminus \pi, p)$$

A policy is *feasible* if the total cost of actions taken is no greater than a given
budget limit $\mathcal{B}$. For a given uncertain attack graph $\mathcal{A}$ with an initial infection $S_0$,
the network hardening problem is to select the optimal policy among all feasible
policies, such that the information gained by the attacker is minimized.

## 3.5   Objective Function

We define two metrics of interest in the network hardening problem: penetrability and expected risk.

**Definition 1. *Penetrability*** *of an uncertain attack graph $\mathcal{A}$ is defined as the expected number of compromised nodes at the end of an attack propagation process that starts at an initial foothold, $S_0$.*

$$\mathcal{P}(\mathcal{A}) = E[S_\infty] \tag{1}$$

*where $S_\infty$ is the number of nodes that are compromised at the end of attack propagation.*

Penetrability is analogous to reachability in stochastic reliability graphs. In a deterministic attack graph, penetrability is defined as the number of nodes that can be reached from a source set $S_0$.

We recall that the uncertain attack graph $\mathcal{A}$ is a probabilistic graph. A probabilistic graph can be a generator for $2^{|E|}$-many deterministic graphs, based on the presence or absence of an edge. In the case of uncertain attack graph $\mathcal{A}$, each deterministic graph it generates is called an attack scenario. Each attack scenario is a possible set of nodes that an attacker has visited using the set of compromised edges. Note that not all of the $2^{|E|}$-many attack scenarios may be plausible, and thus can be excluded from the search space.

**Definition 2. *Risk*** *Consider an attack scenario $A \in \Omega_\mathcal{A}$, where $\Omega_\mathcal{A}$ is the space of attack scenarios that can be generated from uncertain attack graph $\mathcal{A}$. Also, a loss function, $L : A \to \mathbb{R}^+$, for attack scenario $A$, $L(A)$. The risk for a network modeled by the uncertain attack graph $\mathcal{A}$ is defined as the expected loss across all possible attack scenarios:*

$$\mathcal{R}(\mathcal{A}) = \sum_{A \in \Omega_\mathcal{A}} Pr[A] L(A) \tag{2}$$

The goal of the loss function is to quantify the direct (monetary loss due to equipment damage and repair) and indirect losses (loss of intellectual property) a critical infrastructure network faces under attack. The loss function is generally monotone non-decreasing: the more nodes an attacker compromises the greater the loss. While, there is more discussion on loss functions in Sect. 7, in this work we only look at monotone loss functions defined as follows:

**Definition 3.** *Given an attack scenario $A \in \Omega_\mathcal{A}$, the loss under such an attack can be computed as:*

$$L(A) = \sum_{i \in P(A)} c_i \tag{3}$$

*where $c_i$ is the dollar cost of losing the knowledge stored in node $i$ to the attacker and $P(A)$ is the set of compromised nodes in $A$.*

We understand that there is uncertainty associated with deriving the value $c_i$. One source of uncertainty is from attributing the loss as either a loss of confidentiality or one of availability. A loss in confidentiality is characterized by the loss of industrial secrets and proprietary technology present on a node. A loss in availability is due to denial of service which can range from monetary loss due to a blackout for few hours to equipment damage. Thus this loss is more of a distribution rather than a number. While understanding this, we defer this to future work and allow for a treatment of $c_i$ as a number which the user can either define as worst case or average monetary loss based on their analysis.

In any network, some nodes (say an authentication server or the data historian) are more valuable than others. The risk metric intrinsically captures the case where nodes are valued differently. Also, note that the risk metric can be used to define *Penetrability*. If $L(v) = 1$ for every node $v \in A$ and $L(v) = 0$ for every other node, the expected loss is the expected number of nodes reached at the end of attack propagation. While the risk metric is a general solution, it requires more information from the defender, thereby making penetrability a viable alternative. Under this alternate definition, penetrability can be formalized as:

$$\mathcal{P}(\mathcal{A}) = \sum_{A \in \Omega_A} Pr[A]r(A) \tag{4}$$

where $r(A)$ is the $\{0, 1\}$ loss function describing the number of nodes that can be reached in an attack scenario $A$.

### 3.6   Constraints

There is a cost associated with each defender action and the security budget must be split such that the defender actions that restrict attacker movement most should be preferred over the rest. In a simple case where all defender actions cost the same, a cardinality constraint can be used (total number of deleted edges is less than or equal to the budget). In most cases however, the cost is more of a knapsack constraint i.e. the sum of costs of each action must be below the budget.

Considering the objective function and the constraints we defined in this section, the network hardening problem can be translated into the following optimization problem:

$$\underset{\pi \subseteq E}{\operatorname{argmin}} \quad \mathcal{Q}(\mathcal{A} \setminus \pi) \quad \text{subject to} \quad \sum_{i \in \pi} k_i \leq \mathcal{B} \tag{5}$$

Where, the function $\mathcal{Q}(\mathcal{A})$ can be substituted for either $\mathcal{P}(\mathcal{A})$ or $\mathcal{R}(\mathcal{A})$ depending on the analysis.

# 4    Analyzing the Objective Function

In this section we analyze some interesting properties of the objective function. Specifically, we show that the objective function is monotonically decreasing and supermodular in the policy $\pi$.

## 4.1    Live Edge Paths

Influence maximization literature shows that an alternate method to compute the influence function under the linear threshold process is using *live-edge* paths in the influence graph. The claim (as proved in [9]) is that given an initial set of nodes, $S$, the distribution over active nodes obtained by running the linear threshold process to completion is the same as the distribution of nodes reachable from $S$ via live-edge paths, where the live-edge paths are selected as described below:

Each node $v \in V$ picks at most one of its incoming edges at random with probability $p_{uv}$ and selects no edge with probability $1 - \sum_{u:(u,v)\in E} p_{uv}$.

Note that the set of nodes in the generated live-edge graph $X$ is equal to $V$ and the set of *live* edges, $E_X$ is a subset of $E$, i.e., $E_X \subseteq E$. Also note that these sampled edges are unweighted and each vertex has at most one incoming edge.

We will use this live-edge graph to inform our proofs of monotonicity and supermodularity. Before we get to the proofs, we make some useful observations. The probability of a node $v \in V$ having the edge configuration as seen in attack scenario $A$ is given by:

$$p(v, A, \mathcal{A}) = \begin{cases} p_{uv} & \text{if } \exists u : (u,v) \in E_A \\ 1 - \sum_{u:(u,v)\in E} p_{uv} & \text{otherwise} \end{cases} \tag{6}$$

where $E_A$ is the set of edges in $A$. Using this, we can get the probability of a scenario graph $A$ as:

$$Pr[A] = \prod_{v \in V} p(v, A, \mathcal{A}) \tag{7}$$

Note that the loss $(L(A))$ term is deterministic given an instance of the scenario graph $A$. The likelihood term is computed as shown in Eq. 7.

## 4.2    Monotonicity

In this section, we prove that the objective function is monotonically decreasing.

**Lemma 1.** *The reachability function $r(A \setminus \pi)$ is a monotonically decreasing function of the policy $\pi$.*

**Proof.** Given a scenario graph $A \in \Omega_A$, we need to show that for any policy $\pi \subseteq E$ and an edge $e = (u,v) \in E \setminus \pi$

$$r(A \setminus \pi) - r(A \setminus \{\pi \cup e\}) \geq 0$$

Since, A is a live-edge graph such that each vertex $v \in V$ has at most one incoming edge it can be seen that removing an edge $e = (u, v)$ to $v$ will make it unreachable from the source $s$. In fact, if $v$ is not a leaf node, then all its children are rendered unreachable by the removal of edge $e$. Thus, the number of nodes reachable in $A \setminus \pi$ is higher than that in $A \setminus \{\pi \cup e\}$     ∎

**Lemma 2.** *The loss function $L(A \setminus \pi)$ is a monotonically decreasing function of the policy $\pi$.*

**Proof.** Similar to Lemma 1 we must show that

$$L(A \setminus \pi) - L(A \setminus \{\pi \cup e\}) \geq 0$$

By definition, the loss function is a linear combination of non-negative costs based on the reachability of nodes. Thus, since the reachability in $A \setminus \pi$ is higher than that in $A \setminus \{\pi \cup e\}$ by at least 1 (due to v), the loss is also higher by at least $c_v$, which is non-negative     ∎

**Theorem 1.** $\mathcal{R}(A \setminus \pi)$ *is a monotonically decreasing function of the policy $\pi$.*

**Proof.** We defer the proof to the appendix for better readability.

**Theorem 2.** $\mathcal{P}(A \setminus \pi)$ *is a monotonically decreasing function of the policy $\pi$.*

**Proof.** This proof follows the same structure as above and uses Lemma 1 in the final step     ∎

### 4.3   Supermodularity

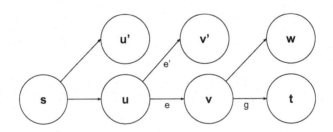

**Fig. 1.** Proof of supermodularity.

**Lemma 3.** *The reachability function $r(A \setminus \pi)$ is a supermodular function of the policy $\pi$.*

**Proof.** In order to prove supermodularity, we must show for an attack graph $\mathcal{A} = (V, E, p)$, a subset of edges to be deleted $\pi \subset E$ and two edges $e, g \in E \setminus \pi$) that

$$r(A \setminus \pi) - r(A \setminus \pi \cup \{e\}) \geq r(A \setminus \pi \cup \{g\}) - r(A \setminus \pi \cup \{e, g\})$$

Figure 1 illustrates our proof. Note that in the live edge graph model, each node has at most one incoming edge as shown in the figure. Consider the scenario shown in the figure to be the $A \setminus \pi$. The difference we are computing is the reduction in reachability when an edge is deleted from this graph. We consider two cases: (1) deleting an edge $e'$ that does not fall in the same path as edge $g$ and (2) deleting an edge $e$ in the same path as edge $g$. In case 1, deleting edge $e'$ from $A \setminus \pi$ leads to the same reduction as in $A \setminus \pi \cup \{g\}$. This is due to the fact that the deletion causes the same set of nodes to become unreachable. Thus, case 1 leads to equality. In case 2, deletion of the edge $e$ from $A \setminus \pi$ disconnects additional nodes which had edge $g$ in their path from the source. Thus in case 2, the reduction in reachable nodes is greater when $e$ is deleted from $A \setminus \pi$ instead of $A \setminus \pi \cup \{g\}$ ∎

**Lemma 4.** *The loss function $L(A\setminus\pi)$ is a supermodular function of the policy $\pi$.*

**Proof.** Similar to Lemma 3, we must show

$$L(A \setminus \pi) - L(A \setminus \pi \cup \{e\}) \geq L(A \setminus \pi \cup \{g\}) - L(A \setminus \pi \cup \{e, g\})$$

Note that in the reachability function, the marginal value after the difference represents the number of nodes that are rendered unreachable after the edge $e$ is deleted. Since the loss function is a linear function of reachability it can be seen that the marginal loss is in fact the sum of costs of nodes that are rendered unreachable by deleting $e$. Thus, it can be seen that for non-negative costs, the marginal loss is greater when $e$ is deleted from $A \setminus \pi$ instead of $A \setminus \pi \cup \{g\}$ ∎

**Lemma 5.** *If $f(S)$ is a supermodular function then its expectation $E[f(S)]$ is a supermodular function of $S$.*

**Proof.** This is a simple extension of the fact that supermodular functions are closed under linear combinations and the expectation function is essentially a non-negative weighted sum of the supermodular function.

**Theorem 3.** *Penetrability is supermodular in the policy $\pi$.*

**Proof.** We note that by the live edge definition in Eq. 4, penetrability is essentially the expected reachability in an attack graph. Hence, by Lemma 3 and Lemma 5 it can be seen that $\mathcal{P}(\mathcal{A} \setminus \pi)$ is supermodular in the policy $\pi$ ∎

**Theorem 4.** *Risk is supermodular in the policy $\pi$.*

**Proof.** Risk is by definition (Eq. 2) expected loss. Hence, by Lemma 4 and Lemma 5 it can be seen that $\mathcal{R}(\mathcal{A} \setminus \pi)$ is supermodular in the policy $\pi$ ∎

## 5 Algorithm

In the previous section we showed our hardening objective functions to be monotone and supermodular under the linear threshold model of attack propagation.

Our optimization problem is one of minimizing a monotone supermodular function under a knapsack constraint. In the case where every defender action has the same cost and we can perform $k$ defender actions, the total possible policies are $\binom{|E|}{k}$. It is combinatorially hard to search this large space for an optimal policy. In fact, the problem of maximizing a monotone submodular function (and in our case a monotone supermodular function) under a cardinality constraint has been shown to be NP-hard [12]. The proof is a reduction of the vertex cover problem [9].

## 5.1    The Greedy Algorithm

A naive approach is to find approximate solutions using the greedy algorithm. The algorithm starts with an empty policy $\pi = \emptyset$ and proceeds in an iterative fashion. There are a total of $k$ iterations (cardinality constraint: $|\pi| \leq k$). In the $j$-th iteration we pick an edge $e$ such that $e = \mathrm{argmax}_{e \in E \setminus \pi} \Delta(e)$, where $\Delta(e)$ is the marginal loss defined as $\mathcal{R}(\mathcal{A} \setminus \pi) - \mathcal{R}(\mathcal{A} \setminus (\pi \cup \{e\}))$. Note that similar steps apply to the Penetrability objective.

A fundamental result by Nemhauser et al. [21] shows that under the cardinality constraint, this greedy algorithm produces a near optimal solution. The greedy algorithm will choose a policy $\pi$ such that $\mathcal{R}(\mathcal{A} \setminus \pi) \geq 1 - \frac{1}{e} \mathcal{R}(\mathcal{A} \setminus \pi^*)$, where $\pi^*$ is the optimal policy for hardening the network.

**Drawbacks of Naive Greedy.** While the greedy algorithm gives a near optimal solution, it needs some tweaks in the practical setting. Note that the $\mathcal{R}(\mathcal{A} \setminus \pi)$ function is evaluated every time we need to find the edge that maximizes $\Delta(e)$. Also, computing this function accurately is #-P hard as shown in [3]. Thus, we need a way to approximate the expected risk in sub-linear time.

## 5.2    Monte Carlo Estimation

In order to find an approximation of risk, we use Monte Carlo estimation. To estimate the risk, the Monte Carlo estimator first draws $N$ possible attack scenarios denoted by $A_0, A_1, ..., A_N$ from the attack graph $\mathcal{A}$ using the live-edge sampling rule stated in Eq. 6. Then, for each possible attack scenario $A_i$, the algorithm evaluates the loss function $L(A_i)$. Finally, the risk estimate (denoted by $\hat{\mathcal{R}}$) is obtained by:

$$\hat{\mathcal{R}}(\mathcal{A}) = \frac{1}{N} \sum_{i \in N} L(A_i) \tag{8}$$

Note that due to the linear combination property of supermodular functions, this estimated expected risk is also monotone and supermodular in the policy $\pi$. Given this estimate, the marginal loss $\Delta(e)$ is given by:

$$\frac{1}{N} \sum_{i \in N} L(A_i \setminus \pi) - L(A_i \setminus \pi \cup \{e\}) \tag{9}$$

## 5.3   Knapsack Budget Constraint

When the defender actions have different costs, the simple greedy algorithm can perform arbitrarily badly. However, an algorithm that defines a loss-cost ratio $\Delta(e)/k_e$ and uses partial enumeration along with the greedy step of picking the edge $e \in E$ that maximizes the loss-cost ratio at every iteration has been shown to achieve an approximation guarantee of $1 - \frac{1}{e}$ [26] [13].

## 5.4   Complexity of the Greedy Algorithm

Note, that in a given iteration we need to compute the marginal loss for every edge $e \in E \setminus \pi$ and every scenario $A_i$. This involves a breadth first search $(O(|V| + |E|))$ to compute reachability which is then used to compute loss. Note that in the live edge graphs, we have at most $|V|$ edges since each node can have only one incoming edge. Thus, the complexity of each iteration is $O(|V|^2 + |V||E|)$. However, we can use the following lemma to improve the time complexity from quadratic to linear.

**Lemma 6.** *The marginal reduction in reachability when an edge $e = (u, v)$ is added to the policy is the same as the size of the sub-tree which is induced by the node $v$.*

**Proof.** We recall that the marginal reduction in reachability, denoted by $r(A \setminus \pi) - r(A \setminus \pi \cup \{e\})$ is the same as the number of nodes rendered unreachable when edge $e$ is deleted from the graph $A \setminus \pi$. Referring to Fig. 1 we see that this is essentially the size of the sub-tree rooted at $v$.

Thus, we can create a tree data structure, which stores at each node the size of the sub-tree it would generate if the incoming edge to it is deleted. Given the data structure, the marginal loss can be computed in a single breadth first search traversal. Also since the number of edges is at most $V$, this takes $O(|V|)$. The total time of the modified greedy algorithm is $O(kN|V|)$ which is linear in the size of the attack graph.

# 6   Related Work

A closely related area is that of influence diffusion through social networks. The question of influence maximization is one of finding the initial set of nodes to infect in order to maximize spread of information over the network [9]. This has been applied to study influence of users on their followers in a social network [2], to examine cascade of news through the web [18] and the propagation of recommendations [17]. There has also been work on inferring the structures of such networks [5,6]. Our work differs from this line of research since our focus is not on empirical analysis of diffusion but rather on modifying a network to minimize the diffusion through it. A similar problem has been addressed in [11] in a different application context, with a slightly different objective function and constraints.

A closer line of inquiry is that of placing sensors to detect changes in the environment. [14, 19] use submodular maximization to maximize the information gained by a sensor placement. There are also actuator placement problems [25] that look to choose a set of actuators to maximize controllability of a control network. Our work differs from this line of inquiry since our study of attack graphs requires analysis of graph modification. Also, the metrics of interest are very different.

## 7    Conclusion and Limitations

Critical infrastructures have complex networks which control unique processes. In order to perform a successful attack on such a network, the adversary must understand the network components they can leverage and the processes they can affect. This requires a long and detailed reconnaissance phase. Thus, the best opportunity for a defender is to harden against such recon activities. This paper provides the defender with a systematic approach to making decisions while limiting the cost of defense. We show that despite imperfect knowledge of the attacker and their actions the defender can still improve their risk posture. While the hardening problem is NP-hard we provide a linear time algorithm with a provable approximation bound of $1 - \frac{1}{e}$.

While this paper provides analysts with an algorithm for hardening their network, there are a few limitations we wish to discuss. First, the loss distribution is often not available to the analysts and while a metric like penetrability can give some insights into the hardening process, in the future we hope to explore other metrics that can capture the loss. Another assumption is the loss function being monotone, however, with the growth of reactive security techniques like honeypots, we will have to look into non-monotone loss in the future. Second, we note that all properties studied in the paper hold only for the linear threshold model and while this is a strong model for reconnaissance we might have to find other models to capture other stages of the attack kill-chain.

## A    Proof of Theorem 1

Given an attack graph $\mathcal{A} = (V, E, p)$ we need to show that for any set $\pi \subseteq E$ and an edge $e = (u, v) \in E \setminus \pi$

$$\mathcal{R}(\mathcal{A} \setminus \pi) - \mathcal{R}(\mathcal{A} \setminus (\pi \cup \{e\})) \geq 0$$

This proof is very similar to the proof in [11] and only differs in our function of interest (attack loss function).

The space of attack scenarios $\Omega_{\mathcal{A}\setminus\pi}$, can be divided into three disjoint partitions based on the edge selected for node $v$. $\Omega^e_{\mathcal{A}\setminus\pi}$ (edge $e = (u, v)$ is chosen), $\Omega^{\bar{e}}_{\mathcal{A}\setminus\pi}$ (a different edge $\bar{e} = (u', v)$ is chosen) and $\Omega^{\emptyset}_{\mathcal{A}\setminus\pi}$ (no incoming edge is selected).

Now for, the space $\Omega_{\mathcal{A}\setminus(\pi\cup\{e\})}$ we note that the space is a subset of $\Omega_{\mathcal{A}\setminus\pi}$ since any scenario graph in the former can be generated in the latter. Also, the only scenarios not present in the former are ones where the edge $e$ is involved. Thus, $\Omega_{\mathcal{A}\setminus(\pi\cup\{e\})}$ can be defined based on two partitions as: $\Omega^{\bar{e}}_{\mathcal{A}\setminus\pi}\cup\Omega^{\emptyset}_{\mathcal{A}\setminus\pi}$.

Using these disjoint partitions, we can write the difference as:

$$\mathcal{R}(\mathcal{A}\setminus\pi)-\mathcal{R}(\mathcal{A}\setminus(\pi\cup\{e\}))$$
$$=\sum_{A\in\Omega^{e}_{\mathcal{A}\setminus\pi}}Pr[A|\mathcal{A}\setminus\pi]L(A)$$
$$+\sum_{A\in\Omega^{\bar{e}}_{\mathcal{A}\setminus\pi}}(Pr[A|\mathcal{A}\setminus\pi]-Pr[A|\mathcal{A}\setminus(\pi\cup e)])L(A)$$
$$+\sum_{A\in\Omega^{\emptyset}_{\mathcal{A}\setminus\pi}}(Pr[A|\mathcal{A}\setminus\pi]-Pr[A|\mathcal{A}\setminus(\pi\cup e)])L(A)$$

For the space $\Omega^{\bar{e}}_{\mathcal{A}\setminus\pi}$ we have $Pr[A|\mathcal{A}\setminus\pi]-Pr[A|\mathcal{A}\setminus(\pi\cup e)]=0$ since from Eq. 6 we have, $p(v,A,\mathcal{A}\setminus\pi)=p(v,A,\mathcal{A}\setminus(\pi\cup\{e\}))=p(\bar{e})$. This is due to the fact that in this space, under both cases, edge $\bar{e}$ is chosen for node $v$.

For the space $\Omega^{\emptyset}_{\mathcal{A}\setminus\pi}$ we have:

$$Pr[A|\mathcal{A}\setminus\pi]-Pr[A|\mathcal{A}\setminus(\pi\cup e)]=-p_e\prod_{v'\neq v}p(v',A,\mathcal{A}\setminus\pi)$$

This stems from the fact that we can rewrite the above difference in terms of the node $v$ and all other nodes $v'\neq v$ as $Pr[A|\mathcal{A}\setminus\pi]-Pr[A|\mathcal{A}\setminus(\pi\cup e)]=\prod_{v'\neq v}p(v',A,\mathcal{A}\setminus\pi)\times[p(v,A,\mathcal{A}\setminus\pi)-p(v,A,\mathcal{A}\setminus(\pi\cup\{e\}))]$.

As for the difference in probabilities when node $v$ has no incoming edge we see that it goes to $-p_e$ due to the fact that $p(v,A,\mathcal{A}\setminus\pi)=1-\sum_{x\in E\setminus\pi}p_x=1-\sum_{x\in E\setminus(\pi\cup\{e\})}p_x-p_e=p(v,A,\mathcal{A}\setminus(\pi\cup\{e\}))-p_e$.

Now consider the following two facts:

- Every graph $A'\in\Omega^{e}_{\mathcal{A}\setminus\pi}$ has a corresponding graph $A\in\Omega^{\emptyset}_{\mathcal{A}\setminus\pi}$ and vice versa where $A'=A\cup\{e\}$ i.e. they differ only in the edge $e$.
- A graph $A'\in\Omega^{e}_{\mathcal{A}\setminus\pi}$ has probability $Pr[A'|\mathcal{A}\setminus\pi]=p_e\prod_{v'\neq v}p(v',A',\mathcal{A}\setminus\pi)$. Note that this is essentially Eq. 7 rewritten in terms of $e$.

Hence:

$$\mathcal{R}(\mathcal{A}\setminus\pi)-\mathcal{R}(\mathcal{A}\setminus(\pi\cup\{e\}))=\sum_{A\in\Omega^{\emptyset}_{\mathcal{A}\setminus\pi}}Pr[A'|\mathcal{A}\setminus\pi][L(A')-L(A)]$$

Since this is a non-negative sum and by Lemma 2 we know that $L(A')-L(A)\geq 0$ we can see that the risk function is monotone decreasing in the policy $\pi$ ∎

# References

1. Anantharaman, P., Palani, K.: What happened when the Kudankulam nuclear plant was hacked - and what real danger did it pose? Scroll.in, 20 November 2019
2. Bakshy, E., Hofman, J.M., Mason, W.A.,Watts, D.J.: Everyone's an influencer: quantifying influence on Twitter. In: Proceedings of the Fourth ACM International Conference on Web Search and Data Mining, pages 65–74 (2011)
3. Chen, W., Wang, C., Wang, Y.: Scalable influence maximization for prevalent viral marketing in large-scale social networks. In: Proceedings of the 16th ACM SIGKDD International Conference on Knowledge Discovery and Data Mining, pp. 1029–1038 (2010)
4. Feige, U.: A threshold of ln n for approximating set cover. J. ACM (JACM) **45**(4), 634–652 (1998)
5. Gomez-Rodriguez, M., Leskovec, J., Krause, A.: Inferring networks of diffusion and influence. ACM Trans. Knowl. Discov. Data (TKDD) **5**(4), 1–37 (2012)
6. Gomez Rodriguez, M., Leskovec, J., Schölkopf, B.: Structure and dynamics of information pathways in online media. In: Proceedings of the Sixth ACM International Conference on Web Search and Data Mining, pp. 23–32 (2013)
7. Granovetter, M., Soong, R.: Threshold models of diffusion and collective behavior. J. Math. Sociol. **9**(3), 165–179 (1983)
8. Jha, S., Sheyner, O., Wing, J.: Two formal analyses of attack graphs. In: Proceedings 15th IEEE Computer Security Foundations Workshop, CSFW-15, pp. 49–63. IEEE (2002)
9. Kempe, D., Kleinberg, J., Tardos, É.: Maximizing the spread of influence through a social network. In: Proceedings of the Ninth ACM SIGKDD International Conference on Knowledge Discovery and Data Mining, pp. 137–146. ACM (2003)
10. Kermack, W.O., McKendrick, A.G.: A contribution to the mathematical theory of epidemics. In: Proceedings of the royal society of London. Series A, Containing Papers of a Mathematical and Physical Character, vol. 115(772), pp. 700–721 (1927)
11. Khalil, E., Dilkina, B., Song, L.: CuttingEdge: influence minimization in networks. In: Proceedings of Workshop on Frontiers of Network Analysis: Methods, Models, and Applications at NIPS (2013)
12. Krause, A., Golovin, D.: Submodular function maximization. In: Tractability: Practical Approaches to Hard Problems, pp. 71–104. Cambridge University Press (2014)
13. Krause, A., Guestrin, C.: A Note on the Budgeted Maximization of Submodular Functions. Carnegie Mellon University, Center for Automated Learning and Discovery (2005)
14. Krause, A., Leskovec, J., Guestrin, C., VanBriesen, J., Faloutsos, C.: Efficient sensor placement optimization for securing large water distribution networks. J. Water Resour. Plann. Manage. **134**(6), 516–526 (2008)
15. Lee, R.: TRISIS malware: analysis of safety system targeted malware. Dragos Inc. (2017)
16. Lee, R., Assante, M., Conway, T.: Analysis of the cyber attack on the Ukrainian power grid. EISAC Technical report (2016)
17. Leskovec, J., Adamic, L.A., Huberman, B.A.: The dynamics of viral marketing. ACM Trans. Web (TWEB) **1**(1), 5es (2007)
18. Leskovec, J., Backstrom, L., Kleinberg, J.: Meme-tracking and the dynamics of the news cycle. In: Proceedings of the 15th ACM SIGKDD International Conference on Knowledge Discovery and Data Mining, pp. 497–506 (2009)

19. Leskovec, J., Krause, A., Guestrin, C., Faloutsos, C., VanBriesen, J., Glance, N.: Cost-effective outbreak detection in networks. In: Proceedings of the 13th ACM SIGKDD International Conference on Knowledge Discovery and Data Mining, pp. 420–429 (2007)
20. Ellen, N.: Russian military was behind NotPetya cyberattack in Ukraine, CIA concludes. The Washington Post, 12 Jan (2018)
21. Nemhauser, G.L., Wolsey, L.A., Fisher, M.L.: An analysis of approximations for maximizing submodular set functions–I. Math. Program. **14**(1), 265–294 (1978)
22. Nguyen, H.H., Palani, K., Nicol, D.M.: An approach to incorporating uncertainty in network security analysis. In: Proceedings of the Hot Topics in Science of Security: Symposium and Bootcamp, pp. 74–84 (2017)
23. Ou, X., Boyer, W.F., McQueen, M.A.: A scalable approach to attack graph generation. In: Proceedings of the 13th ACM Conference on Computer and Communications Security, pp. 336–345 (2006)
24. Sheyner, O., Haines, J., Jha, S., Lippmann, R., Wing, J.M.: Automated generation and analysis of attack graphs. In: Proceedings 2002 IEEE Symposium on Security and Privacy, pp. 273–284. IEEE (2002)
25. Summers, T.H., Cortesi, F.L., Lygeros, J.: On submodularity and controllability in complex dynamical networks. IEEE Trans. Control Netw. Syst. **3**(1), 91–101 (2015)
26. Sviridenko, M.: A note on maximizing a submodular set function subject to a knapsack constraint. Oper. Res. Lett. **32**(1), 41–43 (2004)

# Sensitivity Analysis and Uncertainty Quantification of State-Based Discrete-Event Simulation Models Through a Stacked Ensemble of Metamodels

Michael Rausch[1]([✉]) and William H. Sanders[2]

[1] University of Illinois at Urbana-Champaign, Urbana, IL, USA
mjrausc2@illinois.edu
[2] Carnegie Mellon University, Pittsburgh, PA, USA
sanders@cmu.edu

**Abstract.** Realistic state-based discrete-event simulation models are often quite complex. The complexity frequently manifests in models that (a) contain a large number of input variables whose values are difficult to determine precisely, and (b) take a relatively long time to solve.

Traditionally, models that have a large number of input variables whose values are not well-known are understood through the use of sensitivity analysis (SA) and uncertainty quantification (UQ). However, it can be prohibitively time consuming to perform SA and UQ.

In this work, we present a novel approach we developed for performing fast and thorough SA and UQ on a metamodel composed of a stacked ensemble of regressors that emulates the behavior of the base model. We demonstrate the approach using a previously published botnet model as a test case, showing that the metamodel approach is several orders of magnitude faster than the base model, more accurate than existing approaches, and amenable to SA and UQ.

**Keywords:** Metamodels · Surrogate models · Emulators · Security models · Reliability models · Sensitivity analysis · Uncertainty quantification · Optimization

## 1 Introduction

Many state-based discrete-event simulation models of real-world systems are complex, large, and contain uncertain input parameters. It is challenging to make realistic quantitative models smaller and simpler (and thus faster to execute) because the world is large and complex. It is also very difficult to remove uncertainty in the model input values. Obtaining precise, certain input values in many domains may be prohibitively expensive or even impossible. Special approaches must be developed and used to make effective use of such models, given the issues of long run times and uncertain input values.

© Springer Nature Switzerland AG 2020
M. Gribaudo et al. (Eds.): QEST 2020, LNCS 12289, pp. 276–293, 2020.
https://doi.org/10.1007/978-3-030-59854-9_20

The traditional way of handling uncertain input parameter values is to perform (a) sensitivity analysis (SA) to determine the most sensitive inputs, and (b) uncertainty quantification (UQ) to determine how the uncertainty in the inputs propagates to uncertainty in the model output. Both SA and UQ typically require that models be solved many times, with the input variable values being varied each time. If the calculation of the model's metrics could be done quickly, comprehensive SA and UQ could be accomplished with a reasonable computation and time budget. If the model input values were known with certainty, it would be unnecessary to execute the model multiple times to perform SA and UQ, so the time and computation required to obtain a single model solution would be less of a concern. However, the twin issues of long solution times and uncertain model input values in complex state-based models present a significant challenge to modelers.

We propose the use of *metamodels* (also known as *emulators* or *surrogate models*) to address the two issues. Metamodels are models of the original base model that attempt to approximate the relationship between the base model's inputs and outputs, and can generally be executed much more quickly than the base model. With an acceptably accurate metamodel, fast and comprehensive sensitivity analysis and uncertainty quantification can be performed on the metamodel in place of the original base model. While metamodels can be constructed by hand, normally they are automatically constructed using machine learning techniques (e.g. Gaussian process regressors, multilayer perceptrons, and random forests).

A chief concern with metamodeling is the choice of an appropriate machine learning technique, as each has its own strengths and weaknesses. While most related work arbitrarily chooses a particular machine learning technique, or evaluates a small handful of different techniques and chooses the strongest, in this work we use an ensemble of heterogeneous regressors in an effort to benefit from the strengths of each approach while mitigating the weaknesses. We structure the ensemble using a custom stacking approach. Stacking is a cutting-edge technique used by the winners of some recent machine learning competitions [15], but we adapt it for use on state-based discrete-event simulation models. We use a sophisticated botnet model [16] as a test case and have performed sensitivity analysis and uncertainty quantification on the model, using the metamodeling approach.

To the best of our knowledge, we are the first to propose and demonstrate a metamodeling-based approach to the analysis of complex quantitative security models, and the first to use a stacking-based approach to perform SA and UQ on real-world quantitative models. We show that our stacked metamodels are several orders of magnitude faster than the original models, are more accurate than traditional metamodels, and are amenable to SA and UQ that could not be performed on the original base model within a reasonable time budget. We have demonstrated the approach with the aid of a pre-existing published peer-to-peer botnet security model [16], which we use as a test case.

The rest of this paper is organized as follows. In Sect. 2, we explain the process by which we use stacking to construct an accurate metamodel. In Sect. 3, we describe the sensitivity analysis and uncertainty quantification techniques we employ to analyze the metamodel. In Sect. 4, we briefly explain the botnet model we used as a test case to demonstrate and evaluate the approach. We show the results of our analysis of the botnet model in Sect. 5. We discuss limitations, use cases, and future directions for the approach in Sect. 6, review related work in Sect. 7, and conclude in Sect. 8.

## 2    Approach

In our context, a *metamodel*, also known as a *model surrogate* or *emulator*, is a model of a model (which we will refer to in this work as the *base model*) that attempts, given a particular vector of input variables (which we shall call an *input*), to produce an output that matches as closely as possible the output that the base model would produce given the same input, for all inputs. In this work we consider quantitative outputs (metrics), though the approach could be extended to consider qualitative outputs (i.e., by using classifiers rather than regressors). Metamodels can rarely achieve perfect accuracy in emulating the base model, but they often run much faster than the base models. The long time needed to run the base model, together with the need to run the base model many times to conduct sensitivity analysis, uncertainty quantification, and optimization, provides the motivation to find fast metamodels that can emulate the base model with acceptable accuracy.

While high-quality metamodels can be constructed manually by an expert familiar with the base model, it is often easier and faster to build the metamodel automatically. At a high level, the metamodeling process is conducted in three stages:

1. Data for training and testing are acquired by generating a number of different model inputs and running the base model with those inputs to observe the resulting outputs.
2. The training data are used by a machine learning algorithm to train a metamodel.
3. The test data are used to assess the quality of the trained metamodel.

To begin, in the first stage, data for training and testing must be acquired. Time and computation constraints restrict the maximum number of inputs that can be run on the base model. We can imagine that an $n$-dimensional input vector describes a point in the $n$-dimensional input space. The metamodel will benefit from high-quality training data that gives the most complete view of the input space possible, given the limited number of samples. For example, if all the training inputs are clustered closely together in the input space, the trained metamodel may be accurate only in that limited region of the input space. The input space should be explored as efficiently as possible. The most important decision at this first stage is the choice of strategy for input selection. We consider three input selection strategies in this paper: random sampling, Latin hypercube sampling, and Sobol sequence sampling.

In the second stage, the training data gathered in the previous stage are used to train a metamodel: a model of the original base model that attempts to produce the same output the base model would produce if it were given the same input. As mentioned previously, we consider only quantitative model output in this work, so the metamodels will be regressors. One can choose from a variety of machine learning regressors, including, e.g., kriging (Gaussian process regressors), random forest regressors, support vector machine regressors, and k-nearest neighbors (KNN) regressors. The most important decision at this stage is the selection of the machine learning technique that will be able to produce the most accurate metamodel given the training data collected in the first stage. Instead of selecting one regressor, we use the predictions from multiple heterogeneous regressors through stacking.

In the third stage, the test data are used to evaluate the accuracy of the trained metamodel. Given each input in the test data, the metamodel will produce an output, and the metamodel's output is compared to the base model's output. The absolute value of the difference between the two outputs quantifies the error of the metamodel. If test inputs are generated randomly and independently, one can use the standard statistical methods to determine the average error and associated confidence interval.

The remainder of this section will be devoted to explaining the procedures for obtaining the training data and constructing the metamodel.

## 2.1  Sampling

Acquiring appropriate data for training and testing is vitally important for constructing metamodels and evaluating how well they emulate the base model. The general idea is to choose an input (in other words, a vector that contains a specific value for each input variable), execute the base model with that input, observe the resulting model output or metric, and then record the input and output by adding them to the list of previously observed input-output pairs. This process is repeated multiple times until some stopping criterion is reached, such as the exhaustion of the allocated time budgeted for acquiring training data. The choice of input at each iterative stage is very important. If all the inputs in the training data are "close" to one another in the input space, the metamodel trained on that data may not be able to accurately emulate the base model in other regions. In this paper, we consider three ways of exploring the input space: random sampling, Latin hypercube sampling, and Sobol sequences. The differences between the three methods are illustrated in the two-dimensional case in Figs. 1, 2, and 3.

**Random Sampling.** Random sampling is the simplest sampling strategy that we consider. Random sampling is conducted by selecting a value for each input variable from the range of possible values at random, independently of any prior sample. Random sampling has a chance of exploring any region in the input space, unlike deterministic sampling, but random sampling can have the unfortunate side effect that samples may cluster in regions of the input space that have

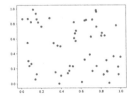 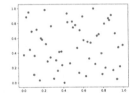 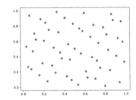

**Fig. 1.** 55 Random samples.

**Fig. 2.** 55 Latin hypercube samples.

**Fig. 3.** 55 Sobol samples drawn from a uniform distribution.

already been well explored while ignoring regions that have not been explored at all. The clustering effect can easily be observed in the two dimensional case shown in Fig. 1.

**Latin Hypercube.** The Latin hypercube sampling (LHS) [5,8] strategy attempts to explore the space more evenly than random sampling. It is inspired by the Latin square puzzle, which consists of an $n$ by $n$ array of $n$ unique symbols such that a particular symbol will appear exactly once in every row and every column.[1] In the two-dimensional case, Latin hypercube sampling generates $N$ samples by dividing the range in the x and y dimensions into $N$ equally likely intervals, forming a grid of $N$ rows and $N$ columns, and choosing a sample such that each row and each column is sampled exactly once.[2] Similarly, in the case of a finite number of inputs, LHS generates $N$ samples by dividing each input variable into $N$ equally probable intervals and choosing exactly one sample from each interval. LHS provides stronger guarantees of coverage than random sampling, because each interval is sampled once, while a particular interval may not be sampled at all with random sampling. However, LHS may not evenly cover the space very well. Indeed, there exist pathological cases in which Latin hypercube sampling leaves large regions of the input space totally unexplored.[3] An illustration of 55 samples chosen via Latin Hypercube sampling is shown in Fig. 2. Notice that the sampling approach produces some pairs that are close to one another in the input space.

**Sobol Sequences.** The Sobol method, in contrast to random sampling and LHS, generates low-discrepancy sequences. Informally, a low-discrepancy sequence will sample the input region relatively evenly. For a formal definition of discrepancy and technical descriptions of the Sobol generation procedure, please see [17]. An illustration in the 2-dimensional case can be seen in Fig. 3.

---

[1] A solved Sudoku puzzle is an example of a Latin square.

[2] This is similar to the classic "8 rooks" problem in chess.

[3] For example, taking a sample at every cell along the diagonal is a valid Latin hypercube sample sequence in two dimensions.

Notice how evenly the Sobol method samples the region of interest compared to the random sampling method and LHS.

## 2.2 Metamodeling

A machine learning model can be trained to emulate the base model, producing outputs that are as close as possible to the base model outputs, given the same input. If the base model output is qualitative, a machine learning classifier can be trained as the metamodel. If, on the other hand, the base model output is a quantitative metric, the metamodel will be a machine learning regressor. In this work, we only consider base models that produce quantitative metrics, but the techniques we utilize could be naturally extended to study base models that produce qualitative results. A wide variety of regressors exist, each with its own strengths. Unfortunately, in general, it is not possible to know a priori which particular machine learning technique will produce the most accurate metamodel for a given black box base model.

We consider a number of regressors in our analysis. Each regressor has its own strengths, weaknesses, and assumptions about the underlying data. We consider seven major types of regressors: random forest (RF) regressors, multilayer perceptrons (MLP), gradient-boosting machines (GBM), the RidgeCV regressor, k-nearest neighbors (KNN) regressors, Gaussian process (kriging) regressors, and stochastic gradient descent (SGD) regressors. Many of the regressors have hyperparameters that change their behavior. For example, the number of neighbors used in the k-nearest neighbors regressor can be any positive integer; different solvers and activation functions can be used by the multilayer perceptron; and the loss function used by the gradient-boosting machine can be changed. In this work we consider twenty-five different regressors of the seven types mentioned above: one random forest, seven different multilayer perceptrons, four different gradient-boosting machines, one Ridge regressor, ten different k-nearest neighbor regressors, one Gaussian process regressor, and one stochastic gradient descent regressor.

One could simply choose a particular regressor to serve as the metamodel (based on intuition or subject matter expertise), or one could train several different candidate regressors, compare their levels of accuracy, and select the one with the best performance to serve as the metamodel. However, the best regressor alone may not perform as well as several regressors working together. The simplest way to use multiple regressors together is to establish a voting committee of regressors, where the regressor predictions are averaged to produce a combined prediction. A drawback of that approach is that poorly performing, inaccurate regressors have the same vote as the most accurate regressor. It would be beneficial to weight the votes, such that the more accurate regressors have more say in the final prediction than the less accurate regressors. Stacking is one way to accomplish such weighting.

*Stacking* is a machine learning technique that accomplishes the weighting by training another regressor (or committee of regressors) on the original training data plus the predictions that the original regressors made [21]. We shall refer to the regressors trained solely on the training data as *layer-1 regressors*, and the regressors that were trained on the training data combined with the predictions from the layer-1 regressors as *layer-2 regressors*. We do not know a priori which regressor will perform the best in the second layer, so we train all twenty-five regressors as layer-2 regressors. We then take the average prediction of the layer-2 regressors as the final metamodel prediction.

In practice, we find that some of the layer-1 regressors produce predictions that are so inaccurate that they significantly degrade the performance of some of the layer-2 regressors. Therefore, we filter the layer-2 training data so they include only the predictions from the most accurate, best-performing layer-1 regressors. Similarly, some of the layer-2 regressors are so inaccurate that their predictions would significantly affect the quality of the final vote, so we filter the predictions of those layer-2 regressors as well. The filtering threshold is set to remove the predictions of any regressor whose average prediction error is more than 25% worse than that of the most accurate regressor at that layer. We found that this filtering strategy was effective in the evaluation of our case study model, but the threshold may have to be adjusted for other models, or a completely different strategy could be used (e.g., using the best $k$ regressors out of the set of $n$, where $k < n$). We plan to evaluate other filtering strategies in future work.

An illustration of the approach can be found in Fig. 4. Once trained, the stacked ensemble of regressors can form an accurate metamodel of the base model. Input analysis (such as sensitivity analysis and uncertainty quantification) techniques can then be applied to the metamodel in place of the base model.

## 3    Analysis Techniques

Uncertainty quantification and sensitivity analysis can help a modeler gain a deeper understanding of the model's input variables. Uncertainty quantification determines the likelihood of different outcomes given the uncertainty in the values of the input variables, and can be conducted in a straightforward manner using a Monte Carlo method [2]. Unfortunately, sensitivity analysis techniques are more complicated, and the different techniques may give differing answers. Sensitivity analysis attempts to determine the degree to which an input variable influences the output. A number of different techniques can be used to perform sensitivity analysis. We briefly describe three that we use in this work: Sobol sensitivity analysis, the Morris method, and the feature importance method. Sensitivity analysis can be used in a variety of ways. For example, a modeler may be able to dedicate only a limited amount of time and resources to reducing the uncertainty in particular model inputs (e.g., by performing experiments or seeking the opinions of experts). If that is the case, the modeler would like to know the input variables whose values have the greatest effect on the model output, so the uncertainty-reduction efforts can be focused on those variables.

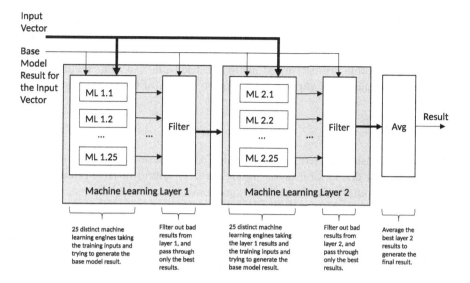

**Fig. 4.** Overview of the metamodel construction approach.

**Sobol Sensitivity Analysis:** Sobol sensitivity analysis [18] is a method for performing global sensitivity analysis. The method calculates the total order index for each value, which measures its total contribution to the variance in the output. The indices can be used to rank the input variables to determine the most and least sensitive inputs.

**Morris Method:** The Morris method is used to perform global sensitivity analysis; for details, consult [1,9]. The method varies one input variable at a time. First, it selects a point in the input space and runs the model with that input to determine the resulting output, which is used as a baseline. Next, each one of the input variables is assigned a new value, one at a time, while the others are held at the original baseline value, and the model is run to determine the magnitude of the difference between the resulting output and the baseline output. Once every input variable has been modified in turn, a completely new input is chosen (i.e., all the values for the input variables are changed at once), and the process repeats several times. The Morris method can use the collected data to perform a global sensitivity analysis, calculating a value $\mu^*$ for each input variable that quantifies its impact on the model output. By comparing the $\mu^*$ values, one can order the input variables based on their impact on the model output.

**Feature Importance Method:** The feature importance method [6,14] can be used as a way to rank the influence each input has on the model output. We shall describe the method at a high level. First, a metamodel is trained on the training data, and its baseline accuracy is recorded. Then, all the values of one input variable in the training data are perturbed through addition of noise. The metamodel is retrained with the modified training data, and the model's accuracy is compared with the accuracy of the baseline metamodel. If the input value has a large impact on the output value, we would expect a relatively large decrease in the accuracy of the new metamodel. If the input value has no impact on the output value, we would expect no appreciable drop in accuracy. The values of each input variable are perturbed in turn in the same way. The technique can be used to rank the input variables from most impactful to least impactful.

## 4   Test Case

Our test case considers a stochastic model of the growth of botnets in different conditions. A full description of the model may be found in [16]. A botnet is a collection of computers that have been compromised and hijacked by a malicious actor. A botnet can grow or shrink in size. The model gives an estimate of the size of the botnet at the end of one week, given a number of assumptions, including the rate at which bots are removed from the botnet by defenders. We imagine that defenders may have a choice among several different methods for removing the bots from the botnet. Methods that remove the bots faster may cost more or have deleterious side effects on the system's performance. In this imagined scenario, the defender would like to know the slowest rate at which the bots may be removed while ensuring that the botnet does not grow above a certain fixed size. The defenders must take into account their lack of knowledge of the precise values of all the input variables. In addition, the defenders would like to know which input variables have the largest effects on the model output, so that they may obtain the most accurate value estimates possible for the sensitive input variables.

We will give a brief overview of the pertinent details. The eleven input variables used in the model are listed in Table 1. In [16], all the input variables were assigned baseline values based on suggestions from subject matter experts. In our analysis, we assume that all the inputs are uncertain, and the uncertainty range is $\pm 50\%$ of the baseline values given in the original paper, with the following exceptions: $ProbConnectToPeers$ and $Prob2ndInjctnSuccessful$ are probabilities and thus may not be greater than 1, so we assign $[0.25, 1]$ as a reasonable range of uncertainty; and we increased the range of uncertainty for $RateConnectBotToPeers$, $RateOfAttack$, and $RateSecondaryInjection$ so that the rates fell between once every 10 s to once every hour, which we believed to be realistic assumptions.

## 5   Evaluation

We used Python to write our sampling, metamodeling, and analysis scripts. The construction of metamodels was accomplished with the aid of the scikit-learn Python package [10], and the SALib package was used to perform Sobol and Morris method sensitivity analysis [4]. All the experiments were run on a machine with an Intel i7-5829K processor and 32 GB of RAM.

**Table 1.** List of inputs used in the botnet test case.

| Variable name | Domain |
|---|---|
| ProbConnectToPeers | [0.25, 1] |
| ProbPropagationBot | [0.05, 0.15] |
| ProbInstallInitialInfection | [0.05, 0.15] |
| Prob2ndInjctnSuccessful | [0.25, 1] |
| RateConnectBotToPeers | [0.0166, 6] |
| RateOfAttack | [0.0166, 6] |
| RateSecondaryInjection | [0.0166, 6] |
| RateBotSleeps | [0.05, 0.15] |
| RateBotWakens | [0.0005, 0.0015] |
| RateActiveBotRemoved | [0.05, 0.15] |
| RateInactiveBotRemoved | [0.00005, 0.00015] |

### 5.1   Speed Comparison

All reported times were rounded to the nearest tenth of a second. The base model was run one thousand times with random inputs, and it took 7,246.1 s (over 2 h) to obtain the corresponding one thousand model outputs. That works out to just over 7 s per input, with a standard deviation of 24.8 s. The longest and shortest times it took to calculate an individual output were 510.6 and 0.8 s, respectively. The metamodel was also run one thousand times with random inputs, and it took a total of 1.9 s to obtain the corresponding one thousand model outputs. It follows that the metamodel can run several thousand times faster than the base model.

We found that the time needed to train the metamodel was correlated with the size of the dataset. Training of the stacked metamodel took 5 min and 19.1 s with the dataset that contained 4,000 random inputs, 1 min 36.0 s with the dataset that contained 1,000 random inputs, and 32.0 s with the dataset that contained 250 random inputs. The time it takes to collect the training data is significantly greater than the time needed to train the metamodel. Recall that in our approach, only a limited number of samples can be collected and used to train the metamodel, because it takes such a long time to execute the base model. The training datasets will therefore be relatively small, leading to relatively fast training times.

### 5.2   Metamodel Accuracy

Having established the speed with which the metamodel can be executed, we turn to an evaluation of the metamodel's accuracy. Recall that the metamodel attempts to produce an estimate of the final size of the botnet after one week. The estimation error is the absolute value of the difference between the metamodel's estimate and the base model's estimate, given a particular input. The errors reported are the average absolute difference between the metamodel's predictions and the corresponding base model's outputs across all the data in the test set. 2,500 randomly generated inputs (and associated base model outputs) comprised the test set. The minimum and maximum botnet sizes recorded in the test set were 0 and 37,143, respectively.

First, we consider the accuracy of the metamodel given the three different input sampling strategies and three different training dataset sizes. The results can be seen in Fig. 5. Unsurprisingly, we found that the metamodel error decreased when the training set contained more data. We also found that the Sobol sequence was the best-performing sampling strategy as the number of samples grew: as the number of samples grows, the effect of the low-discrepancy attribute of the Sobol sampling sequence becomes more obvious. Encouragingly, it appears that the metamodel can perform well even with a relatively low number of training samples: the metamodel trained with 250 random samples, the least accurate of the nine shown in Fig. 5, had an average estimation error that was less than 1% of the range found in the test data.

Next, we show that using our stacking approach is better than simply using the predictions from the best-performing of the twenty-five regressors we consider (which we call the *Best of Many* metamodel), and that the predictions of both methods perform better than a naive metamodel. Our naive metamodel calculates the average value of the outputs in the training dataset, and gives that average as its prediction regardless of the model input. Therefore, any well-performing regressor should produce more accurate predictions than the naive metamodel. We compare the errors of the naive metamodel, the most accurate single regressor (the Best of Many metamodel), and the stacked metamodel, given different training data. The results can be seen in Table 2. We see that the Best of Many metamodel has about half the average prediction error as the naive metamodel. In the worst case, the metamodel composed of stacked regressors has a 10% average error reduction compared to the best single regressor, and in the best case, it achieves a 32% average error reduction. This analysis shows that regressor stacking can lead to a significantly more accurate metamodel for our cybersecurity model compared to simply using a single regressor that is the best among many candidate regressors.

### 5.3   Uncertainty Quantification: Determination of Optimal Removal Rate

Assume that a defender has the ability to remove nodes from the botnet at a specific rate, but a faster rate costs the defender more than a slower rate. That may be the case when the defender can implement more effective but more expensive countermeasures to respond to the attack. The defender may wish to know how quickly the botnet can be expected to grow given different removal rates, and uncertainty in the other input parameters.

We conducted an experiment in which we used our stacked metamodel trained on the dataset consisting of four thousand Sobol samples. For each

**Table 2.** Average metamodel prediction error (lower is better).

| Training data | Naive metamodel error | Best of many metamodel error | Stacked metamodel error | Error reduction stacked vs. best of many |
|---|---|---|---|---|
| Random250 | 691 | 338 | 281 | 17% |
| Random1000 | 589 | 285 | 230 | 20% |
| Random4000 | 973 | 245 | 192 | 22% |
| LHS250 | 646 | 330 | 260 | 21% |
| LHS1000 | 696 | 296 | 222 | 25% |
| LHS4000 | 848 | 243 | 219 | 10% |
| Sobol250 | 761 | 371 | 282 | 24% |
| Sobol1000 | 825 | 321 | 230 | 28% |
| Sobol4000 | 704 | 234 | 161 | 32% |

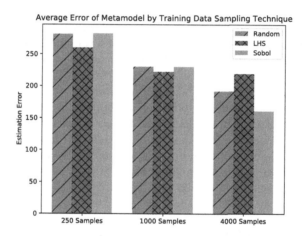

**Fig. 5.** Comparison of regressors trained on data obtained from random sampling, Latin hypercube sampling, and Sobol sequence sampling, respectively.

experiment, we fixed the value of the *RateActiveBotRemoved* variable. We then generated ten thousand Sobol samples for the other input variables in the ranges given by Table 1, and ran the regressor with those inputs to observe the predicted botnet size given the conditions described by the input variables. For each value of *RateActiveBotRemoved* we tested, we found the average botnet size, the 95th percentile, the 99th percentile, and the largest recorded botnet. That information can help a defender determine the slowest permissible removal rate given uncertainty in the input parameters. The results of our analysis can be found in Table 3.

**Table 3.** Estimated botnet size given removal rate.

| Removal rate | Average | 95% | 99% | Largest |
|---|---|---|---|---|
| 0.05 | 170 | 614 | 2436 | 9277 |
| 0.04 | 264 | 984 | 3861 | 11568 |
| 0.03 | 493 | 1970 | 6330 | 30004 |
| 0.02 | 1000 | 4229 | 10033 | 31364 |
| 0.01 | 1661 | 7206 | 27029 | 35593 |
| 0.001 | 2163 | 8594 | 27573 | 48152 |
| 0.0001 | 2207 | 8650 | 27965 | 48152 |

## 5.4 Sensitivity Analysis

We performed a sensitivity analysis to determine the degree to which model inputs impact the value of the output. As discussed in Sect. 2, traditional SA techniques often cannot be applied directly to the base model because of the long model run times. To overcome that issue, we used several different SA techniques and compared the results to determine the highly sensitive and highly insensitive model inputs. We conducted sensitivity analysis through the Morris method, the feature importance method, and the Sobol method.

The results of the analysis can be found in Table 4. Each of the three methods has two columns; the left column gives the score calculated by the method ($M_{\mu}^{*}$ for the Morris method, increase in metamodel error rate for the feature importance method, and the total order indices for the Sobol method).[4] Finally, in the rightmost two columns, we show the average ranking across the three methods for each input variable and the standard deviation. We will comment on the results of each SA method in turn, and then discuss how they may be interpreted when taken together.

First, the Morris method returns a $\mu^{*}$ value for each parameter, and since higher values indicate higher sensitivities, we can rank the input parameters relative to one another by using this value. The Morris method indicates that the model output is most sensitive to the *RateOfAttack* input variable, and least sensitive to the *RateBotSleeps* variable.

Second, the feature importance method determines how much the average error of the regressor's prediction increases (or conversely, how much its accuracy decreases) when a particular input variable's value is distorted with noise. The regressor is least accurate when the *RateOfAttack* input variable's value is corrupted with noise, indicating that variable's importance. The negative error calculated for the *RateSecondaryInjection* and *RateConnectBotToPeers* input variables indicates that the regressor does not make much use of these values when performing the regression.

---

[4] The $\mu^{*}$ values and the feature importance errors were rounded to the nearest integer, and all other values in the table were rounded to the nearest hundredth.

**Table 4.** Inputs variables ranked from most to least sensitive by taking the average of the rankings provided by the Morris, feature importance, and Sobol methods.

| Input name | Morris | | F. I. | | Sobol | | Combined | |
|---|---|---|---|---|---|---|---|---|
| | $\mu^*$ | Rank | Error | Rank | Total ord. ind. | Rank | Avg. rank | Std. dev. |
| ROfAttack | 3620 | 1 | 355 | 1 | 3.54 | 1 | 1 | 0 |
| RActiveBRemoved | 2648 | 3 | 303 | 3 | 3.53 | 2 | 2.67 | 0.58 |
| PPropagationB | 2492 | 4 | 307 | 2 | 1.25 | 5 | 3.67 | 1.53 |
| PInstall1stInfection | 2675 | 2 | 74 | 7 | 1.20 | 6 | 5 | 2.65 |
| RBWakens | 1311 | 8 | 120 | 4 | 1.69 | 4 | 5.33 | 2.31 |
| PConnectToPeers | 2041 | 6 | 13 | 9 | 1.83 | 3 | 6 | 3 |
| RInactiveBRemoved | 2373 | 5 | 77 | 6 | 1.09 | 10 | 7 | 2.65 |
| RBSleeps | 0 | 11 | 103 | 5 | 1.13 | 7 | 7.67 | 3.06 |
| PSecondaryInjection | 1979 | 7 | 74 | 8 | 1.09 | 9 | 8 | 1 |
| RSecondaryInjection | 1113 | 9 | −20 | 11 | 1.13 | 8 | 9.33 | 1.53 |
| RConnectBToPeers | 1073 | 10 | −19 | 10 | 1.06 | 11 | 10.33 | 0.58 |

Third, the Sobol method calculates total-order indices for the input variables, with higher values indicating more sensitivity. We can see that the Sobol method is in agreement with the other two methods in finding that the model output is most sensitive to the value of the *RateOfAttack* variable. The output is also quite sensitive to the *RateActiveBotRemoved* variable.

Finally, when they are taken together, it can be seen that there is broad agreement among the three methods in their ranking of the inputs. In Table 4, we show the combined average rank for each input, which we calculated by summing the input variable's sensitivity ranks as determined by the three SA methods and dividing by the number of SA methods. We also calculated the standard deviation of the three rankings; a low standard deviation shows that the methods are in agreement about the sensitivity rank of an input variable, while a high standard deviation shows disagreement. For example, each method ranked the *RateOfAttack* variable as the most sensitive, so it is given an average rank of $(1 + 1 + 1)/3 = 1$ and a standard deviation of 0, which shows perfect agreement. On the other hand, the three methods disagreed the most on the relative sensitivity ranking of the *RateBotSleeps* variable. The Morris method ranked it as the least sensitive, while the feature importance method ranked it as the fifth most sensitive, and the Sobol method ranked it as the seventh most sensitive, for a combined average rank of $(11 + 5 + 7)/3 = 7.67$, and a standard deviation of 3.06. In the absence of ground truth from the base model, it is encouraging to find that multiple SA methods largely agree on the rankings.

## 6   Discussion

**Limitations.** We believe that the metamodeling approach outlined in this work can aid in the evaluation and validation of realistic, complex, long-running computer security models. However, there are limitations to the approach. First, the approach generally sacrifices some accuracy for speed, since usually the meta-model cannot perfectly emulate the behavior of the base model. If the base model can be run very quickly, or if the modeler can afford to wait for the base model to run, the speedup achieved through the approach described in this work may not justify the decrease in accuracy. In practice, we believe that there are many realistic models that cannot be solved quickly, and that it would be prohibitively expensive to obtain enough computational resources to sufficiently decrease model execution times. On the other hand, the model may run so slowly that it is infeasible to obtain enough training samples to train a sufficiently accurate metamodel. If enough training samples cannot be obtained, we recommend modifying the base model so it runs more quickly, or building a metamodel by hand. A further limitation is that the metamodels may be accurate for most inputs, but inaccurate in a region that the modeler considers particularly interesting. In that case, we recommend that the modeler use adaptive sampling to drive the sampling towards the interesting region [7]. Finally, changing the base model invalidates the training data and the metamodel constructed using the training data, so the process would need to be repeated for every change to the base model.

**Use Cases.** We believe that our approach can help architects design better, more secure systems by helping them understand how a system might perform given different conditions and assumptions. The approach can be applied (1) to simulation and analytical models, as this paper demonstrates; (2) to long-running emulation experiments, e.g., experiments that use virtual machines and virtual switches to emulate a real network with real hosts cheaply and realistically, and (3) to prototypes of a system.

In the cybersecurity context, many realistic security models run too slowly to be used in an online fashion, so they can be used only at design time. However, a metamodel may be able to give helpful, near real-time guidance to human defenders responding to an ongoing cyber attack. An advanced intrusion prevention system (IPS) may also be able to use information from the metamodel in an online fashion (along with normal monitoring data, such as alerts from intrusion detection systems (IDS)), to thwart attacks more effectively.

**Future Work.** We are in the process of applying the metamodeling-based analysis approach to other published security models, particularly, to an advanced metering infrastructure (AMI) cybersecurity model [11,13] and a password cybersecurity model [12], to see whether the results generalize. We intend to explore whether more accurate metamodels can be constructed if adaptive sampling is used to obtain the training data [7]. We also intend to further explore different ways to structure the ensemble stack, for example, by using more than two layers.

## 7   Related Work

To the best of our knowledge, no prior work exists that uses metamodels in place of complex cybersecurity models. In addition, we know of no published work on the application of model stacking to state-based discrete-event simulation models. For a thorough introduction to the topic of metamodeling, see the excellent survey papers on the topic [7,20]. Xiao, Zuo, and Zhou propose an adaptive sampling approach that can be used to construct metamodels for reliability analysis, and demonstrate its use on four small reliability model problems [22]. Eisenhower et al. demonstrated a methodology that uses a support vector machine with a Gaussian kernel as a metamodel to perform an optimization for a building energy model with over 1,000 parameters [3]. Tenne proposes an approach that uses multiple metamodels and optimizers [19], but in a one-at-a-time fashion; the metamodel predictions are never fused or combined, whereas in our work the predictions of multiple regressors are taken together to form a better overall prediction. Zhou et al. like us, propose to use an ensemble of metamodels together, rather than train several metamodels and pick the one that performs best [23]. However, their work uses a recursive arithmetic average method to combine the predictions from multiple regressors, while we use stacking. Many papers have been published on sensitivity analysis in general. An overview of methods for global sensitivity analysis (including metamodeling-based approaches) can be found in [6].

# 8    Conclusion

In this work, we introduced a metamodeling-based approach for performing sensitivity analysis and uncertainty quantification on complex real world models. We demonstrated the approach on a published cybersecurity model. Our stacked metamodel is orders of magnitude faster than the base model, and more accurate than common existing approaches to metamodeling. The metamodeling-based approach outlined in this paper can be used by system architects to evaluate and validate realistic long-running models with a number of input variables whose values are not known with certainty.

**Acknowledgements.** The authors would like to thank Jenny Applequist, Lowell Rausch, and the reviewers for their feedback on the paper. This material is based upon work supported by the Maryland Procurement Office under Contract No. H98230-18-D-0007. Any opinions, findings and conclusions or recommendations expressed in this material are those of the author(s) and do not necessarily reflect the views of the Maryland Procurement Office.

# References

1. Campolongo, F., Cariboni, J., Saltelli, A.: An effective screening design for sensitivity analysis of large models. Environ. Model. Softw. **22**(10), 1509–1518 (2007)
2. Cunha, A., Nasser, R., Sampaio, R., Lopes, H., Breitman, K.: Uncertainty quantification through the Monte Carlo method in a cloud computing setting. Comput. Phys. Commun. **185**(5), 1355–1363 (2014)
3. Eisenhower, B., O'Neill, Z., Narayanan, S., Fonoberov, V.A., Mezić, I.: A methodology for meta-model based optimization in building energy models. Energy Build. **47**, 292–301 (2012)

4. Herman, J., Usher, W.: SALib: an open-source Python library for sensitivity analysis. J. Open Source Softw. **2**(9) (2017). https://doi.org/10.21105/joss.00097

5. Iman, R.L., Helton, J.C., Campbell, J.E.: An approach to sensitivity analysis of computer models: Part I - introduction, input variable selection and preliminary variable assessment. J. Qual. Technol. **13**(3), 174–183 (1981)

6. Iooss, B., Lemaître, P.: A review on global sensitivity analysis methods. In: Dellino, G., Meloni, C. (eds.) Uncertainty Management in Simulation-Optimization of Complex Systems: Algorithms and Applications, pp. 101–122. Springer, Boston (2015). https://doi.org/10.1007/978-1-4899-7547-8_5

7. Liu, H., Ong, Y.-S., Cai, J.: A survey of adaptive sampling for global metamodeling in support of simulation-based complex engineering design. Struct. Multi. Optim. **57**(1), 393–416 (2017). https://doi.org/10.1007/s00158-017-1739-8

8. McKay, M.D., Beckman, R.J., Conover, W.J.: Comparison of three methods for selecting values of input variables in the analysis of output from a computer code. Technometrics **21**(2), 239–245 (1979)

9. Morris, M.D.: Factorial sampling plans for preliminary computational experiments. Technometrics **33**(2), 161–174 (1991)

10. Pedregosa, F., et al.: Scikit-learn: machine learning in Python. J. Mach. Learn. Res. **12**, 2825–2830 (2011)

11. Rausch, M.: Determining cost-effective intrusion detection approaches for an advanced metering infrastructure deployment using ADVISE. Master's thesis, University of Illinois at Urbana-Champaign (2016)

12. Rausch, M., Fawaz, A., Keefe, K., Sanders, W.H.: Modeling humans: a general agent model for the evaluation of security. In: McIver, A., Horvath, A. (eds.) Quantitative Evaluation of Systems. Proceedings of International Conference on Quantitative Evaluation of Systems, pp. 373–388. Springer, Cham (2018). https://doi. org/10.1007/978-3-319-99154-2_23

13. Rausch, M., Feddersen, B., Keefe, K., Sanders, W.H.: A comparison of different intrusion detection approaches in an advanced metering infrastructure network using ADVISE. In: Agha, G., Van Houdt, B. (eds.) QEST 2016. LNCS, vol. 9826, pp. 279–294. Springer, Cham (2016). https://doi.org/10.1007/978-3-319-43425-4_19

14. Razmjoo, A., Xanthopoulos, P., Zheng, Q.P.: Online feature importance ranking based on sensitivity analysis. Expert Syst. Appl. **85**, 397–406 (2017)

15. Risdal, M.: Stacking made easy: an introduction to StackNet by competitions grandmaster Marios Michailidis (KazAnova). http://blog.kaggle.com/2017/06/15/stacking-made-easy-an-introduction-to-stacknet-by-competitions-grandmaster-marios-michailidis-kazanova/. Accessed 13 Dec 2019

16. Ruitenbeek, E.V., Sanders, W.H.: Modeling peer-to-peer botnets. In: Proceedings of 2008 Fifth International Conference on Quantitative Evaluation of Systems, pp. 307–316, September 2008

17. Sobol, I.: On the distribution of points in a cube and the approximate evaluation of integrals. USSR Comput. Math. Math. Phys. **7**(4), 86–112 (1967)

18. Sobol, I.: Global sensitivity indices for nonlinear mathematical models and their Monte Carlo estimates. Math. Comput. Simul. **55**(1), 271–280 (2001)

19. Tenne, Y.: An optimization algorithm employing multiple metamodels and optimizers. Int. J. Autom. Comput. **10**(3), 227–241 (2013)

20. Viana, F., Gogu, C., Haftka, R.: Making the most out of surrogate models: tricks of the trade. In: Proceedings of the ASME Design Engineering Technical Conference, vol. 1, pp. 587–598 (2010)

21. Wolpert, D.H.: Stacked generalization. Neural Netw. **5**(2), 241–259 (1992)
22. Xiao, N.C., Zuo, M.J., Zhou, C.: A new adaptive sequential sampling method to construct surrogate models for efficient reliability analysis. Reliab. Eng. Syst. Saf. **169**, 330–338 (2018)
23. Zhou, X.J., Ma, Y.Z., Li, X.F.: Ensemble of surrogates with recursive arithmetic average. Struct. Mult. Optim. **44**(5), 651–671 (2011)

# Correction to: The Dynamic Fault Tree Rare Event Simulator

Carlos E. Budde⑩, Enno Ruijters⑩, and Mariëlle Stoelinga⑩

## Correction to:
## Chapter "The Dynamic Fault Tree Rare Event Simulator"
## in: M. Gribaudo et al. (Eds.): *Quantitative Evaluation*
## *of Systems*, LNCS 12289,
## https://doi.org/10.1007/978-3-030-59854-9_17

In the original version of this chapter Reference 5 was published incorrectly. Reference 5 has now been corrected.

The updated version of this chapter can be found at
https://doi.org/10.1007/978-3-030-59854-9_17

# Tutorials

# Flexible Nets

Jorge Júlvez[(✉)] [iD]

Department of Computer Science and Systems Engineering,
University of Zaragoza, Zaragoza, Spain
`julvez@unizar.es`

This tutorial introduces Flexible Nets, a novel modelling formalism for dynamic systems that can account for a number of parameter uncertainties and that facilitates the performance evaluation, optimization and control of the modelled systems [1]. A Flexible Net is composed of two nets, an event net that captures how the processes of the system produce changes in its state, and an intensity net that models how the state induces speeds in the processes. These nets have three types of vertices: places (that model the state), transitions (that model processes), and handlers (that model the relationships between the state and the processes). Handlers can be equipped with inequalities in order to model system uncertainties, and optionally with piecewise linear functions to account for nonlinear dynamics [3]. After introducing the main features of Flexible Nets, several net examples, including a resource allocation system, a partially observed system, and a biological system [2], will be presented together with some of the analysis, optimization and control possibilities that can be used [4]. The last part of the tutorial will introduce the open-source Python tool fnyzer (https://fnyzer.readthedocs.io/) for the analysis of Flexible Nets.

**Acknowledgments.** This work was supported by the Spanish Ministry of Science, Innovation and Universities [ref. Medrese-RTI2018-098543-B-I00].

## References

1. Júlvez, J., Oliver, S.G.: Flexible nets: a modeling formalism for dynamic systems with uncertain parameters. Discrete Event Dyn. Syst. **29**(3), 367–392 (2019). https://doi.org/10.1007/s10626-019-00287-9
2. Júlvez, J., Dikicioglu, D., Oliver, S.G.: Handling variability and incompleteness of biological data by flexible nets: a case study for Wilson disease. Syst. Biol. Appl. **4**(1), 7 (2018). https://doi.org/10.1038/s41540-017-0044-x
3. Júlvez, J., Oliver, S.G.: Modeling, analyzing and controlling hybrid systems by guarded flexible nets. Nonlinear Anal. Hybrid Syst. **32**, 131–146 (2019). https://doi.org/10.1016/j.nahs.2018.11.004
4. Júlvez, J., Oliver, S.G.: Steady state analysis of flexible nets. IEEE Trans. Autom. Control **65**(6), 2510–2525 (2020). https://doi.org/10.1109/TAC.2019.2931836

# Verifying Probabilistic Programs

Benjamin Kaminski[1], Joost-Pieter Katoen[2(✉)], and Christoph Matheja[3]

[1] University College London, London, UK
b.kaminski@ucl.ac.uk
[2] RWTH Aachen University, Aachen, Germany
katoen@cs.rwth-aachen.de
[3] ETH Zürich, Zürich, Switzerland
contact@cmath.eu

Probabilistic programs enrich computation with randomization by allowing ordinary programs to (a) sample from probability distributions and (b) condition such distributions on some observed evidence. They are finite representations of potentially infinite-state Markov chains.

Randomization in computation has been around at least since Rabin introduced probabilistic automata in the early 1960s. Its importance has been ever-growing since then. Randomization is an important tool for the design and analysis of efficient algorithms, serves as a tie-breaker in distributed protocols, and is ubiquitous in cybersecurity. At the moment, probabilistic programs are receiving fast-growing attention in artificial intelligence, where they serve as a powerful modeling formalism that is both more expressive and more accessible than classical graphical models.

In light of the increasing deployment of probabilistic systems, their correctness is paramount. Establishing the correctness of probabilistic systems, however, is notoriously difficult. Even the notion of correctness itself becomes blurred: A program that produces correct results with high probability may be perfectly adequate.

In this tutorial, we will give an in-depth introduction to the foundations of quantitative verification of probabilistic programs:

(a) We present deductive techniques for verifying quantitative properties, such as both correctness and termination probabilities or expected runtimes. These techniques work directly on source-code level without explicitly constructing any Markov chain.

(b) We discuss invariant-style reasoning for (potentially unbounded) probabilistic loops – the main cause of infinite state spaces.

(c) We apply our techniques to automatically analyze expected simulation times of Bayesian networks – a popular graphical model in artificial intelligence.

© Springer Nature Switzerland AG 2020
M. Gribaudo et al. (Eds.): QEST 2020, LNCS 12289, p. 298, 2020.
https://doi.org/10.1007/978-3-030-59854-9

# Author Index

Printed in the United States
By Bookmasters